Ioan Merches
Dorian Tatomir
Roxana E. Lupu

Basics of

Quantum Electrodynamics

CISP

CRC Press
Taylor & Francis Group
Boca Raton London New York

CRC Press is an imprint of the
Taylor & Francis Group, an **informa** business

CRC Press
Taylor & Francis Group
6000 Broken Sound Parkway NW, Suite 300
Boca Raton, FL 33487-2742

First issued in paperback 2019

ISBN-13: 978-1-4665-8037-4 (hbk)
ISBN-13: 978-0-367-38055-7 (pbk)

Visit the Taylor & Francis Web site at
http://www.taylorandfrancis.com

and the CRC Press Web site at
http://www.crcpress.com

We physicists are always checking to see if there is something the matter with the theory. That's the game, because if there is something the matter, it's interesting! But so far, we have found nothing wrong with the theory of quantum electrodynamics. It is, therefore, I would say, THE JEWEL OF PHYSICS - our proudest possession.

Richard Phillips Feynman

To the memory of our adviser, collaborator and friend,
Professor Emeritus Ioan Gottlieb

PREFACE

Quantum Field Theory was born only several years after Quantum Mechanics, more precisely in 1927, when P.A.M.Dirac performed the quantization of the electromagnetic radiation (Proc.Roy.Soc.A, vol. 114, 1927, pp.243 and 710). Unlike Quantum Mechanics, which became a well established chapter of Theoretical Physics in less than a decade, Quantum Field Theory (QFT) needed two decades to become sufficiently well developed. The explanation is that there were no applications by that time. The crucial role in the development of QFT was played by Bethe's paper (Phys.Rev. vol.72, 1947, p.339) concerning the so-called *Lamb shift* (or *displacement*), which is a small difference in energy between two energy levels $2S_{1/2}$ and $2P_{1/2}$ (in terms of symbolic notation) of the hydrogen atom. Hans Bethe was the first to explain the Lamb shift in the hydrogen spectrum, and he thus laid the foundation for the development of modern quantum electrodynamics. The Lamb shift currently provides a measurement of the fine-structure constant α to better than one part per million, allowing a precision test of quantum electrodynamics.

By Quantum Electrodynamics (QED) one understands the *relativistic quantum field theory of electrodynamics*. It describes all phenomena that involve interacting charged particles, and it can be viewed as a perturbation theory of the electromagnetic quantum vacuum.

Currently, QED represents a vast discipline in the family of quantum theories, and a very efficient instrument for approaching problems in most chapters of physics: elementary particles, atomic nucleus, solid state, etc. For this reason, almost all modern Quantum Mechanics textbooks end with a chapter dedicated to the quantum aspects of electromagnetic interactions.

The first two chapters of this textbook deal with the general theory of free fields. Here some results encountered within other disciplines are synthesized, generalized and systematized. The reader is also told about the necessity of field quantization. As a matter of fact, the necessity of field quantization can be fully understood only *a posteriori*, through a set of consequences which can be experimentally verified.

The next three chapters of the book are concerned with the quantization of the scalar, electromagnetic, and spinorial fields. This has the purpose to prepare the reader for understanding field interactions. We presented here the quantization of the scalar field for two reasons: this is the most simple case, and some of the results (operator products, vacuum state, Wick's theorems) will also be used for all interaction processes.

The sixth chapter presents some general problems of field interactions. Here we introduce the *scattering matrix* (or *S-matrix*), relating the initial and final states of a physical system undergoing a scattering process, the *Feynman-Dyson diagrams* or *Dyson graphs*, and the *effective scattering cross section*.

The next chapter studies the divergence-free second order processes. As two main examples we consider *Compton scattering*, and *Møller scattering*. In both cases the effective cross sections and *transition probabilities* are calculated.

The eighth (and last) chapter deals with divergent second-order processes. Knowledge of the photon self-energy diagram makes it possible to study the *vacuum polarization* as well as the *mass* and *charge renormalization*.

Keeping in mind that this textbook is dedicated to both students and their instructors, to facilitate the application of the theory, we have also included a chapter containing several solved problems.

The book ends with a useful Appendix on the theory of distributions. Starting with the *Dirac delta function*, some functions connected with the delta function, known as retarded and advanced functions, as well as some causal functions, are also discussed.

There are several excellent books on QED. Our belief is that a textbook on this subject has to present the intimate connection between Quantum Mechanics and QED, at an appropriate level, in two general steps: the quantization of free fields, followed by their interactions with applications.

It is a fact that no advanced physical theory can be approached without possessing a good background regarding QFT. We feel that the detailed presentation of the quantization of various fields, the original presentation of the theory and the detailed study of the most important interaction processes in QED make this textbook a useful guide for those studying physics, at both graduate and undergraduate level.

Iasi, Mountain View, April 2012   The Authors

CONTENTS

CHAPTER I
GENERAL FIELD THEORY

Anyone who is not shocked by the quantum theory has not understood it.

Niels Bohr

The aim of the present chapter is to establish a general method to derive the fundamental quantization laws valid for *any* field. As soon as we are prepared with this general theory, we shall apply it to the quantization of the scalar, electromagnetic, and spinorial field, considered as free fields. To do this, we shall use the methods and principles of Analytical Mechanics.

1.1. Basic field equations

We begin our investigation with the fundamental postulate that any free field obeys Hamilton's principle

$$\delta S = 0, \tag{1.1.1}$$

where the *action* S is given by the quadruple integral

$$S = \int \mathcal{L} \, dx \, dy \, dz \, dt. \tag{1.1.2}$$

Here $\mathcal{L}$ stands for the *Lagrangian density* of the field. To write the action S in a convenient form, we shall define the elementary four-volume

$$d\Omega = dx_1 dx_2 dx_3 dx_4$$

in the four-dimensional complex Euclidean space E_4, defined by $x_1 = x, x_2 = y, x_3 = z, x_4 = ict$. As known, $d\Omega$ is a relativistic invariant. We then can write

$$S = \frac{1}{ic} \int \mathcal{L} \, d\Omega. \tag{1.1.3}$$

1

General field theory

It is worthwhile mentioning that $\mathcal{L}$ and S are the Lagrangian density and the action for some arbitrary field, respectively, without any specification about its concrete form.

But, while the Lagrangian of a charged particle moving in an external electromagnetic field writes

$$L = \frac{1}{2}mv^2 - eV + e\mathbf{v} \cdot \mathbf{A}, \qquad (1.1.4)$$

where the Cartesian coordinates x, y, z, t play the role of generalized coordinates, and v_x, v_y, v_z are the generalized velocities, the Lagrangian density of the free electromagnetic field (the Larmor invariant) is

$$\mathcal{L} = \frac{1}{2\mu_0}B^2 - \frac{1}{2}\epsilon_0 E^2 \qquad (1.1.5)$$

with $\mathbf{E} = -\nabla V - \frac{\partial \mathbf{A}}{\partial t}$; $\mathbf{B} = \nabla \times \mathbf{A}$. This time the dependent variables are the electromagnetic potentials $\mathbf{A}$ and V, each being functions of the independent variables x, y, z, t. We also observe that the Lagrangian density (1.1.5) depends on the partial derivatives of *all* the field variables $U_{(r)}$ with respect to *all* the independent variables x_k; $k = \overline{1,4}$, while the Lagrangian (1.1.4) only depends on the derivatives of x, y, z with respect to a *single* independent variable, the time t.

The explanation is obvious: a field cannot be concentrated in one point, that is the Lagrangian density cannot depend on the "velocity" of that point, because it doesn't exist.

Therefore, we can state that, in general, the Lagrangian density $\mathcal{L}$ of a field depends on the independent variables x_k, on some *field functions* $U_{(r)}$, also called *field variables*, as well as on the partial derivatives of the field variables with respect to x_k as well, i.e.

$$\mathcal{L} = \mathcal{L}\big[x_k, U_{(r)}, U_{(r),k}\big], \qquad (1.1.6)$$

where

$$U_{(r),k} = \frac{\partial U_{(r)}}{\partial x_k}. \qquad (1.1.7)$$

We denoted by (r) a group of indices with the following properties:
- does not exist in the case of the zero-order tensor field (scalar field): $U_{(r)} \equiv U$;
- reduces to a single index (without parentheses) for a first-order tensor (vector) field, and spinorial field as well: $U(r) \equiv U_r$;
- consists of two indices (without parentheses) in the case of a second-order tensor field: $U_{(r)} \equiv U_{ik}$.

2

Basic field equations

The examples (1.1.4) and (1.1.5) show that the Lagrangian depends on the coordinates x_k only if there exist point charges interacting with the field. In this respect, one can observe that the Lagrangian density (1.1.5) does not explicitly depend on x_k. Consequently, since a free field does not have any concentration centers in space, the Lagrangian density of a free field does not explicitly depend on x_k. The first five chapters of this book are concerned with free fields.

We mention that, in general, the field variables $U_{(r)}$ can be complex quantities. Consequently, the number of field variables becomes twice as big. Furthermore, the field functions $U_{(r)}$ can be expressed as matrices. These reasons motivate choosing as field variables in the Lagrangian density the quantities $U_{(r)}$, as well as their Hermitian conjugates $U_{(r)}^+$, independent of each other.

In addition to the field variables $U_{(r)}$ and $U_{(r)}^+$, the Lagrangian density must also contain the first derivatives of these functions with respect to x_k, that is $U_{(r),k}$ and $U_{(r),k}^+$. The second (third, etc.) derivatives cannot appear in the Lagrangian density $\mathcal{L}$, because the usual equations of physics (wave propagation equation, Schrödinger equation, etc.) are second-order, partial differential equations.

Consequently, we choose the Lagrangian density as

$$\mathcal{L} = \mathcal{L}\big[U_{(r)}, U_{(r)}^+, U_{(r),k}, U_{(r),k}^+\big]. \tag{1.1.8}$$

Obviously, if the field variables $U_{(r)}$ are common functions, then the Hermitian conjugate $U_{(r)}^+$ reduces to complex conjugate $U_{(r)}^*$.

Using the action (1.1.3) and the Lagrangian density (1.1.8), Hamilton's principle (1.1.1) leads to a system of second-order, partial differential equations for $U_{(r)}$ and $U_{(r)}^+$. Simplifying by ic, we have:

$$\delta \int_\Omega \mathcal{L}\, d\Omega = \int_\Omega \Big[\frac{\partial \mathcal{L}}{\partial U_{(r)}} \delta U_{(r)} + \delta U_{(r)}^+ \frac{\partial \mathcal{L}}{\partial U_{(r)}^+}$$

$$+ \frac{\partial \mathcal{L}}{\partial U_{(r),k}} \delta U_{(r),k} + \delta U_{(r),k}^+ \frac{\partial \mathcal{L}}{\partial U_{(r),k}^+} \Big]\, d\Omega = 0, \tag{1.1.9}$$

where the summation convention has been used. Integrating by parts the last two terms, we arrive at

$$\int_\Omega \frac{\partial \mathcal{L}}{\partial U_{(r),k}} \delta U_{(r),k}\, d\Omega = \int_\Omega \frac{\partial \mathcal{L}}{\partial U_{(r),k}} \cdot \frac{\partial}{\partial x_k}(\delta U_{(r)})\, d\Omega$$

$$= \int_\Omega \frac{\partial \mathcal{L}}{\partial x_k}\Big[\frac{\partial \mathcal{L}}{\partial U_{(r),k}} \delta U_{(r)}\Big]\, d\Omega - \int_\Omega \frac{\partial \mathcal{L}}{\partial x_k}\Big[\frac{\partial \mathcal{L}}{\partial U_{(r),k}}\Big] \delta U_{(r)}\, d\Omega$$

$$= \int_S \frac{\partial \mathcal{L}}{\partial U_{(r),k}} \delta U_{(r)} \, dS_k - \int_\Omega \frac{\partial \mathcal{L}}{\partial x_k} \Big[\frac{\partial \mathcal{L}}{\partial U_{(r),k}} \Big] \delta U_{(r)} \, d\Omega, \qquad (1.1.10)$$

as well as

$$\int_\Omega \delta U^+_{(r),k} \frac{\partial \mathcal{L}}{\partial U^+_{(r)}} \, d\Omega = \int_S \delta U^+_{(r)} \frac{\partial \mathcal{L}}{\partial U^+_{(r),k}} \, dS_k$$

$$- \int_\Omega \delta U^+_{(r)} \frac{\partial}{\partial x_k} \Big[\frac{\partial \mathcal{L}}{\partial U^+_{(r),k}} \Big] \, d\Omega. \qquad (1.1.11)$$

We have denoted by S the three-dimensional hypersurface bounding the domain Ω, while dS_k is the hypersurface element orthogonal to dx_k, namely

$$d\Omega = dS_k \, dx_k \quad \text{(no summation).} \qquad (1.1.12)$$

Let us assume that the field variables $U_{(r)}$ and $U^+_{(r)}$ are given on the frontier S of the domain Ω. In this case

$$\delta U_{(r)}|_S = 0; \quad \delta U^+_{(r)}|_S = 0, \qquad (1.1.13)$$

and the three-dimensional integrals in (1.1.10) and (1.1.11) vanish. Eq. (1.1.9) yields

$$\int_\Omega \Big[\frac{\partial \mathcal{L}}{\partial U_{(r)}} - \frac{\partial}{\partial x_k} \Big(\frac{\partial \mathcal{L}}{\partial U_{(r),k}} \Big) \Big] \delta U_{(r)} \, d\Omega$$

$$+ \int_\Omega \delta U^+_{(r)} \Big[\frac{\partial \mathcal{L}}{\partial U^+_{(r)}} - \frac{\partial}{\partial x_k} \Big(\frac{\partial \mathcal{L}}{\partial U^+_{(r),k}} \Big) \Big] \, d\Omega = 0. \qquad (1.1.14)$$

Since the elementary variations $\delta U_{(r)}$ and $\delta U^+_{(r)}$ are arbitrary, we arrive at the well-known Euler-Lagrange equations

$$\frac{\partial \mathcal{L}}{\partial U_{(r)}} - \frac{\partial}{\partial x_k} \Big(\frac{\partial \mathcal{L}}{\partial U_{(r),k}} \Big) = 0; \quad \frac{\partial \mathcal{L}}{\partial U^+_{(r)}} - \frac{\partial}{\partial x_k} \Big(\frac{\partial \mathcal{L}}{\partial U^+_{(r),k}} \Big) = 0. \quad (1.1.15)$$

We shall call them *basic field equations*.

As known, the system of (say, s) second-order, partial differential equations (1.15) is equivalent to a system of $2s$ first-order partial differential equations. If we define the *generalized momentum densities* as

$$\pi^+_{k(r)} = \frac{\partial \mathcal{L}}{\partial U_{(r),k}}; \quad \pi_{k(r)} = \frac{\partial \mathcal{L}}{\partial U^+_{(r),k}}, \qquad (1.1.16)$$

then the system of equations (1.1.15) writes

$$\frac{\partial \mathcal{L}}{\partial U_{(r)}} = \pi^+_{k(r),k}; \qquad \frac{\partial \mathcal{L}}{\partial U^+_{(r)}} = \pi_{k(r),k}. \qquad (1.1.17)$$

Equations (1.1.16) are justified by the fact that the Lagrangian density $\mathcal{L}$ must be a real function, so it must contain products like $U_{(r),k} U^+_{(r),k}$, and the derivation is straightforward.

Equations (1.1.16), together with (1.1.17), are equivalent to Eqs. (1.1.15). These equations are essential in derivation of the fundamental notions and conservation laws in QFT.

1.2. Infinitesimal Lorentz transformation

We call a *Lorentz transformation* any rotation of the four-dimensional, Euclidean complex frame $x_1 = x, x_2 = y, x_3 = z, x_4 = ict$, expressed by the linear transformation

$$x'_i = a_{ik} x_k; \qquad i, k = \overline{1, 4}. \qquad (1.2.1)$$

Imposing the condition that x_i squared, i.e. $x_i x_i$, remains unchanged with respect to the transformation (1.2.1), we immediately get

$$a_{ik} a_{il} = \delta_{kl}, \qquad (1.2.2)$$

which is the *orthogonality condition* for the coefficients a_{ik}. Since the transformation (1.2.1) is orthogonal and taking into account the fact that the four-dimensional space is Euclidean, there is no distinction between contravariant and covariant vectors, and we are allowed to use lower indices only.

Let us consider the infinitesimal transformation

$$x'_i = x_i - \omega_{ik} x_k, \qquad (1.2.3)$$

or

$$\delta x_i = x'_i - x_i = -\omega_{ik} x_k, \qquad (1.2.3')$$

where ω_{ik} are the components of a second-order infinitesimal tensor (i.e. one can neglect its higher powers). The transformation (1.2.3) can also be written as

$$x'_i = (\delta_{ik} - \omega_{ik}) x_k, \qquad (1.2.4)$$

which yields

$$a_{ik} = \delta_{ik} - \omega_{ik} \qquad (1.2.6)$$

and (1.2.2) becomes

$$\delta_{kl} = a_{ik}\, a_{il} = (\delta_{ik} - \omega_{ik})\,(\delta_{il} - \omega_{il}) \approx \delta_{kl} - \omega_{kl} - \omega_{lk},$$

showing that

$$\omega_{ik} + \omega_{ki} = 0. \qquad (1.2.6)$$

This result says that (1.2.3) is an infinitesimal Lorentz transformation only if the tensor ω_{ik} is antisymmetric.

The transformation (1.2.3) can be easily inverted. Since

$$x_i = x_i' + \omega_{ik}\, x_k = x_i' + \omega_{ik}(x_k' + \omega_{kl}\, x_l),$$

the *inverse transformation* writes

$$x_i = x_i' + \omega_{ik}\, x_k'. \qquad (1.2.7)$$

We define *Lorentz invariance* as the invariance with respect to the infinitesimal Lorentz transformation (1.2.3). The Lorentz invariance of the Lagrangian density $\mathcal{L}$ and the action S leads to some fundamental laws in QFT. To this end, we have to know the transformation relations of the field variables $U_{(r)}$ and $U_{(r),k}$ under the Lorentz transformation (1.2.3).

Whatever the tensor rank of $U_{(r)}$ may be, the most general form for the transformation of the field quantities writes

$$U_{(r)}' = U_{(r)} - \omega_{ik}\, S_{(r)(p)ik}\, U_{(p)}. \qquad (1.2.8)$$

This transformation is *infinitesimal* due to ω_{ik}, while the coefficients $S_{(r)(p)ik}\, U_{(p)}$ depend *only* on the tensor character of $U_{(r)}$.

Let us now derive the transformation laws of the infinitesimal quantities

$$\delta U_{(r)} = U_{(r)}' - U_{(r)} \quad ; \quad \delta U_{(r),l} = U_{(r),l}' - U_{(r),l} \qquad (1.2.9)$$

with respect to the infinitesimal Lorentz transformation (1.2.3). To do this, let us take the derivative of (1.2.8) with respect to x_l'. The result is

$$U_{(r),l}' = \frac{\partial U_{(r)}'}{\partial x_l'} = \frac{\partial U_{(r)}}{\partial x_s}\frac{\partial x_s}{\partial x_l'} - \omega_{ik}\, S_{(r)(p)ik}\frac{\partial U_{(p)}}{\partial x_s}\frac{\partial x_s}{\partial x_l'}.$$

Transformation of the quantities $U_{(r)}$ in particular cases

According to (1.2.7), we have

$$x_s = x'_s + \omega_{sl}\, x'_l = (\delta_{sl} + \omega_{sl})\, x'_l; \qquad \frac{\partial x_s}{\partial x'_l} = \delta_{sl} + \omega_{sl},$$

so that

$$U'_{(r),l} = U_{(r),s}(\delta_{sl} + \omega_{sl}) - \omega_{ik}\, S_{(r)(p)ik}\, U_{(p),s}\, (\delta_{sl} + \omega_{sl}). \qquad (1.2.10)$$

Neglecting the product $\omega_{ik}\, \omega_{sl}$, and using (1.2.8), (1.2.9), and (1.2.10), we finally obtain

$$\delta U_{(r)} = -\omega_{ik}\, S_{(r)(p)ik}\, U_{(p)}, \qquad (1.2.11)$$

$$\delta U_{(r),l} = -\omega_{ls}\, U_{(r),s} - \omega_{ik}\, S_{(r)(p)ik}\, U_{(p),l}. \qquad (1.2.12)$$

Similarly, we have:

$$\delta U^+_{(r)} = -\omega_{ik}\, U^+_p\, S^+_{(r)(p)ik}; \qquad (1.2.13)$$

$$\delta U^+_{(r),l} = -\omega_{ls}\, U^+_{(r),s} - \omega_{ik}\, U^+_{(p),l} S^+_{(r)(p)ik}. \qquad (1.2.14)$$

1.3. Transformation of the quantities $U_{(r)}$ in particular cases

We are now prepared to write the actual expressions for the quantities $\delta U_{(r)}$ and $\delta U_{(r),l}$ and, consequently, the coefficients $S_{(r)(p)ik}$, for some well-known fields.

a) **Scalar field.** In this case $U_{(r)}$ is a scalar: $U_{(r)} \equiv U$. Being a scalar, U is invariant with respect to the coordinate transformation (1.2.3), that is $U'(x'_i) = U(x_i)$, or

$$\delta U = 0. \qquad (1.2.1)$$

A solution of (1.2.11) then is

$$S_{(r)(p)ik} = 0. \qquad (1.3.2)$$

According to (1.2.12), we also have

$$\delta U_{,l} = -\omega_{ls}\, U_{,s}. \qquad (1.3.3)$$

b) **Vector field.** Since $U_{(r)} \equiv U_r$ is a vector, it transforms according to (1.2.3), namely

$$\delta U_r = -\omega_{rk}\, U_k. \qquad (1.3.4)$$

General field theory

Using the Kronecker symbol δ_{ik} we can also write

$$\delta U_r = -\omega_{ik}\,\delta_{ir}\,\delta_{kp}\,U_p, \tag{1.3.5}$$

and comparing (1.3.5) with (1.2.11) one obtains

$$S_{rpik} = \delta_{ir}\,\delta_{kp}, \tag{1.3.6}$$

while (1.2.12) yields

$$\delta U_{r,l} = -\omega_{ls}\,U_{r,s} - \omega_{rk}\,U_{k,l}. \tag{1.3.7}$$

c) **Second-order tensor field.** As discussed in Section 1.1, in this case (r) is a group of two indices: $U_r \equiv U_{gh}$. This is a second-order orthogonal affine tensor, whose components transform according to

$$U'_{gh} = a_{gf}\,a_{hq}\,U_{fq}. \tag{1.3.8}$$

In view of (1.2.5), we have:

$$U'_{gh} = (\delta_{gf} - \omega_{gf})(\delta_{hq} - \omega_{hq})\,U_{fq} \approx U_{gh} - (\omega_{gf}\delta_{hq} + \omega_{hq}\delta_{gf})\,U_{fq}$$

$$= U_{gh} - \omega_{ik}\,(\delta_{ig}\delta_{kf}\delta_{hq} + \delta_{ih}\delta_{kq}\delta_{gf})\,U_{fq},$$

or, equivalently,

$$\delta U_{gh} = -\omega_{ik}\,S_{ghfqik}\,U_{fq} \tag{1.3.9}$$

where

$$S_{ghfqik} = \delta_{ig}\delta_{kf}\delta_{hq} + \delta_{ih}\delta_{kq}\delta_{gf}. \tag{1.3.10}$$

Following (1.2.12), we then have

$$\delta U_{gh,l} = -\omega_{ls}\,U_{gh,s} - \omega_{ik}\,S_{ghfqik}\,U_{fq,l},$$

or, in view of (1.3.9)

$$\delta U_{gh,l} = -\omega_{ls}\,U_{gh,s} - \omega_{gf}\,U_{fh,l} - \omega_{hq}\,U_{gq,l}. \tag{1.3.11}$$

d) **Spinorial field.** This time, the quantities $U_{(r)} \equiv U_r$ $(r = \overline{1,4})$ form a spinor. It is more convenient to use U and U^+, without indices, where by U and U^+ we mean the one column and, respectively, one row matrices

$$U = \begin{pmatrix} U_1 \\ U_2 \\ U_3 \\ U_4 \end{pmatrix}; \qquad U^+ = \begin{pmatrix} U_1^* & U_2^* & U_3^* & U_4^* \end{pmatrix}. \tag{1.3.12}$$

Transformation of the quantities $U_{(r)}$ in particular cases

As previously mentioned, U must be given as a linear combination of the field variables. Since this time U is a vector, we choose the linear combination

$$\delta U = -AU, \qquad (1.3.13)$$

where A is an infinitesimal matrix whose explicit form is going to be determined. Obviously, we also have

$$\delta U^+ = -U^+ A^+. \qquad (1.3.14)$$

Anticipating, one can show that it is more convenient to use

$$\overline{U} = U^+ \gamma_4, \qquad (1.3.15)$$

instead of U^+, where

$$\gamma_4 = \begin{pmatrix} \epsilon & 0 \\ 0 & -\epsilon \end{pmatrix} \qquad (1.3.16)$$

while ϵ is the second-order unit matrix

$$\epsilon = \begin{pmatrix} 1 & 0 \\ 0 & 1 \end{pmatrix}.$$

For a primer on $\gamma-$ matrices we refer the reader to Section 5.1. The explicit form of these matrices, following the conventions used in this book, is given by the expressions (5.3.5).

Multiplying (1.3.14) on the right by γ_4, we have

$$(\delta U^+)\gamma_4 = \delta(U^+\gamma_4) = -U^+ A^+\gamma_4 = -(U^+\gamma_4)(\gamma_4 A^+\gamma_4).$$

Denoting

$$\overline{A} = \gamma_4 A^+ \gamma_4, \qquad (1.3.17)$$

we can write

$$\delta\overline{U} = -\overline{U}\,\overline{A}. \qquad (1.3.18)$$

In light of the above considerations, let us now determine the quantities δU and $\delta U_{,l}$. Combining the equations (1.2.7) and (1.3.13), we successively have:

$$U' = U + \delta U = U - AU;$$

$$U'_{,l} = \frac{\partial U'}{\partial x'_l} = \frac{\partial U}{\partial x_s}\frac{\partial x_s}{\partial x'_l} - A\frac{\partial U}{\partial x_s}\frac{\partial x_s}{\partial x'_l}$$

$$= U_{,s}(\delta_{sl} + \omega_{sl}) - AU_{,s}(\delta_{sl} + \omega_{sl}) = U_{,l} + \omega_{sl}U_{,s} - AU_{,l},$$

9

leading to

$$\delta U_{,l} = \omega_{sl} U_{,s} - A U_{,l}.\tag{1.3.19}$$

Taking the Hermitian conjugate of (1.3.19) and multiplying by γ_4 on the right, one obtains

$$\delta \overline{U}_{,l} = -\omega_{ls} \overline{U}_{,s} - \overline{u}_{,l} \overline{A}.\tag{1.3.20}$$

Using the above considerations, we are now able to find the explicit form of the infinitesimal matrix A. To do this, we impose the following two conditions: (a) the product $U\overline{U}$ is an invariant; (b) the product $\overline{U}\gamma_r U$, where γ_r are the Dirac matrices, is a four-vector. Under these assumptions, we have

$$\delta(\overline{U}U) = 0;\tag{1.3.21}$$

$$\delta(\overline{U}\gamma_r U) = -\omega_{rk} \overline{U}\gamma_k U.\tag{1.3.22}$$

Relation (1.3.21) then yields

$$\overline{U}\delta U + (\delta \overline{U})U = -\left(\overline{U}(AU) + (\overline{U}\,\overline{A})U\right) = -\overline{U}(A + \overline{A})U = 0,$$

that is, for arbitrary U and $\overline{U}$, we have

$$\overline{A} = -A.\tag{1.3.23}$$

On the other hand, condition (1.3.22) gives

$$\overline{U}\,\overline{A}\gamma_r U - \overline{U}\gamma_r A U = -\omega_{rk} \overline{U}\gamma_k U,$$

or, in view of (1.3.23)

$$\overline{U}(\gamma_r A - A\gamma_r) U = \omega_{rk} \overline{U}\gamma_k U.$$

We then have

$$\gamma_r A - A\gamma_r = \omega_{rk}\,\gamma_k.\tag{1.3.24}$$

Let us now multiply (1.3.24) by γ_r on the right, and remember that $\gamma_r \gamma_r = \gamma_1^2 + \gamma_2^2 + \gamma_3^2 + \gamma_4^2 = 4$. The result is

$$A = \frac{1}{4}\omega_{rk}\gamma_r\gamma_k + B,\tag{1.3.25}$$

where

$$B = \frac{1}{4}\gamma_r A \gamma_r.\tag{1.3.26}$$

Transformation of the quantities $U_{(r)}$ in particular cases

Since B contains A, it is difficult to solve (1.3.26). However, if we take $A = \frac{1}{4} w_{rk} \gamma_r \gamma_k$ in B, one obtains $B = 0$, meaning that this choice makes the two equations $A = \frac{1}{4} w_{rk} \gamma_r \gamma_k$ and $B = 0$ compatible with each other. Indeed, from equation (1.3.26) and the anti-commutation rule for Dirac matrices

$$\gamma_i \gamma_k + \gamma_k \gamma_i = 2 \delta_{ik}, \tag{1.3.27}$$

we have

$$B = \frac{1}{4} \gamma_r \left(\frac{1}{4} w_{ik} \gamma_i \gamma_k \right) \gamma_r = \frac{1}{16} w_{ik} (2 \delta_{ir} - \gamma_i \gamma_r)(2 \delta_{rk} - \gamma_r \gamma_k)$$

$$= \frac{1}{4} w_{ik} \gamma_i \gamma_k = 0,$$

where we applied the product rule between a symmetric and an anti-symmetric tensor. We can then write

$$A = \frac{1}{4} w_{ik} \gamma_i \gamma_k. \tag{1.3.28}$$

By replacing (1.3.28) into (1.3.13), we get

$$\delta U = -\frac{1}{4} w_{ik} \gamma_i \gamma_k \, U, \tag{1.3.29}$$

or, if we go back to index notation,

$$\delta U_r = -\frac{1}{4} w_{ik} \left(\gamma_i \gamma_k \right)_{rp} U_p. \tag{1.3.30}$$

If we now compare (1.2.11) with (1.3.30), we find

$$S_{rpik} = \frac{1}{4} \left(\gamma_i \gamma_k \right)_{rp}, \tag{1.3.31}$$

and (1.2.12) leads to

$$\delta U_{r,l} = -w_{ls} \, U_{r,s} - \frac{1}{4} w_{ik} \left(\gamma_i \gamma_k \right)_{rp} U_{p,l}. \tag{1.3.32}$$

In view of (1.3.18), and (1.3.23), we also have

$$\delta \overline{U}_r = \frac{1}{4} w_{ik} \overline{U}_p \left(\gamma_i \gamma_k \right)_{pr}. \tag{1.3.33}$$

11

General field theory

But

$$S^+_{rpik} = -\frac{1}{4}(\gamma_i\gamma_k)_{pr} = \frac{1}{4}(\gamma_k\gamma_i)_{pr}; \quad (i \neq k)$$

and (1.2.14) yields

$$\delta\overline{U}_{r,l} = -\omega_{ls}\,\overline{U}_{r,s} + \frac{1}{4}\,\omega_{ik}\,\overline{U}_{p,l}\,(\gamma_i\gamma_k)_{pr}. \qquad (1.3.34)$$

Observation. We did not consider the quantities δU^+_r and $\delta U_{r,l}$ in the case of the scalar, vector and second-order tensor fields. This is an easy task and we leave it to the reader.

1.4. Invariance of the Lagrangian density under infinitesimal Lorentz transform

Starting from equation (1.1.8) and using the same procedure as in (1.1.9), let us perform an arbitrary variation of the Lagrangian density $\mathcal{L}$:

$$\delta\mathcal{L} = \frac{\partial\mathcal{L}}{\partial U_{(r)}}\delta U_{(r)} + \frac{\partial\mathcal{L}}{\partial U_{(r),l}}\delta U_{(r),l}$$

$$+\delta U^+_{(r)}\frac{\partial\mathcal{L}}{\partial U^+_{(r)}} + \delta U^+_{(r),l}\frac{\partial\mathcal{L}}{\partial U^+_{(r),l}}, \qquad (1.4.1)$$

or, in view of (1.1.16) and (1.1.17)

$$\delta\mathcal{L} = \pi^+_{l(r),l}\,\delta U_{(r)} + \pi^+_{l(r)}\,\delta U_{(r),l} + \delta U^+_{(r)}\pi_{l(r),l} + \delta U^+_{(r),l}\pi_{l(r)}. \qquad (1.4.2)$$

We next impose the condition that $\mathcal{L}$ is invariant under the infinitesimal Lorentz transformation (1.2.3). Practically, this means that we have to replace $\delta U(r)$, $\delta U_{(r),l}$, $\delta U^+_{(r)}$, $\delta U^+_{(r),l}$ given by (1.2.11) - (1.2.14) into (1.4.2), and equate the result to zero. We then have

$$\omega_{ik}\pi^+_{l(r),l}S_{(r)(p)ik}\,U_{(p)} + \omega_{li}\,\pi^+_{l(r)}U_{(r),i} + \omega_{ik}\pi^+_{l(r)}S_{(r)(p)ik}\,U_{(p),l}$$

$$+\omega_{ik}\,U^+_{(p)}S_{prik}\,\pi_{l(r),l} + \omega_{li}\,U^+_{(r),i}\pi_{l(r)}$$

$$+\omega_{ik}\,U^+_{(p),l}S^+_{(r)(p)ik}\,\pi_{l(r)} = 0. \qquad (1.4.3)$$

Let us change the summation index l with k in the terms 2 and 5 of (1.4.3), and use the fact that ω_{ik} is antisymmetric. Then we can write

$$\omega_{ik}\left[\pi^+_{k(r)}U_{(r),i} + U^+_{(r),i}\pi_{k(r)}\right.$$

12

$$-\frac{\partial}{\partial x_l}\left(\pi^+_{l(r)}S_{(r)(p)ik}U_{(p)} + U(p)^+ S^+_{(r)(p)ik}\pi_{l(r)}\right)\Big]$$

$$= \omega_{ik}\, t_{ik} = 0, \tag{1.4.4}$$

where

$$t_{ik} = \pi^+_{k(r)}U_{(r),i} + U^+_{(r),i}\pi_{k(r)}$$

$$-\frac{\partial}{\partial x_l}\left(\pi^+_{l(r)}S_{(r)(p)ik}U_{(p)} + U^+_{(p)}S^+_{(r)(p)ik}\pi_{l(r)}\right). \tag{1.4.5}$$

Since the tensor ω_{ik} is antisymmetric and arbitrary, (1.4.4) shows that t_{ik} given by (1.4.5) is a symmetric tensor. This can be written in a more compact form by denoting

$$g_{ikl} = -\left(\pi^+_{l(r)}S_{(r)(p)ik}U_{(p)} + U(p)^+ S^+_{(r)(p)ik}\pi_{l(r)}\right). \tag{1.4.6}$$

With this notation, t_{ik} writes

$$t_{ik} = \pi^+_{k(r)}U_{(r),i} + U^+_{(r),i}\pi_{k(r)} + g_{ikl,l}. \tag{1.4.7}$$

We mention that, except for the spinorial field,

$$S^+_{(r)(p)ik} = S_{(r)(p)ik}.$$

To conclude, the Lorentz invariance of the Lagrangian density $\mathcal{L}$ leads to the definition of a second-order symmetric tensor t_{ik}. As we shall show in the following investigation, this tensor plays an essential role in the symmetrization of some fundamental quantities involved in the study of QFT.

1.5. The energy-momentum tensor of a field

One of the basic problems arising in QFT is establishing the general laws describing the space and time variation of some fundamental quantities: energy, momentum, charge, etc. As we shall see, each of these laws of conservation is written as a vanishing four-divergence.

Let us first define the *energy-momentum tensor* of the field. Its conservation law is obtained using the Lagrangian density $\mathcal{L}$, and the basic field equations (1.1.8). Taking the partial derivative of $\mathcal{L}$ with respect to x_i, we have:

$$\frac{\partial \mathcal{L}}{\partial x_i} = \frac{\partial \mathcal{L}}{\partial U_{(r)}}U_{(r),i} + \frac{\partial \mathcal{L}}{\partial U_{(r),k}}U_{(r),ki} + U^+_{(r),i}\frac{\partial \mathcal{L}}{\partial U^+_{(r)}} + U^+_{(r),ki}\frac{\partial \mathcal{L}}{\partial U_{(r),k}}$$

$$= \pi^+_{k(r),k} U_{(r),i} + \pi^+_{k(r)} U_{(r),ki} + U^+_{(r),i}\pi_{k(r),k} + U^+_{(r),ki}\pi_{k(r)}$$

$$= \frac{\partial}{\partial x_k}\left[\pi^+_{k(r)} U_{(r),i} + U^+_{(r),i}\pi_{k(r)}\right]. \tag{1.5.1}$$

On the other hand, one can write

$$\frac{\partial \mathcal{L}}{\partial x_i} = \frac{\partial}{\partial x_k}(\mathcal{L}\delta_{ik}). \tag{1.5.2}$$

Comparing Eqs. (1.5.1) and (1.5.2) we find the equation of conservation

$$\frac{\partial}{\partial x_k}\left(T^{(c)}_{ik}\right) = 0, \tag{1.5.3}$$

with

$$T^{(c)}_{ik} = \pi^+_{k(r)} U_{(r),i} + U^+_{(r),i}\pi_{k(r)} - \mathcal{L}\delta_{ik}. \tag{1.5.4}$$

The tensor defined by (1.5.4) is named *canonical energy-momentum tensor*. To determine the physical significance of its components, let us choose, for example, $i = k = 4$. In this case

$$T^{(c)}_{44} = \pi^+_{4(r)} U_{(r),4} + U^+_{(r),4}\pi_{4(r)} - \mathcal{L}. \tag{1.5.5}$$

Comparing (1.5.5) with the Hamiltonian function defined in analytical mechanics

$$H = \sum_j p_j \dot{q}_j - L,$$

it follows that $T^{(c)}_{44}$ represents the energy density of the field. By analogy with electrodynamics, we can assert that equation (1.5.3) expresses the *law of conservation of energy and momentum of the field*. We also observe that the role of $\dot{q}_j$ is now played by $\partial U_{(r)}/\partial t$ and $\partial U^+_{(r)}/\partial t$, which means that the components of the momentum density are given by (see 1.1.16)

$$\pi^+_{0(r)} = \frac{\partial \mathcal{L}}{\partial(\partial U_{(r)}/\partial t)}, \quad \text{and} \quad \pi_{0(r)} = \frac{\partial \mathcal{L}}{\partial(\partial U^+_{(r)}/\partial t)}, \tag{1.5.6}$$

respectively.

As we also know, the tensor $T^{(c)}_{ik}$ serves to define the momentum four-vector

$$P_i = \frac{i}{c}\int T^{(c)}_{ik}\, dS_k. \tag{1.5.7}$$

14

The energy-momentum tensor of a field

One can observe that the tensor $T_{ik}^{(c)}$ is not uniquely determined by equations (1.5.3) and (1.5.7). Indeed, if we choose

$$T_{ik}'^{(c)} = T_{ik}^{(c)} + f_{ikl,l}', \qquad (1.5.8)$$

where f_{ikl}' is a tensor antisymmetric in the last two indices

$$f_{ikl}' = - f_{ilk}', \qquad (1.5.9)$$

then the new tensor $T_{ik}'^{(c)}$ satisfies the same conservation law (1.5.3), and leads to the same four-momentum (1.5.7). Among all tensors f_{ikl}' satisfying (1.5.9), we are interested only in that tensor (say, f_{ikl}) which makes $T_{ik}'^{(c)}$ symmetric. Denoting this symmetrized tensor by $T_{ik}^{(m)}$, called *metric energy-momentum tensor*, we have

$$T_{ik}^{(m)} = T_{ik}^{(c)} + f_{ikl,l}. \qquad (1.5.10)$$

The symmetrization of the tensor $T_{ik}^{(c)}$ has a special physical importance, because the equation

$$\frac{\partial T_{ik}^{(m)}}{\partial x_k} = 0 \qquad (1.5.11)$$

expresses the law of conservation of *energy, momentum, and angular momentum of the field*.

In view of (1.4.7) and (1.5.4), we may write

$$t_{ik} = T_{ik}^{(c)} + \mathcal{L}\delta_{ik} + g_{ikl,l}, \qquad (1.5.12)$$

where t_{ik} is a symmetric tensor. In this case, the tensor

$$\tau_{ik} = T_{ik}^{(c)} + g_{ikl,l} = t_{ik} - \mathcal{L}\delta_{ik} \qquad (1.5.13)$$

is also symmetric.

Comparing (1.5.10) and (1.5.13) we realize that τ_{ik} would coincide with $T_{ik}^{(m)}$ if g_{ikl} were antisymmetric in the last two indices. Since the definition (1.4.6) of g_{ikl} does not ensure this property, in general τ_{ik} is not the metric energy-momentum tensor, but we can exploit the symmetry of τ_{ik} in order to construct the tensor f_{ikl} with the aforementioned properties, i.e. : 1) $f_{ikl} = -f_{ilk}$, and 2) $T_{ik}^{(m)}$ defined by (1.5.10) is also symmetric.

To this end, let us consider three tensors, antisymmetric in the last two indices, defined as

$$g_{ikl}^{(1)} = g_{ikl} - g_{ilk};$$

$$g_{ikl}^{(2)} = g_{lki} - g_{kli}; \tag{1.5.14}$$

$$g_{ikl}^{(3)} = g_{lik} - g_{kil}.$$

The next step is to look for f_{ikl} as a linear combination of $g_{ikl}^{(\alpha)}$, written as

$$f_{ikl} = a_\alpha \, g_{ikl}^{(\alpha)}; \quad \alpha = 1, 2, 3. \tag{1.5.15}$$

Subtracting (1.5.13) from (1.5.10), we obtain

$$T_{ik}^{(m)} - \tau_{ik} = a_\alpha \, g_{ikl,l}^{(\alpha)} - g_{ikl,l} = \tau'_{ik}. \tag{1.5.16}$$

To solve this problem, we therefore have to find the coefficients a_α so that the tensor τ'_{ik} defined by (1.5.16) becomes symmetric. This means that the coefficients a_α will be found from the condition

$$\tau'_{ik} = \tau'_{ki}. \tag{1.5.17}$$

In view of (1.5.14), (1.5.16), and (1.5.17), we then have:

$$a_1(g_{ikl,l} - g_{ilk,l}) + a_2(g_{lki,l} - g_{kli,l}) + a_3(g_{lik,l} - g_{kil,l}) - g_{ikl,l}$$

$$= a_1(g_{kil,l} - g_{kli,l}) + a_2(g_{lik,l} - g_{ilk,l}) + a_3(g_{lki,l} - g_{ikl,l}) - g_{kil,l},$$

which gives $a_1 = a_2 = a_3 = \frac{1}{2}$, so that

$$f_{ikl} = \frac{1}{2}(g_{ikl} - g_{ilk} + g_{lki} - g_{kli} + g_{lik} - g_{kil}). \tag{1.5.18}$$

Using f_{ikl} given by (1.5.18), the tensor $T_{ik}^{(m)}$ can be symmetrized. If g_{ikl} is antisymmetric in the last two indices, then (1.5.18) gives $f_{ikl} = g_{ikl}$. Indeed,

$$f_{ikl} = \frac{1}{2}(g_{ikl} + g_{ikl} + g_{lki} - g_{kli} - g_{lki} - g_{kli}) = g_{ikl}. \tag{1.5.19}$$

1.6. The angular momentum tensor of a field

Going now back to our discussion in the previous section regarding the conservation of the angular momentum, let us first define some essential quantities. Using (1.5.7), let

$$dP_i = \frac{i}{c} T_{il} \, dS_l \tag{1.6.1}$$

be the components of the infinitesimal momentum. The corresponding infinitesimal angular momentum is then

$$dI_{ik} = x_i \, dP_k - x_k \, dP_i, \tag{1.6.2}$$

or, if we substitute equation (1.6.1),

$$dI_{ik} = \frac{i}{c} \left(x_i \, T_{kl} - x_k \, T_{il} \right) dS_l = J_{ikl} \, dS_l, \tag{1.6.3}$$

where

$$J_{ikl} = \frac{i}{c} \left(x_i \, T_{kl} - x_k \, T_{il} \right) \tag{1.6.4}$$

stands for the *total angular momentum density*.

The law of conservation of angular momentum demands

$$J_{ikl,l} = 0. \tag{1.6.5}$$

In view of (1.6.4) and (1.6.5), we then have

$$\delta_{il} T_{kl} + x_i \, T_{kl,l} - \delta_{kl} T_{il} - x_k T_{il,l} = 0.$$

But the second and fourth terms in this relation cancel due to the equation of conservation (1.5.3), and we are left with $T_{ik} = T_{ki}$. Consequently, the angular momentum is conserved only if T_{ik} is symmetric, i.e. it coincides with the metric tensor $T_{ik}^{(m)}$. This way we have justified the statement that equation (1.5.11) summarizes the laws of conservation of energy, momentum and angular momentum of the field. We therefore can write

$$J_{ikl} = \frac{i}{c} \left(x_i \, T_{kl}^{(m)} - x_k \, T_{il}^{(m)} \right),$$

or, in view of (1.5.10) and (1.5.19):

$$J_{ikl} = \frac{i}{c} \left(x_i \, T_{kl}^{(c)} - x_k \, T_{il}^{(c)} + x_i \, f_{kls,s} - x_k \, f_{ils,s} \right). \tag{1.6.6}$$

General field theory

Recalling that J_{ikl} is the angular momentum density, the angular momentum is found by performing the integral

$$I_{ik} = \int J_{ikl} \, dS_l. \tag{1.6.7}$$

To this end, one observes that the last two terms of (1.6.6) can be written as

$$\begin{cases} x_i f_{kls,s} = \frac{\partial}{\partial x_s}(x_i f_{kls}) - \delta_{is} f_{kls} = \frac{\partial}{\partial x_s}(x_i f_{kls}) - f_{kli} \ ; \\ x_k f_{ils,s} = \frac{\partial}{\partial x_s}(x_k f_{ils}) - \delta_{sk} f_{iks} = \frac{\partial}{\partial x_s}(x_k f_{ils}) - f_{ilk}. \end{cases} \tag{1.6.8}$$

The integration of the four-divergences gives zero

$$\int \frac{\partial}{\partial x_k}(x_i f_{kls}) \, dS_l = \int x_i f_{kls} \, d\sigma_{ls} = 0, \tag{1.6.9}$$

because the integration domain is the whole space, which means that the last integral extends over the bi-dimensional surface situated at infinity. Here

$$dS_l = d\sigma_{ls} \, dx_s : \quad \text{(no summation over s)} . \tag{1.6.10}$$

If we take into account (1.6.8) and (1.6.9), then (1.6.6) yields

$$J_{ikl} = S_{ikl} + M_{ikl}, \tag{1.6.11}$$

where we denoted

$$S_{ikl} = \frac{i}{c}\left(f_{ilk} - f_{kli}\right) = \frac{i}{c}\left(g_{kil} - g_{ikl}\right); \tag{1.6.12}$$

$$M_{ikl} = \frac{i}{c}\left(x_i T_{kl}^{(c)} - x_k T_{il}^{(c)}\right). \tag{1.6.13}$$

According to (1.6.11) - (1.6.13) it follows that we have obtained the total angular momentum density J_{ikl} as a sum of two terms. The first term S_{ikl} does not explicitly depend on the coordinates, but depends on the tensorial nature of the field. It is called *proper angular momentum density*, or *spin*. The second term M_{ikl} does not depend on the tensorial nature of the field, but depends on the coordinates, i.e. on the motion of the field. We shall call it *orbital angular momentum density*, or *current momentum density*.

In view of (1.4.6), relation (1.5.12) yields:

$$S_{ikl} = \frac{i}{c}\left[\pi_{l(r)}^+ S_{(r)(p)ik} \, U_{(p)} + U_{(p)}^+ S_{(r)(p)ik}^+ \, \pi_{l(r)}\right.$$

$$-\pi^{+}_{l(r)} S_{(r)(p)ki}\, U_{(p)} - U^{+}_{(p)} S^{+}_{(r)(p)ki}\, \pi_{l(r)}\big] \qquad (1.6.14)$$

$$= \frac{i}{c}\big[\pi^{+}_{l(r)} \tilde{S}_{(r)(p)ik}\, U_{(p)} + U^{+}_{(p)} \tilde{S}^{+}_{(r)(p)ik}\, \pi_{l(r)}\big],$$

where

$$\begin{cases} \tilde{S}_{(r)(p)ik} = S_{(r)(p)ik} - S_{(r)(p)ki}\ ; \\ \tilde{S}^{+}_{(r)(p)ik} = S^{+}_{(r)(p)ik} - S^{+}_{(r)(p)ki}. \end{cases} \qquad (1.6.15)$$

Next, to write the orbital angular momentum density, we use the definition (1.5.4) of the canonical energy-momentum tensor. After a convenient rearrangement of terms, we get:

$$\begin{cases} T^{(c)}_{kl} = \pi^{+}_{l(r)} U_{(r),k} + U^{+}_{(r),k} \pi_{l(r)} - \mathcal{L}\delta_{kl}, \\ T^{(c)}_{il} = \pi^{+}_{l(r)} U_{(r),i} + U^{+}_{(r),i} \pi_{l(r)} - \mathcal{L}\delta_{il}. \end{cases} \qquad (1.6.16)$$

Substituting (1.6.16) into (1.6.13) and grouping the terms, we get:

$$M_{ikl} = \frac{i}{c}\big[\pi^{+}_{l(r)} \Big(x_i \frac{\partial}{\partial x_k} - x_k \frac{\partial}{\partial x_i}\Big) U_{(r)}$$

$$+ \Big(x_i \frac{\partial}{\partial x_k} - x_k \frac{\partial}{\partial x_i}\Big) U^{+}_{(r)} \pi_{l(r)} - \mathcal{L}\Big(x_i \frac{\partial}{\partial x_k} - x_k \frac{\partial}{\partial x_i}\Big)\big] \qquad (1.6.17)$$

$$= \frac{i}{c}\big[\pi^{+}_{l(r)} \vec{D}_{ik} U_{(r)} + U^{+}_{(r)} \overleftarrow{D}_{ik} \pi_{l(r)} - \mathcal{L}\vec{D}_{ik}\, x_l\big],$$

where

$$D_{ik} = x_i \frac{\partial}{\partial x_k} - x_k \frac{\partial}{\partial x_i} \qquad (1.6.18)$$

is named *the rotation operator*. The direction of the arrow above D_{ik} shows on which quantity the operator acts.

Replacing (1.6.14) and (1.6.17) in the expression for the total angular momentum density, we obtain:

$$J_{ikl} = \frac{i}{c}\Big\{\pi^{+}_{l(r)} \big[\vec{D}_{ik}\delta_{(r),(p)} + \tilde{S}_{(r)(p)ik}\big] U_{(p)}$$

$$+ U^{+}_{(p)} \big[\overleftarrow{D}_{ik}\delta_{(r)(p)} + \tilde{S}^{+}_{(r)(p)ik}\big] \pi_{l(r)} - \mathcal{L}\vec{D}_{ik}\, x_l\Big\} \qquad (1.6.19)$$

$$= \frac{i}{c}\Big\{\pi^{+}_{l(r)} \vec{D}_{(r)(p)ik} U_{(p)} + U^{+}_{(p)} \overleftarrow{D}_{(r)(p)ik} \pi_{l(r)} - \mathcal{L}\vec{D}_{ik}\, x_l\Big\},$$

where

$$D_{(r)(p)ik} = D_{ik}\delta_{(r)(p)} + \tilde{S}_{(r)(p)ik}. \qquad (1.6.20)$$

To display the physical significance of this derivation, let us give an example. As we have seen, the spin S_{ikl} is connected to the tensor rank of the field. In the case of a scalar field, for instance, we have $S_{(r)(p)ik} = 0$, and according to (1.4.6), this gives $g_{ikl} = 0$, meaning that $T_{ik}^{(c)}$ is always symmetric. It also follows, in view of (1.6.14), that the quanta of a scalar field have null spin, and vice-versa. One can then say that the tensor order of a field determines the spin of the quanta of that field. Moreover, the tensor order of a field equals (in units of $\bar{h}$) the spin of the quanta of this field (Marius Fierz, Helvetica Physica Acta, Vol.12, 1939, p.3). Indeed, the quanta of the electromagnetic field, which is a first-order tensor field, are characterized by spin 1, while the gravitational (second-order tensor) field has spin 2, etc.

1.7. Symmetry transformations

In the study of field theory, a central role is played by those transformations which leave the differential equations of motion (Euler-Lagrange equations (1.1.15)) invariant. These transformations are called *symmetry transformations*. In this category fall the space-time transformations (infinitesimal translations and rotations), gauge transformations, etc.

Let us show that the conservation laws obtained in the previous sections can also be established as a consequence of the invariance of the Lagrangian density $\mathcal{L}$ with respect to the space-time symmetry transformation group. This result follows from a very important theorem, which can be stated as follows: *any invariance with respect to a group of transformations represents a conservation law.* This theorem belongs to the German mathematician Emmy Noether (Nachr. Kgl. Wiss. Götingen, 1918, p.235).

To start with, we shall prove that both the existence of $T_{ik}^{(c)}$ and the vanishing of its four-divergence appear as a result of the invariance of $\mathcal{L}$ with respect to an infinitesimal transformation. Consider, in this respect, the infinitesimal transformation

$$x_i \to x_i' = x_i + \xi_i, \tag{1.7.1}$$

or

$$\delta x_i = \xi_i. \tag{1.7.2}$$

To be invariant with respect to (1.7.1), the Lagrangian density must obey

$$\mathcal{L}'(x_i') = \mathcal{L}(x_i), \tag{1.7.3}$$

or, if we take into account (1.7.2) and perform a series expansion about x_i',

$$\mathcal{L}'(x_i') = \mathcal{L}(x_i' - \xi_i) \approx \mathcal{L}(x_i') - \xi_i \frac{\partial \mathcal{L}}{\partial x_i'}, \qquad (1.7.4)$$

where we neglected higher order terms. Since (1.7.4) is valid for any field point, we can abandon the index "prime" and get

$$\delta \mathcal{L} + \xi_i \frac{\partial \mathcal{L}}{\partial x_i} = 0, \qquad (1.7.5)$$

where

$$\delta \mathcal{L} = \mathcal{L}'(x_i) - \mathcal{L}(x_i). \qquad (1.7.6)$$

Noting that the translation (1.7.2) does not imply tensor transformations of the field variables, we can write

$$U'_{(r)}(x_i') = U_{(r)}(x_i), \qquad (1.7.7)$$

or, if a series expansion is performed,

$$\delta U_{(r)} = - U_{(r),i}\, \xi_i; \quad \delta U_{(r),l} = - U_{(r),il}\, \xi_i. \qquad (1.7.8)$$

By virtue of (1.4.2) and (1.7.8), we then have:

$$\left(\pi^+_{l(r),l} U_{(r),i} + \pi^+_{l(r)} U_{(r),il} + U^+_{(r),i} \pi_{l(r),l} + U^+_{(r),il} \pi_{l(r)} - \frac{\partial \mathcal{L}}{\partial x_i} \right) \xi_i = 0. \qquad (1.7.9)$$

Since the infinitesimal quantities ξ_i are arbitrary, we are left with

$$\frac{\partial}{\partial x_l} \left(\pi^+_{l(r)} U_{(r),i} + U^+_{(r),i} \pi_{l(r)} - \mathcal{L}\delta_{il} \right) = 0, \qquad (1.7.10)$$

which is precisely the equation of conservation for the canonic energy-momentum tensor (1.5.3).

Next, we shall prove that both the existence of the angular momentum density J_{ikl} and its conservation appear as a result of the invariance of $\mathcal{L}$ with respect to an infinitesimal rotation of the field. Consider the infinitesimal rotation

$$x_i' = x_i + \omega_{ik}\, x_k; \qquad \delta x_i = \omega_{ik}\, x_k. \qquad (1.7.11)$$

The invariance of the Lagrangian density $\mathcal{L}$ with respect to (1.7.11) demands

$$\mathcal{L}'(x_i') = \mathcal{L}(x_i), \qquad (1.7.12)$$

or, if we use a series expansion, keeping only terms linear in ω_{ik}, and taking into account (1.7.11)

$$\mathcal{L}'(x_i') = \mathcal{L}(x_i') - \omega_{ik}\, x_k' \frac{\partial \mathcal{L}}{\partial x_i'}. \tag{1.7.13}$$

Since (1.7.13) is valid for any field point, we may write

$$\mathcal{L}'(x_i) = \mathcal{L}(x_i) - \omega_{ik}\, x_k \frac{\partial \mathcal{L}}{\partial x_i},$$

or

$$\delta\mathcal{L} = -\omega_{ik}\, x_k \frac{\partial \mathcal{L}}{\partial x_i}. \tag{1.7.14}$$

Unlike translation, the rotation (1.7.11) implies tensor transformations of the field variables. In view of (1.2.8), we have

$$U'_{(r)}(x_j') = U_{(r)}(x_j) + \omega_{ik}\, S_{(r)(p)ik} U_{(p)}(x_j). \tag{1.7.15}$$

Expanding in series, we can write:

$$U'_{(r)}(x_j') = U_{(r)}(x_j' - \omega_{jk}\, x_k') + \omega_{ik} S_{(r)(p)ik} U_{(p)}(x_j')$$

$$= U_{(r)}(x_j') - \omega_{jk}\, x_k' \frac{\partial U_{(r)}}{\partial x_j'} + \omega_{ik} S_{(r)(p)ik} U_{(p)}(x_j'),$$

which means that the variation of the field variables $U_{(r)}$ with regard to the rotation (1.7.11) is

$$\delta U_{(r)} = -\omega_{ik}\left(x_k U_{(r),i} - S_{(r)(p)ik} U_{(p)} \right), \tag{1.7.16}$$

where we dropped the "prime" index of x_j. We have

$$\delta U_{(r),l} = -\omega_{ik}\left[\frac{\partial}{\partial x_l}(x_k\, U_{(r),i}) - S_{(r)(p)ik} U_{(p),l} \right], \tag{1.7.17}$$

as well as

$$\delta U_{(r)}^+ = -\omega_{ik}\left[x_k U_{(r),i}^+ - U_{(p)}^+ S_{(r)(p)ik}^+ \right]; \tag{1.7.18}$$

$$\delta U_{(r),l}^+ = -\omega_{ik}\left[\frac{\partial}{\partial x_l}(x_k U_{(r),i}^+) - U_{(p),l}^+ S_{(r)(p)ik}^+ \right]. \tag{1.7.19}$$

Making allowance for (1.7.16) - (1.7.19), we have:

$$\delta\mathcal{L} = -\omega_{ik}\Big[\pi^+_{l(r),l}x_kU_{(r),i} - \pi^+_{l(r),l}S_{(r)(p)ik}U_{(p)}$$

$$+\pi^+_{l(r)}\frac{\partial}{\partial x_l}\big(x_kU_{(r),i}\big) - \pi^+_{l(r)}S_{(r)(p)ik}U_{(p),l}$$

$$+x_kU^+_{(r),i}\pi_{l(r),l} - U^+_{(p)}S^+_{(r)(p)ik}\pi_{l(r),l}$$

$$+\frac{\partial}{\partial x_l}\big(x_kU^+_{(r),i}\big)\pi_{l(r)} - U^+_{(p),l}S^+_{(r)(p)ik}\pi_{l(r)}\Big] \qquad (1.7.20)$$

$$= -\omega_{ik}\frac{\partial}{\partial x_l}\Big[\pi^+_{l(r)}x_kU_{(r),i} - \pi^+_{l(r)}S_{(r)(p)ik}U_{(p)}$$

$$+x_kU^+_{(r),i}\pi_{l(r)} - U^+_{(p)}S^+_{(r)(p)ik}\pi_{l(r)}\Big].$$

In order to compare (1.7.20) to (1.7.14), it is necessary to write (1.7.14) in a convenient form. One observes that

$$x_k\frac{\partial\mathcal{L}}{\partial x_i} = \frac{\partial}{\partial x_i}(\mathcal{L}x_k) - \mathcal{L}\delta_{ik} = \frac{\partial}{\partial x_l}(\mathcal{L}x_k\delta_{il}) - \mathcal{L}\delta_{ik}. \qquad (1.7.21)$$

Substituting (1.7.21) into (1.7.14), then subtracting the result from (1.7.20), we have:

$$\omega_{ik}\frac{\partial}{\partial x_l}\Big[\pi^+_{l(r)}x_kU_{(r),i} - \pi^+_{l(r)}S_{(r)(p)ik}U_{(p)} + x_kU^+_{(r),i}\pi_{l(r)}$$

$$-U^+_{(p)}S^+_{(r)(p)ik}\pi_{l(r)} - \mathcal{L}x_k\delta_{il}\Big] = \omega_{ik}A_{ik} = 0, \qquad (1.7.22)$$

where we took into account that $\omega_{ik}\delta_{ik} = 0$. Relation (1.7.22) shows that A_{ik} is a symmetric tensor: $A_{ik} = A_{ki}$. We then have

$$\frac{\partial}{\partial x_l}\Big\{\pi^+_{l(r)}\big(x_k\frac{\partial}{\partial x_i} - x_i\frac{\partial}{\partial x_k}\big)U_{(r)} - \pi^+_{l(r)}\big(S_{(r)(p)ik} - S_{(r)(p)ki}\big)U_{(p)}$$

$$+\big(x_k\frac{\partial}{\partial x_i} - x_i\frac{\partial}{\partial x_k}\big)U^+_{(r)}\pi_{l(r)} - U^+_{(p)}\big(S^+_{(r)(p)ik} - S^+_{(r)(p)ki}\big)\pi_{l(r)}$$

$$-\mathcal{L}\big(x_k\frac{\partial}{\partial x_i} - x_i\frac{\partial}{\partial x_k}\big)x_l\Big\} = 0,$$

or, by means of notations (1.6.15), (1.6.18), (1.6.20), and (1.6.21),

$$\frac{\partial}{\partial x_l}\Big\{\pi^+_{l(r)}\overrightarrow{D}_{(r)(p)ik}U_{(p)} + U^+_{(p)}\overleftarrow{D}_{(r)(p)ik}\pi_{l(r)} - \mathcal{L}\overrightarrow{D}_{ik}x_l\Big\} = 0. \quad (1.7.23)$$

We recognize, between the curly brackets, the expression for the angular momentum density tensor (up to the constant factor i/c). We therefore retrieved the conservation law of angular momentum (1.6.5).

1.8. Phase transformations

As mentioned in Section 1.1, the Lagrangian density $\mathcal{L}$ must be a real function. To this end, the field quantities $U_{(r)}, U_{(r)}^+$ on the one hand, and $U_{(r),l}, U_{(r),l}^+$, on the other, have to enter the Lagrangian density only in combinations of the form

$$U_{(r)}^+ U_{(r)} \; ;$$

$$U_{(r),l}^+ U_{(r),l}.$$

This condition is also fulfilled if $U_{(r)}$ and $U_{(r)}^+$ undergo transformations of the form

$$\overline{U}_{(r)} = e^{i\alpha_{(r)}} U_{(r)}, \tag{1.8.1}$$

$$\overline{U}_{(r)}^+ = e^{-i\alpha_{(r)}} U_{(r)}^+, \tag{1.8.1}$$

where $\alpha_{(r)}$ are some arbitrary real constants.

The transformations (1.8.1) and (1.8.2) are called *phase transformations*, while the invariance of the Lagrangian density $\mathcal{L}$ with respect to these transformations is termed *phase invariance*. We mention that there is no summation over (r) in (1.8.1) and (1.8.2).

In what follows we shall study the invariance of the Lagrangian density $\mathcal{L}$ with respect to the infinitesimal phase transformations

$$\overline{U}_{(r)} = U_{(r)} + i\alpha_{(r)} U_{(r)}; \qquad \delta U_{(r)} = i\alpha_{(r)} U_{(r)}, \tag{1.8.3}$$

$$\overline{U}_{(r)}^+ = U_{(r)}^+ - i\alpha_{(r)} U_{(r)}^+; \qquad \delta U_{(r)}^+ = -i\alpha_{(r)} U_{(r)}^+. \tag{1.8.4}$$

These relations are obtained from (1.8.1) and (1.8.2) by performing series expansions on $e^{i\alpha_{(r)}}$ and $e^{-i\alpha_{(r)}}$, and keeping only the terms linear in $\alpha_{(r)}$, e.g.

$$e^{i\alpha_{(r)}} \simeq 1 + i\alpha_{(r)}.$$

The invariance of $\mathcal{L}$ with respect to (1.8.3) and (1.8.4) demands

$$\delta\mathcal{L} = i\alpha_{(r)} \left\{ \pi_{l(r),l}^+ U_{(r)} + \pi_{l(r)}^+ U_{(r),l} - U_{(r)}^+ \pi_{l(r),l} - U_{(r),l}^+ \pi_{l(r)} \right\}$$

$$= i\alpha_{(r)}\frac{\partial}{\partial x_l}\left[\pi^+_{l(r)}U_{(r)} - U^+_{(r)}\pi_{l(r)}\right] = 0.$$

Denoting

$$s_{l(r)} = i\left[\pi^+_{l(r)}U_{(r)} - U^+_{(r)}\pi_{l(r)}\right], \tag{1.8.5}$$

we have

$$s_{l(r),l} = 0. \tag{1.8.6}$$

This equation expresses the conservation law of the quantities

$$Q_{(r)} = \int s_{l(r)}\ dS_l. \tag{1.8.7}$$

If the index (r) takes n values, then equation (1.8.6) expresses the conservation of the n quantities (1.8.7). Performing summation over r in (1.8.5), we get

$$s_l = i\left[\pi^+_{l(r)}U_{(r)} - U^+_{(r)}\pi_{l(r)}\right], \tag{1.8.8}$$

while (1.8.6) yields

$$s_{l,l} = 0. \tag{1.8.9}$$

which is an *equation of continuity*. In this case the quantities s_l signify the components of a *current density*, while

$$Q = \int s_l\ dS_l \tag{1.8.10}$$

represents *the total charge of the field*.

Equation (1.8.8) leads to a very important conclusion. In the case of real fields, the Hermitian conjugate of a quantity coincides with the quantity itself, which means that $s_l = 0$. Consequently, *real fields are chargeless*. This theory predicts, for example, that the quanta of the electromagnetic field, which is a real field, are chargeless. The experiment shows that photons have indeed no charge.

* * *

At the end of this chapter we remind the reader that during our investigation we chose some particular solutions of (1.2.11), in agreement with the tensor order of the studied field. Indeed, we obtained $S_{(r)(p)ik}$, step by step, as being given by (1.3.2), (1.3.6), (1.3.10), and (1.3.31) for each type of field, respectively. One notes that to each of these expressions we can add any tensor $t_{(r)(p)ik}$, symmetric in i and

k (symmetry being demanded by $\omega_{ik}t_{(r)(p)ik} = 0$), but arbitrary in (r) and (p). As easily seen, this choice maintains the validity of (1.2.11). Consequently, the general solution (1.2.11) could be written as

$$S'_{(r)(p)ik} = S_{(r)(p)ik} + t_{(r)(p)ik}. \qquad (1.8.11)$$

By substituting (1.8.11) into (1.6.15), we find

$$\tilde{S}'_{(r)(p)ik} = \tilde{S}_{(r)(p)ik},$$

meaning that the tensor $t_{(r)(p)ik}$ plays no role in the expression of the field angular, and it can be dropped.

CHAPTER II
GENERAL PROBLEMS OF FIELD QUANTIZATION

> *Every act of creation is first an act of destruction.*
>
> Pablo Picasso

2.1. Necessity of field quantization

With all its remarkable achievements, Quantum Mechanics as established in the fundamental papers written by Heisenberg, de Broglie, Schrödinger, Dirac, etc. contains some inconsistencies and internal contradictions. Quantum mechanics deals with quantum processes in which *the number* and *the type* of particles do not vary during their interaction, but only their energy and momentum states. The formalism of Quantum Mechanics is not capable of describing neither the variation of the number of particles, nor the transformation of one type of particle into another type. Even if in some cases the number of particles is allowed to change, this is somewhat forced and inconsistent. For example, Quantum Mechanics is not able to give a satisfactory explanation of the spontaneous emission, in which an electron situated on an excited energy level emits a quanta γ and decays to a lower energy level.

Therefore, it became apparent that a new formalism was necessary, able to explain the transformation of particles. This formalism consists in the quantization of the de Broglie waves associated to particles and it is called *Quantum Field Theory* (QFT).

Let us consider a system of n particles situated in a non-stationary quantum state described by the ket vector $|\psi>$, and let $|\phi_i>$ be the complete orthonormal set of the eigenvectors of the operator $\hat{A}$, associated to a physical observable with a discreet spectrum A. Then we have

$$|\psi> = \sum_i c_i |\phi_i>, \qquad (2.1.1)$$

where c_i are the coefficients of the series expansion of $|\psi>$ in terms of the eigenvectors $|\phi_i >$. According to the usual theory, $c_i^* c_i$ is the probability that, at the moment t, a number n_i of particles out of the total number n are in the state $|\phi_i >$. Since, in view of our hypothesis the state is not stationary, this probability is changing during the process. That is, at least one particle initially in state $|\phi_i >$, after a time interval Δt, passes to another quantum state $|\phi_k >$. The n-norm of the state function $|\psi >$ is

$$< \psi|\psi >= n, \tag{2.1.2}$$

or

$$\sum_i < \phi_i|c_i^* c_i|\phi_i >= \sum_i n_i = n. \tag{2.1.3}$$

Assuming that $|\phi_i >$ is normalized to unity, (2.1.3) yields

$$\sum_i |c_i|^2 = n, \tag{2.1.4}$$

as well as

$$|c_i|^2 = n_i. \tag{2.1.5}$$

It then follows that when a particle passes from state $|\phi_i >$ to $|\phi_k >$, the number of particles $n_i = c_i^* c_i$ decreases by one, while the number of particles $n_k = c_k^* c_k$ increases by one. In other words, we have a *discrete* variation of c_i, leading to a discontinuous variation of $|\psi >$. But traditional Quantum Mechanics considers both $|\psi >$ and c_i as being continuous and even derivable functions. This way we have arrived at the aforementioned contradiction.

The discrete variation of the state function $|\psi >$ and of the coefficients c_i leads to the necessity of their *quantization*. Therefore, they have to be represented by certain operators. For example, to the number n_i of particles in the quantum state $|\phi_i >$ corresponds the operator

$$c_i^+ c_i = n_i. \tag{2.1.6}$$

The above considerations are ingeniously illustrated in an *imaginary experiment* by Louis de Broglie. Consider a system of identical particles described by the vector ket $|\psi >$ whose norm is n (see 2.1.2). Let us also pick two systems of observable quantities, associated with the quantum state of the particles. Consider two such quantities A and B, with their associated operators $\hat{A}$ and $\hat{B}$. We also suppose that the operators $\hat{A}$ and $\hat{B}$ do not commute, i.e. the two systems of quantities

can only be determined with some characteristic 'uncertainties'. Then
we may write

$$\hat{A}\,|\phi_i> = \alpha_i\,|\phi_i>\,; \qquad (2.1.7)$$

$$\hat{B}\,|\chi_k> = \beta_k\,|\chi_k>\,, \qquad (2.1.8)$$

where we assumed that each set of eigenvectors $|\phi_i>$ and $|\chi_k>$ is
complete orthonormal, that is

$$<\phi_i|\phi_{i'}> = \delta_{ii'}, \qquad (2.1.9)$$

$$<\chi_k|\chi_{k'}> = \delta_{kk'}. \qquad (2.1.10)$$

The series expansion of $|\psi>$ can be accomplished either in terms
of the eigenvectors $|\phi_i>$

$$|\psi> = \sum_i c_i\,|\phi_i>, \qquad (2.1.11)$$

or in terms of the eigenvectors $|\chi_k>$

$$|\psi> = \sum_k d_k\,|\chi_k>, \qquad (2.1.12)$$

depending on which system we want to determine, A or B. According
to (2.1.7) and (2.1.8) it follows that the number of particles for which
α_i is a eigenvalue of $\hat{A}$ is

$$c_i^+ c_i = n_i, \quad (\text{ no summation }) \qquad (2.1.13)$$

while the number of particles for which β_k is a eigenvalue of $\hat{B}$ is

$$d_k^+ d_k = n'_k. \qquad (2.1.14)$$

Obviously,

$$\sum_i n_i = \sum_k n'_k = n. \qquad (2.1.15)$$

Recalling the way we defined the systems A and B, the particle
number n_i (or n_k) is subject to one of the two quantum statistics:
Bose-Einstein, or Fermi-Dirac. In other words, n_i can be any natural
number (including zero) in the case of bosons, and zero or one (due to
the Pauli exclusion principle) in the case of fermions.

The transition from representation $|\phi_i>$ to the representation $|\chi_k>$ is accomplished by the unitary matrix s_{ki}

$$|\phi_i> = \sum_k s_{ki} |\chi_k>, \qquad (2.1.16)$$

where

$$\sum_k s^*_{ki} s_{kj} = \delta_{ij}. \qquad (2.1.17)$$

Then (2.1.11) and (2.1.12) yield

$$|\psi> = \sum_i c_i |\phi_i> = \sum_i \sum_k c_i s_{ki} |\chi_k>$$

$$= \sum_k (\sum_i c_i s_{ki}) |\chi_k> = \sum_k d_k |\chi_k>,$$

leading to

$$d_k = \sum_i c_i s_{ki}. \qquad (2.1.18)$$

Suppose that the system of n particles is contained in a finite domain R which moves in a given direction. In the way of this particle beam are placed two devices D_A and D_B (see Fig.2.1) that determine the systems of observables A and B, respectively. The quantities c_i and c_i^+ can be written as

$$c_i = |c_i| e^{i\theta_i}; \quad c_i^+ = |c_i| e^{-i\theta_i}, \qquad (2.1.19)$$

where $|c_i|$ is the amplitude, and θ_i the phase of the quantity c_i. One can discern the following three cases:

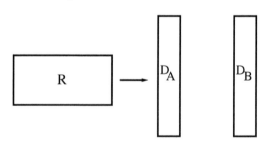

Fig.2.1

a) Only the first device, D_A, is present. In this case we can determine only the number of particles n_i and the amplitudes $|c_i|$ according to (2.1.13)

$$c_i = \sqrt{n_i}, \qquad (2.1.20)$$

but without knowing the phases θ_i.

b) This time only the device D_B is used, and we can determine the number of particles n'_k. In view of (2.1.14), (2.1.18), and (2.1.19), we have:

$$n'_k = \sum_i \sum_j s^*_{ki} s_{kj} c^+_i c_j = \sum_i |s_{ki} c_i|^2 + \sum_{\substack{i,j \\ i \neq j}} s^*_{ki} s_{kj} |c_i| |c_j| e^{i(\theta_j - \theta_i)}.$$

(2.1.21)

Therefore in this case we are able to know the phase difference $\theta_j - \theta_i$, but we cannot know the amplitudes $|c_i|$, because the device D_A is absent.

c) Let us now introduce both devices D_A and D_B, as shown in Fig. 2.1. The device D_A will distinguish n_i particles in the state $|\phi_i >$, described by the vector ket

$$|\psi_i> = c_i |\phi_i> = \sum_k s_{ki} c_i |\chi_k > .$$

(2.1.22)

Among those n_i particles arriving at D_B, there is a number of

$$n'_{ik} = |s_{ki} c_i|^2$$

(2.1.23)

particles having the eigenvalue β_k for B. The total number of particles with eigenvalue β_k is then

$$n''_k = \sum_i n'_{ik} = \sum_i |s_{ki} c_i|^2 = \sum_k |s_{ki}|^2 |c_i|^2 \neq n'_k.$$

(2.1.24)

The total number of particles is

$$\sum_k n''_k = \sum_i \sum_k s^*_{ki} s_{ki} c^+_i c_i = \sum_i c^+_i c_i = \sum_i |c_i|^2 = n.$$

(2.1.25)

Therefore, in this case we are able to determine only the amplitudes.

In a similar way it can be shown that if the two devices are placed in order D_B, D_A, the result is the same: the quantities $|c_i|$ and θ_i cannot be simultaneously determined.

To conclude, the fact that the quantities $|c_i|$ and θ_i cannot be simultaneously determined translates into quantum mechanics language as the non-commutation relation

$$[|c_i|, \theta_i] \neq 0.$$

(2.1.26)

General problems of field quantization

But, if $|c_i|$ and θ_i are operators, then necessarily the state function $|\psi>$ is also an operator. This way, we fall back on the necessity of field quantization.

2.2. Commutation and anti-commutation relations. Emission and absorption operators

For a deeper understanding of the physical meaning of the operator c_i, let us choose the commutator

$$[|c_i|^2, \theta_i] = [n_i, \theta_i] \neq 0. \tag{2.2.1}$$

By virtue of the theoretical knowledge accumulated while studying Quantum Mechanics, we know that if two operators are non-commutative, then one of them could be a differential operator with respect to the quantity expressed by the other. (We recall, for example, Heisenberg's uncertainty relation

$$[p_x, x] = \frac{\hbar}{i},$$

where $p_x \to \frac{\hbar}{i}\frac{\partial}{\partial x}$). Let θ_i be a differential operator, conjugated to the operator n_i. By analogy with the generalized momentum

$$\hat{\vec{P}} = \frac{\hbar}{i}\nabla - e\vec{A} \tag{2.2.2}$$

we shall choose θ_i as

$$\hat{\theta}_i = \frac{\kappa}{i}\frac{\partial}{\partial x_i} + \lambda_i, \tag{2.2.3}$$

where κ is a constant whose physical meaning is going to be found.

Once known the fact that the operators n_i and θ_i do no commute, next step is to set a convenient choice for the order of the modulus and the argument of the operators c_i and c_i^+, so that the relation

$$c_i^+ c_i = |c_i|^2 = n_i$$

can be satisfied. To this end, instead of (2.1.19) we shall choose

$$c_i = e^{i\theta_i}|c_i|; \qquad c_i^+ = |c_i|e^{-i\theta_i}. \tag{2.2.4}$$

Substituting (2.2.3) into (2.2.4), we get

$$c_i = e^{\kappa\frac{\partial}{\partial n_i}}\mu_i|c_i|; \qquad c_i^+ = |c_i|\mu_i^+ e^{-\kappa\frac{\partial}{\partial n_i}}, \tag{2.2.5}$$

32

where

$$\mu_i = e^{i\lambda_i}; \quad \mu_i^+ = e^{-i\lambda_i} \tag{2.2.6}$$

with $\mu_i^+\mu_i = \mu_i\mu_i^+ = 1$.

Let $f(n_i)$ be a function which depends only on the number of particles. Noting that the operator $e^{\kappa\frac{\partial}{\partial n_i}}$ acts only on the quantities with this property, we have:

$$e^{\kappa\frac{\partial}{\partial n_i}} f(n_i) = \Big(1 + \frac{\kappa}{1!}\frac{\partial}{\partial n_i} + \frac{\kappa^2}{2!}\frac{\partial^2}{\partial n_i^2} + \ldots$$

$$+ \frac{\kappa^q}{q!}\frac{\partial^q}{\partial n_i^q} + \ldots\Big) f(n_i) = f(n_i + \kappa). \tag{2.2.7}$$

In a similar way it can be proven that $e^{-\kappa\frac{\partial}{\partial n_i}}$ applied on $f(n_i)$ results in $f(n_i - \kappa)$. Synthesizing these results, we may write

$$e^{\pm\kappa\frac{\partial}{\partial n_i}} f(n_i) = f(n_i \pm \kappa). \tag{2.2.8}$$

Since n_i signifies the number of particles, κ is necessarily *an integer*.

The operator c^+ is called *creation operator*, while c is named *annihilation operator*. Sometimes names like *emission* and *absorption* operators, respectively, are used instead. These names will be justified in what follows.

First, we shall derive the commutation relations satisfied by the creation and annihilation operators for bosons and fermions.

a) If the particles are *bosons*, then n_i can be any integer, including zero. This condition is fulfilled by the choice $\kappa = 1$. Then , using (2.1.20), we have:

$$c_i = e^{\frac{\partial}{\partial n_i}} \mu_i \sqrt{n_i}; \quad c_i^+ = \sqrt{n_i}\mu_i^+ e^{-\frac{\partial}{\partial n_i}}. \tag{2.2.9}$$

Then, in view of (2.2.6) and (2.1.20), we may write

$$c_i c_i^+ = e^{\frac{\partial}{\partial n_i}}(n_i)e^{-\frac{\partial}{\partial n_i}} = n_1 + 1; \quad c_i^+ c_i = \sqrt{n_i}.\sqrt{n_i} = n_i,$$

that is

$$[c_i, c_i^+] = 1, \tag{2.2.10}$$

where 1 is the unit operator. If c_i and c_k^+ are two independent operators, corresponding to two different states, or two different categories of particles, then

$$[c_i, c_k^+] = 0. \tag{2.2.11}$$

The commutation relations (2.2.10) and (2.2.11) can be compressed as

$$[c_i, c_k^+] = \delta_{ik}. \tag{2.2.12}$$

Obviously, we may also write

$$[c_i, c_k] = [c_i^+, c_k^+] = 0. \tag{2.2.13}$$

To obtain (2.2.13) we assumed that μ_i and n_k are independent, and their commutator is zero. Since the only condition satisfied by μ_i is $\mu_i^+ \mu_i = 1$, we may choose $\mu_i = 1$, and then (2.2.6) yields $\lambda_i = 0$.

b) Consider, now, that our particles are *fermions*. Since this type of particles obey Pauli's exclusion principle, the number of particles n_i may take only the values 0 and 1. It consequently results from $(2.2.5)_1$ that, in the case of operator c_i, if $n_i = 1$, only the operator $e^{-\frac{\partial}{\partial n_i}}$ can act, meaning that $\kappa = -1$; if $n_i = 0$, the only operator which can act is $e^{\frac{\partial}{\partial n_i}}$, that is $\kappa = +1$. Synthesizing these two possibilities, we can express κ as a linear function of n_i

$$\kappa = a n_i + b.$$

Taking into account the aforementioned conditions, we must choose $a = -2, b = 1$, that is

$$\kappa = 1 - 2 n_i = \begin{cases} -1 & \text{for} \quad n_i = 1; \\ +1 & \text{for} \quad n_i = 0. \end{cases}$$

Passing now to the operator c_i^+, and applying again Pauli's principle, we observe that, if $n_i = 1$, we must set $-\kappa = -1$, while if $n_i = 0$, we have to set $-\kappa = +1$. Summarizing, we can write

$$-\kappa = \begin{cases} -1 & \text{for} \quad n_i = 1; \\ +1 & \text{for} \quad n_i = 0. \end{cases}$$

This means that in $(2.2.5)_2$ we shall use

$$-\kappa = 1 - 2 n_i.$$

The operators c_i and c_i^+ therefore are

$$\begin{cases} c_i = e^{(1-2n_i)\frac{\partial}{\partial n_i}} \mu_i \sqrt{n_i}; \\ c_i^+ = \sqrt{n_i}\, \mu_i^+ e^{(1-2n_i)\frac{\partial}{\partial n_i}}. \end{cases} \tag{2.2.14}$$

Commutation and anti-commutation relations

Before calculating the products $c_i c_i^+$ and $c_i^+ c_i$, one observes that, in view of (2.2.8), we successively have:

$$\left[e^{(1-2n_i)\frac{\partial}{\partial n_i}} \right]^2 f(n_i) = e^{(1-2n_i)\frac{\partial}{\partial n_i}} \, e^{(1-2n_i)\frac{\partial}{\partial n_i}} \, f(n_i)$$

$$= e^{(1-2n_i)\frac{\partial}{\partial n_i}} \, f(1 - n_i) = f(n_i).$$

But

$$\frac{\partial}{\partial n_i} = \frac{\partial}{\partial(1-n_i)} \frac{\partial(1-n_i)}{\partial n_i} = -\frac{\partial}{\partial(1-n_i)} :$$

that is

$$e^{-(1-2n_i)\frac{\partial}{\partial(1-n_i)}} \, f(1 - n_i) = f[1 - n_i - (1 - 2n_i)] = f(n_i),$$

so that we finally have

$$\left[e^{(1-2n_i)\frac{\partial}{\partial n_i}} \right]^2 = 1. \tag{2.2.15}$$

By virtue of (2.2.6), (2.2.14) and (2.2.15), we can write

$$c_i c_i^+ = e^{(1-2n_i)\frac{\partial}{\partial n_i}} \, \mu_i \, (n_i) \, \mu_i^+ e^{(1-2n_i)\frac{\partial}{\partial n_i}} = n_i + (1 - 2n_i) = 1 - n_i,$$

$$c_i^+ c_i = \sqrt{n_i} . \sqrt{n_i} = n_i,$$

which gives

$$c_i^+ c_i + c_i c_i^+ = \{c_i, c_i^+\} = 1. \tag{2.2.16}$$

The braces of (2.2.16) define an *anti-commutator*. As in the case of bosons, this relation can be generalized for two independent operators c_i and c_k^+:

$$\{c_i, c_k^+\} = 0. \tag{2.2.17}$$

Putting together the anti-commutation rules (2.2.16) and (2.2.17), we have

$$\{c_i, c_k^+\} = \delta_{ik}, \tag{2.2.18}$$

as well as

$$\{c_i, c_k\} = \{c_i^+, c_k^+\} = 0. \tag{2.2.19}$$

In order to satisfy the anti-commutation relations (2.2.17) and (2.2.19), we shall use the idea of P.Jordan and E.P.Wigner (Zeitschift für Physik, vol.47, 1928, p.631) who defined the function μ_i as

$$\mu_i = (-1)^{n_1 + n_2 + \ldots + n_i} = (-1)^{\sum_{l=1}^{i} n_l}. \tag{2.2.20}$$

This choice gives

$$c_i c_i^+ = \left[e^{(1-2n_i)\frac{\partial}{\partial n_i}} \mu_i \sqrt{n_i} \right] \left[\sqrt{n_k} \mu_k^+ e^{(1-2n_k)\frac{\partial}{\partial n_k}} \right]. \qquad (2.2.21)$$

Consider, for example, $i < k$. Since $(-1)^{-2n_i} = 1$ for any n_i, we can write

$$e^{(1-2n_i)\frac{\partial}{\partial n_i}} \mu_k = (-1)^{n_1+...+n_{i-1}+[n_i+(1-2n_i)]+n_{i+1}+...+n_k} = -\mu_k. \qquad (2.2.22)$$

Since $k > i$, the operator $\frac{\partial}{\partial n_k}$ does not act on μ_i. This allows us to change the order of some factors in (2.2.21), to give

$$c_i c_k^+ = \left[-\sqrt{n_k} \mu_k^+ e^{(1-2n_k)\frac{\partial}{\partial n_k}} \right] \left[e^{(1-2n_i)\frac{\partial}{\partial n_i}} \mu_i \sqrt{n_i} \right] = -c_i^+ c_i, \qquad (2.2.23)$$

which is precisely the commutation rule (2.1.18). We also have

$$c_i c_i = \left[e^{(1-2n_i)\frac{\partial}{\partial n_i}} \mu_i \sqrt{n_i} \right] \left[e^{(1-2n_i)\frac{\partial}{\partial n_i}} \mu_i \sqrt{n_i} \right]$$

$$= \left[-\mu_i \sqrt{1-n_i} e^{(1-2n_i)\frac{\partial}{\partial n_i}} \right] \left[e^{(1-2n_i)\frac{\partial}{\partial n_i}} \mu_i \sqrt{n_i} \right] = -\sqrt{n_i(1-n_i)}, \qquad (2.2.24)$$

where Eq.(2.2.15) and relation $\mu_i \mu_i = 1$ have been used.

Examining (2.2.24), we observe that always $c_i c_i = 0$, because n_i can be either 0 or 1; that is, we shall always have $\sqrt{n_i(1-n_i)} = 0$. This way, we proved that the anti-commutation relation (2.2.19) is also satisfied.

2.3. Commutation relations and the Bose-Einstein statistics

As seen in the previous paragraph, in the case of a field whose quanta obey the Bose-Einstein statistics the quantities c_i and c_i^+ are operators and satisfy certain commutation rules. In the following investigation we shall reverse the way of our reasoning: show that the commutations relations lead - without any supplementary hypotheses - to the Bose-Einstein quantum statistics. This method offers a deeper understanding of the operators c_i and c_i^+ and their proper functions and eigenvalues and, more importantly, introduces their matrix representation.

Let us write the commutation relation (2.2.10) as

$$c c^+ - c^+ c = 1, \qquad (2.3.1)$$

Commutation relations and the Bose-Einstein statistics

where, for convenience, we dropped the index i. Multiplying (2.3.1) on the right by c and grouping the terms, we have

$$(c^+ c)\, c = c\,(c^+ c - 1). \qquad (2.3.2)$$

Suppose now that $|\alpha >$ is an eigenvector of the Hermitian operator $c^+ c$, corresponding to the eigenvalue α, that is

$$(c^+ c)\,|\alpha >= \alpha\,|\alpha > . \qquad (2.3.3)$$

Applying (2.3.2) on the eigenvector $|\alpha >$ and taking into account (2.3.3), we have

$$(c^+ c)(c\,|\alpha >) = c\,(c^+ c - 1)\,|\alpha >= (\alpha - 1)\,(c\,|\alpha >). \qquad (2.3.4)$$

The last relation shows that, if $|\alpha >$ is a eigenvector of the operator $c^+ c$ corresponding to the eigenvalue α, then $c\,|\alpha >$ is also a eigenvector of the same operator, belonging to the eigenvalue $(\alpha - 1)$. In other words, $c\,|\alpha >$ is a multiple of the vector $(\alpha - 1)$:

$$c\,|\alpha >= k\,|\alpha - 1 >, \qquad (2.3.5)$$

where k is a constant. This can easily be shown by introducing (2.3.5) into (2.3.4):

$$(c^+ c)\,|\alpha - 1 >= (\alpha - 1)\,|\alpha - 1 > .$$

To determine k we take the Hermitian conjugate of (2.3.5)

$$< \alpha|c^+ =< \alpha - 1|\,k^*, \qquad (2.3.5')$$

and write the scalar product of (2.3.5) and (2.3.5'):

$$< \alpha|c^+ c|\alpha >= |k|^2 \,< \alpha - 1|\alpha - 1 > .$$

If the eigenvectors $|\alpha >$ are orthonormal, taking into account (2.3.3) we find

$$|k|^2 = \alpha. \qquad (2.3.6)$$

Since no other conditions are imposed, we may choose the constant k as being real, and (2.3.5) gives

$$c\,|\alpha >= \sqrt{\alpha}\,|\alpha - 1 > . \qquad (2.3.7)$$

Let us now multiply (2.3.1), on the left, by the operator c^+. Grouping the terms, we obtain

$$(c^+c)\,c^+ = c^+\,(c^+c+1). \qquad (2.3.8)$$

Apply now (2.3.8) to the eigenvector $|\alpha>$ and make use of (2.2.2). The result is:

$$(c^+c)\,(c^+\,|\alpha>) = (c^+\,c+1)\,|\alpha>= (\alpha+1)\,(c^+\,|\alpha>), \qquad (2.3.9)$$

meaning that $c^+\,|\alpha>$ is also a eigenvector of the operator $c^+\,c$, corresponding to the eigenvalue $\alpha+1$. Then we can write

$$c^+\,|\alpha>= k'|\alpha+1>, \qquad (2.3.10)$$

where k' is, again, a constant. To determine k' we take the adjoint of (2.3.10)

$$<\alpha|\,c =<\alpha+1|k'^{*},$$

and write the scalar product

$$<\alpha|c\,c^+|\alpha>= |k'|^2\,<\alpha+1|\alpha+1>,$$

or, in view of (2.3.1)

$$<\alpha|(1+c^+\,c)|\alpha>= (\alpha+1)<\alpha|\alpha>$$

leading to

$$|k'|^2 = \alpha+1. \qquad (2.3.11)$$

By means of (2.3.11), the relation (2.3.10) writes

$$c^+\,|\alpha>= \sqrt{\alpha+1}|\alpha+1> . \qquad (2.3.12)$$

Following (2.3.3), (2.3.4), and (2.3.9) it then follows that $\alpha -1, \alpha, \alpha+1$ are eigenvalues of the operator c^+c, meaning that α *can vary only by unity.* Relation (2.3.6) still shows that always $\alpha \geq 0$, i.e. $c^+\,c$ cannot have negative eigenvalues. There arises the following question: which is the smallest eigenvalue α of the operator c^+c?

In this respect, we remind the reader that, *by definition*, $c|0>= 0$, where $|0>$ is the vacuum state operator. On the other hand, the basic vectors on the Hilbert space can be obtained from the vacuum state $|0>$ by a successive action of the operator c^+:

$$c^+|0>= \sqrt{1}\,|1>,$$

....

$$c^+|\alpha> = (\alpha+1)|\alpha+1>, \quad \text{etc.}$$

Since $\alpha \geq 0$, let us first try the minimum possible value, i.e. $\alpha = 0$. In this case (2.3.1), by means of (2.3.7) and (2.3.12), yields

$$(c c^+ - c^+ c)|0> = c(c^+|0>) - c^+(c|0>) = \sqrt{1}\,c\,|1> -0 = |0>,$$

meaning that the commutation relation (2.3.1) is satisfied by the eigen-value $\alpha = 0$. The eigenvalues taken by α are then

$$\alpha = 0, 1, 2, 3, ...$$

showing that α can be any natural number, including zero. This fact straightforwardly suggests that α is *a number of particles*, $\alpha = n$. The value $\alpha = 0$ corresponds to the vacuum state. The equations (2.3.7) and (2.3.12) then become

$$c\,|n> = \sqrt{n}\,|n-1>, \tag{2.3.13}$$

$$c^+|n> = \sqrt{n+1}\,|n+1>. \tag{2.3.14}$$

One can easily verify that the commutation rule (2.3.1) is satisfied by (2.3.13) and (2.3.14). Indeed,

$$(c c^+ - c^+ c)|n> = \sqrt{n+1}\,c\,|n+1> - \sqrt{n}\,c^+\,|n-1>$$

$$= (\sqrt{n+1})^2|n> -(\sqrt{n})^2\,|n> = |n>.$$

Summarizing, we realize that the commutation rule (2.3.1), among other things, leads to the definition of the vector ket $|n>$, which de-scribes the state of a system of n particles. The action of the operators c and c^+ on the vector $|n>$ leads to the decrease or increase of the number of particles, respectively. For this reason, c^+ can be called *emission operator*, and c - *absorption operator*. Since the number of particles n in a certain quantum state is arbitrary, we conclude that the *commutation rule (2.3.1) necessarily leads to the Bose-Einstein statistics*.

If we consider, for example, a process of transformation of one type of particles into another type, we first have to apply the corre-sponding absorption operator to the state describing the initial type of particles, followed by the second operator applied to the second kind of particles. The emission and absorption operators are usually named *creation* and *annihilation* operators, respectively.

Matrix representation. Consider the complete orthonormal set of vectors $|n>$ and take it as the *basis* for the matrix representation of the operators c^+c, cc^+, c and c^+.

We first note that the operator c^+c is in its proper representation, namely a diagonal matrix, with the diagonal elements being the eigenvalues 0, 1, 2, ...:

$$(c^+c) = \begin{pmatrix} 0 & 0 & 0 & 0 & ... \\ 0 & 1 & 0 & 0 & ... \\ 0 & 0 & 2 & 0 & ... \\ . & . & . & . & ... \end{pmatrix}. \qquad (2.3.15)$$

A matrix element $(n'n)$ of the operator c, in view of (2.3.13), writes

$$(c)_{n'n} = < n'|c|n> = \sqrt{n} <n'|n-1> = \sqrt{n}\, \delta_{n',n-1}. \qquad (2.3.16)$$

As one can see, the elements of the matrix (c) are arranged on a line parallel to the principal diagonal:

$$(c) = \begin{pmatrix} 0 & \sqrt{1} & 0 & 0 & 0 & ... \\ 0 & 0 & \sqrt{2} & 0 & 0 & ... \\ 0 & 0 & 0 & \sqrt{3} & 0 & ... \\ . & . & . & . & . & ... \end{pmatrix}. \qquad (2.3.17)$$

Similarly, in view of (2.3.13), the elements of (c^+) are

$$(c^+)_{n'n} = < n'|c^+|n> = \sqrt{n+1} <n'|n+1> = \sqrt{n+1}\, \delta_{n',n+1}, \qquad (2.3.18)$$

that is

$$(c^+) = \begin{pmatrix} 0 & 0 & 0 & 0 & 0 & ... \\ \sqrt{1} & 0 & 0 & 0 & 0 & ... \\ 0 & \sqrt{2} & 0 & 0 & 0 & ... \\ 0 & 0 & \sqrt{3} & 0 & 0 & ... \\ . & . & . & . & . & ... \end{pmatrix}. \qquad (2.3.19)$$

According to (2.3.17) and (2.3.19), we have

$$(cc^+) = \begin{pmatrix} 1 & 0 & 0 & 0 & 0 & ... \\ 0 & 2 & 0 & 0 & 0 & ... \\ 0 & 0 & 3 & 0 & 0 & ... \\ . & . & . & . & . & ... \end{pmatrix}. \qquad (2.3.20)$$

Subtracting (2.3.15) from (2.3.20), we finally obtain

$$(cc^+) - (c^+c) = \begin{pmatrix} 1 & 0 & 0 & 0 & 0 & \cdots \\ 0 & 1 & 0 & 0 & 0 & \cdots \\ 0 & 0 & 1 & 0 & 0 & \cdots \\ \cdot & \cdot & \cdot & \cdot & & \cdots \end{pmatrix} = I, \qquad (2.3.21)$$

where I is the unit matrix. Therefore, the commutation rule (2.3.1) is satisfied.

2.4. Anti-commutation relations and the Fermi-Dirac statistics

Using the same procedure as in the previous paragraph, let us multiply the anti-commutation relation

$$cc^+ + c^+c = I, \qquad (2.4.1)$$

on the right, by c (see 2.2.16). A convenient arrangements of terms gives

$$(c^+c)c = c(1 - c^+c). \qquad (2.4.2)$$

If $|\alpha >$ is a eigenvector of the operator c^+c, corresponding to a eigenvalue α, then

$$(c^+c)|\alpha >= \alpha|\alpha > . \qquad (2.4.3)$$

Apply now (2.4.2) to $|\alpha >$, and get

$$(c^+c)(c|\alpha >) = c(1 - c^+c)|\alpha >= (1 - \alpha)(c|\alpha > . \qquad (2.4.4)$$

This means that, if $|\alpha >$ is a eigenvector of the operator c^+c, corresponding to the eigenvalue α, then $c|\alpha >$ is also a eigenvector of the same operator, corresponding to the eigenvalue $1 - \alpha$. Then we may write

$$c|\alpha >= k|1 - \alpha > . \qquad (2.4.5)$$

To determine the constant k, we take the adjoint of (2.4.5) and perform the scalar product

$$< \alpha|c^+c|\alpha >= k^2 < 1 - \alpha|1 - \alpha >,$$

which gives

$$|k|^2 = \alpha, \qquad (2.4.6)$$

and (2.4.5) becomes

$$c \, |\alpha >= \sqrt{\alpha} \, |1 - \alpha > . \tag{2.4.7}$$

Next, let us multiply (2.4.1), on the left, by c^+. The result can be written as

$$(c^+ c) c^+ = c^+ (1 - c^+ c). \tag{2.4.8}$$

Apply now (2.4.8) to the eigenvector $|\alpha >$:

$$(c^+ c)(c^+ |\alpha >) = c^+ (1 - c^+ c) |\alpha >= (1 - \alpha)(c^+ |\alpha >),$$

because $c^+ |\alpha >$ is a eigenvector of the operator $c^+ c$, corresponding to the eigenvalue $(1 - \alpha)$. Then we can write

$$c^+ |\alpha >= k' \, |1 - \alpha > . \tag{2.4.9}$$

The constant k' is determined using the already known procedure, taking the scalar product

$$< \alpha | c c^+ |\alpha >=< \alpha | (1 - c^+ c) |\alpha >= |k'|^2 < 1 - \alpha | 1 - \alpha >,$$

which gives

$$|k'|^2 = 1 - \alpha, \tag{2.4.10}$$

and (2.4.9) reads

$$c^+ |\alpha >= \sqrt{1 - \alpha} |1 - \alpha > . \tag{2.4.11}$$

According to (2.4.6) and (2.4.10), $\alpha \geq 0$, and $1 - \alpha \geq 0$, which means

$$0 \leq \alpha \leq 1. \tag{2.4.12}$$

One can observe that the anti-commutation relation

$$(c c^+ + c^+ c) |\alpha >= |\alpha > \tag{2.4.13}$$

is satisfied only for $\alpha = 0$ and $\alpha = 1$, in agreement with (2.4.12). The two eigenvalues 0 and 1 show that $\alpha = n$ is a number of particles and, consequently, *only one particle can exist in a quantum state.*

Using (2.4.7) and (2.4.10), we can still write

$$c \, |n >= \sqrt{n} |1 - n > ; \tag{2.4.14}$$

$$c^+ |n >= \sqrt{1 - n} |1 - n > . \tag{2.4.15}$$

If n takes the values 0 and 1, we find

$$c\,|0>=0; \quad c\,|1>=0; \quad c^+|0>=1; \quad c^+|1>=0.$$

As easily seen, the anti-commutation relation (2.4.13) is verified by (2.4.14) and (2.4.15). Indeed,

$$(cc^+ + c^+c)|n>=\sqrt{1-n}\,c\,|1-n>+\sqrt{n}c^+|1-n>$$

$$=(\sqrt{1-n})^2|n>+(\sqrt{n})^2|n>=|n>.$$

To conclude, the fact that there cannot be more than one particle in a quantum state shows that *the anti-commutation rule (2.4.1) necessarily leads to the Fermi-Dirac statistics*.

Matrix representation. This time, the complete orthonormal set of vectors $|n>$, which we take as the *basis* for the matrix representation of the operators c^+c, cc^+, c and c^+, is formed by two vectors only: $|0>$ and $|1>$. We then have:

$$\begin{cases} (c)_{n'n} =< n'|c|n >= \sqrt{n} < n'|1-n >= \sqrt{n}\,\delta_{n',1-n}; \\ (c^+)_{n'n} =< n'|c^+|n >= \sqrt{1-n} < n'|1-n >= \sqrt{1-n}\,\delta_{n',1-n}, \end{cases}$$
$$(2.4.16)$$

which shows that (c) and (c^+) are second-order square matrices

$$c = \begin{pmatrix} 0 & 1 \\ 0 & 0 \end{pmatrix} \quad ; \quad c^+ = \begin{pmatrix} 0 & 0 \\ 1 & 0 \end{pmatrix}. \qquad (2.4.17)$$

We have

$$(c\,c^+) = \begin{pmatrix} 1 & 0 \\ 0 & 0 \end{pmatrix} \quad ; \quad (c^+c) = \begin{pmatrix} 0 & 0 \\ 0 & 1 \end{pmatrix}.$$

Obviously, the anti-commutation rule (2.4.1) is satisfied:

$$(c\,c^+) + (c^+c) = \begin{pmatrix} 1 & 0 \\ 0 & 1 \end{pmatrix}. \qquad (2.4.18)$$

2.5. Alternative methods of field quantization

In the previous sections of this chapter we justified the necessity of field quantization, by means of the expansion of the state function $|\psi>$ in terms of the eigenfunctions $|\phi_i>$. This way we ascertained that the coefficients c_i intervening in the expansion of $|\psi>$ are operators, that obey a commutation rule for bosons, and an anti-commutation rule for fermions, depending on which category the state vector $|\psi>$ stands for.

The method we have mentioned is not the only possible way of field quantization. In the following, we present two more procedures of field quantization, as generalizations of some methods from Quantum Mechanics. These two alternative methods are, as we shall see, in full agreement with the one already presented. These considerations will support, again, the necessity of the field quantization.

a) One of these methods uses the relation for the time-variation of an operator $\hat{F}$

$$\frac{d\hat{F}}{dt} = \frac{\partial \hat{F}}{\partial t} + \frac{i}{\hbar}[\hat{H}, \hat{F}].\tag{2.5.1}$$

Suppose that $\hat{F} = \hat{F}(p_i, q_i, t)$, where the canonical conjugate variables q_i, p_i satisfy Hamilton's canonical equations

$$\dot{q_i} = \frac{\partial H}{\partial p_i}; \qquad \dot{p_i} = -\frac{\partial H}{\partial q_i}.\tag{2.5.2}$$

We then have

$$\frac{d\hat{F}}{dt} = \frac{\partial \hat{F}}{\partial t} + \dot{q_i}\frac{\partial \hat{F}}{\partial q_i} + \dot{p_i}\frac{\partial \hat{F}}{\partial p_i} = \frac{\partial \hat{F}}{\partial t} + \frac{\partial \hat{H}}{\partial p_i}\frac{\partial \hat{F}}{\partial q_i} - \frac{\partial \hat{H}}{\partial q_i}\frac{\partial \hat{F}}{\partial p_i}.\tag{2.5.3}$$

Comparing (2.5.1) and (2.5.3), we can write

$$\frac{\partial \hat{H}}{\partial p_i}\frac{\partial \hat{F}}{\partial q_i} - \frac{\partial \hat{H}}{\partial q_i}\frac{\partial \hat{F}}{\partial p_i} = \frac{i}{\hbar}[\hat{H}, \hat{F}],\tag{2.5.4}$$

where we did not use the symbol for the Poisson bracket in the l.h.s., in order to avoid any possible confusion with the commutator notation.

Choose, for example, $\hat{F} = q_k$. Then (2.5.4) yields

$$\frac{\partial \hat{H}}{\partial p_k} = \frac{i}{\hbar}[\hat{H}, q_k].\tag{2.5.5}$$

Suppose that q_k is a state function of the form

$$q_k = c_k\, e^{-i\omega t}.\tag{2.5.6}$$

If $\hat{H}$ does not explicitly depend on the time, which is the case for a free field, we obtain

$$p_k = -\int \frac{\partial \hat{H}}{\partial q_k}\, dt = -\int \frac{\partial \hat{H}}{\partial c_k} e^{i\omega t}\, dt = \frac{i}{\omega} f_k\, e^{i\omega t},\tag{2.5.7}$$

where

$$f_k = \frac{\partial \hat{H}}{\partial q_k}. \qquad (2.5.8)$$

With the help of (2.5.7), Eq.(2.5.5) then yields

$$\frac{\partial \hat{H}}{\partial p_k} = -i\omega \frac{\partial \hat{H}}{\partial f_k} e^{-i\omega t} = \frac{i}{\hbar}[\hat{H}, c_k] e^{-i\omega t},$$

or

$$\frac{\partial \hat{H}}{\partial f_k} = -\frac{1}{\hbar\omega}[\hat{H}, \hat{c}_k]. \qquad (2.5.9)$$

For any given field, here is a succession of operations we have to trace. First, the component T_{44} of the energy-momentum tensor is determined. Then, by integrating over the whole three-dimensional space, we obtain the Hamiltonian density H. Next, we proceed to the series expansion of the state vector $|\psi>$, associated to the field quanta, as

$$|\psi> = \sum_k \hat{c}_k |\phi_k>,$$

where we take into account the fact that, if $\hat{c}_k$ is an operator, then $|\psi>$ is also an operator that acts on the number of particles. Thereafter, the quantities f_k and $\frac{\partial \hat{H}}{\partial f_k}$ are determined. The last step is to impose the condition that the quantities $\frac{\partial \hat{H}}{\partial f_k}$, $\hat{H}$, and $\hat{c}_k$ satisfy the commutation rule (2.5.9).

b) Another method of field quantization uses *Heisenberg's commutation relation*

$$[p_i, q_k] = \frac{\hbar}{i}\delta_{ik}, \qquad (2.5.10)$$

where p_i are the momentum components at the point of coordinates q_k, both of them taken at the same instance t. Since our system is a field, instead of q_k we choose the field variable $U_{(k)}$:

$$q_k \equiv U_{(k)}. \qquad (2.5.11)$$

On the other hand, using the definition of the generalized momentum density (see Cap.1, §1)

$$\pi^+_{l(i)} = \frac{\partial \mathcal{L}}{\partial U_{(i),l}}, \qquad (2.5.12)$$

and integrating over the hyperplane $t = $ const., we find the momentum four-vector

$$p_i = \frac{1}{ic}\int \pi^+_{4(i)} dS_4 = \int \pi^+_{0(i)} d\vec{x}; \qquad d\vec{x} = dx\ dy\ dz. \qquad (2.5.13)$$

Replacing (2.5.11) and (2.5.13) into (2.5.10), we obtain the following generalization of Heisenberg's relation

$$[\int \pi_{0(i)}^{+}(\vec{x}, t) \, d\vec{x} \, , \, U_{(k)}(\vec{x}', t)] = \frac{\hbar}{i} \delta_{(i)(k)}. \qquad (2.5.14)$$

Since the quantities $\pi_{0(i)}^{+}$ and $U_{(k)}$ are defined at different places at time t, there is no causal connection between them, and the commutator (2.5.14) differs from zero only for $\vec{x}' = \vec{x}$. Then (2.5.14) can be generalized as

$$[\pi_{0(i)}^{+}(\vec{x}, t) \, , \, U_{(k)}(\vec{x}', t)] = \frac{\hbar}{i} \delta_{(i)(k)} \, \delta(\vec{x} - \vec{x}'), \qquad (2.5.15)$$

were $\delta(\vec{x} - \vec{x}')$ is the Dirac delta function. As shown in the Appendix,

$$\int \delta(\vec{x} - \vec{x}') \, d\vec{x} = 1, \qquad (2.5.16)$$

the integral being extended over the whole three-dimensional space. Relation (2.5.15) can also be considered as the starting point for the field quantization.

At the end of this section we mention that various authors use different methods of field quantization. In the next chapters we shall approach the quantization of some particular fields (scalar, electromagnetic, spinorial) and show that there is a perfect correspondence between all these methods.

2.6. Notations and units in QFT

Before approaching the quantization of some specific fields, we have to adopt certain conventions on the notations and units we are going to use.

Notations. We emphasize that the entire theory is exclusively relativistic. The Latin indices will run from 1 to 4, and the Greek ones from 1 to 3. For example, if we denote $x_1 = x$, $x_2 = y$, $x_3 = z$, $x_4 = ict = i x_0$, then the Minkowski metric writes

$$ds^2 = dx_i dx_i = dx_\alpha dx_\alpha + dx_4 dx_4 = dx_\alpha dx_\alpha - dx_0 dx_0. \qquad (2.6.1)$$

Units. Two constants are going to frequently arise in our investigation: the speed of light c and the Planck action constant $\hbar = h/2\pi$.

The first is directly connected to the relativistic nature of the problem, while the second is related to the field quantization. To simplify calculations, we shall choose

$$\hbar = 1 \quad ; \qquad c = 1. \qquad (2.6.2)$$

To rewrite the final relations in SI (International System of Units), we attach to these units suitable dimensions. We define the LSV fundamental system of dimensions (length-action-velocity). One notes that the choice (2.6.2) does not affect the electric quantities, so we have to work on mechanical quantities only. The conversion from the system LSV to the system LMT is

$$L = L \; ; \quad S = ML^2T^{-1} \; ; \quad V = LT^{-1}, \qquad (2.6.3)$$

while the inverse transformation is given by

$$L = L \; ; \quad M = SL^{-1}V^{-1} \; ; \quad T = LV^{-1}. \qquad (2.6.4)$$

The table below shows the most frequently used physical quantities, expressed in LMT and LSV units:

No	PHYSICAL QUANTITY	LMT	LSV
1	LENGTH	L	L
2	MASS	M	$SL^{-1}V^{-1}$
3	TIME	T	SLV^{-1}
4	VELOCITY	LT^{-1}	V
5	ACTION	ML^2T^{-1}	S
6	ENERGY	ML^2T^{-2}	$SL^{-1}V$
7	MOMENTUM	MLT^{-1}	SL^{-1}
8	ANGULAR MOMENTUM	ML^2T^{-1}	S

To exemplify the conversion from one system (LSV) to the other (LMT), let us consider the following two cases:

1. In LSV units the energy-momentum relation writes

$$E^2 = p^2 + m^2. \qquad (2.6.5),$$

where E is the energy of the particle, p - its momentum, and m - its rest mass. This relation can also be written as

$$E^2 = k_1\,p^2 + k_2\,m^2, \qquad (2.6.6)$$

where k_1 and k_2 are two constants having value one, but whose dimensions are different from one. The corresponding dimension equation in LSV is

$$S^2 T^{-2} V^2 = [k_1] S^2 L^{-2} + [k_2] S^2 L^{-2} V^{-2},$$

meaning that

$$[k_1] = V^2 \; ; \quad [K_2] = V^4.$$

But the only quantity in LSV with value one and dimension V is c, therefore the energy-momentum relation reads

$$E^2 = p^2 c^2 + m^2 c^4. \tag{2.6.7}$$

2. In LSV, the Klein-Gordon-Schrödinger equation writes

$$(\Box - m^2)|\psi> = 0, \tag{2.6.8}$$

where $\Box$ is the D'Alembertian operator

$$\Box = \frac{\partial}{\partial x_i} \frac{\partial}{\partial x_i}. \tag{2.6.9}$$

Since $[\Box] = L^{-2}$, we must have

$$[k] \, [m^2] = L^{-2}, \tag{2.6.10}$$

k being a constant whose dimension is now different from 1. Relation (2.6.10) gives

$$[k] = S^{-2} V^2$$

that is

$$k = \frac{c^2}{\hbar^2}, \tag{2.6.11}$$

and the the Klein-Gordon-Schrödinger equation in LMT units reads

$$\left(\Box - \frac{m^2 c^2}{\hbar^2}\right)|\psi> = 0. \tag{2.6.12}$$

CHAPTER III

THE QUANTIZATION OF THE SCALAR FIELD

*The great revelation of the quantum the-
ory was that features of discreteness were
discovered in the Book of Nature...*

Erwin Schrödinger

3.1. The Lagrangian formalism

Based on the results accumulated in the first two chapters, we shall
continue our investigation with the quantization of the *free scalar field*.
More precisely, the scalar field we are going to quantize is the one
described by the Klein-Gordon-Schrödinger equation.

This starting point is suggested by the fact that the quantization
of the scalar field is relatively simple as compared to the other fields.
Moreover, we shall use these results in the context of our further investigations.

As we remember, no scalar field has a proper angular momentum
(see Chap.I, § 6), which means that the quantization process will lead
to *spinless quanta* associated with the field. As any other scalar field,
the quanta associated with the field described by the Klein-Gordon-
Schrödinger equation are spinless. These quanta are named *mesons*.

The Klein-Gordon-Schrödinger equation emerges from the energy-
momentum relation (2.6.7) which, in our conventional unit system
LSV, can be written as

$$E^2 - \vec{p}^2 - m^2 = 0, \qquad (3.1.1)$$

where E, $\vec{p}$, m are the energy, the momentum and the rest mass of the
particle, respectively. Let us now define the momentum four-vector
p_i, as

$$p_1 = p_x, \quad p_2 = p_y, \quad p_3 = p_z, \quad p_4 = ip_0 = iE. \tag{3.1.2}$$

Equation (3.3.3) then becomes

$$\sum_{\alpha=1}^{3} p_\alpha^2 + p_4^2 + m^2 = \sum_{i=1}^{4} p_i^2 + m^2 = 0. \tag{3.1.3}$$

If, instead of $\sum_{i=1}^{4} p_i^2$ we use the notation p^2, the energy-momentum relation (3.1.3) finally reads

$$p^2 + m^2 = 0. \tag{3.1.4}$$

On the other hand, from Quantum Mechanics the reader is familiar with the components of the momentum four-vector operator

$$\hat{p}_\alpha = -i\frac{\partial}{\partial x_\alpha} \; ; \quad \hat{p}_4 = iE = -\frac{\partial}{\partial t} = -i\frac{\partial}{\partial x_4}. \tag{3.1.5}$$

For convenience, in what follows we shall use either Greek indices, or arrows to designate the three-dimensional vectors. We also abandon the notation $\char"5E$ for the operators, and the symbols $<|, |>$ which designate the bra- and ket-vectors, respectively. In view of (3.1.5), we then have

$$p^2 = p_i p_i = -\frac{\partial}{\partial x_\alpha}\frac{\partial}{\partial x_\alpha} - \frac{\partial}{\partial x_4}\frac{\partial}{\partial x_4} = -\frac{\partial}{\partial x_i}\frac{\partial}{\partial x_i} = -\partial_i \partial_i = -\Box. \tag{3.1.6}$$

The Klein-Gordon-Schrödinger equation is then obtained by applying (3.1.4) to the state function ψ:

$$(\Box - m^2)\psi = 0. \tag{3.1.7}$$

Since ψ is a scalar function, its Hermitian conjugate ψ^+ satisfies the same equation

$$(\Box - m^2)\psi^+ = 0. \tag{3.1.8}$$

To find the fundamental properties (energy, momentum, etc.) of the field, it is necessary to construct the Lagrangian density first. If ψ and ψ^+ are taken as independent field variables, then the complex scalar field is described by

$$\mathcal{L} = -\psi_{,i}^+ \psi_{,i} - m^2 \psi^+ \psi, \tag{3.1.9}$$

The Lagrangian formalism

where

$$\psi_{,i} = \frac{\partial \psi}{\partial x_i}. \tag{3.1.10}$$

Before going further, let us verify that the Lagrangian density $\mathcal{L}$ and Euler-Lagrange equations (1.1.15) lead, without any other supplementary conditions, to the Klein-Gordon-Schrödinger equation (3.1.7). Indeed, since

$$\frac{\partial \mathcal{L}}{\partial \psi^+_{,i}} = -\psi_{,i} \; ; \quad \frac{\partial}{\partial x_i}\left(\frac{\partial \mathcal{L}}{\partial \psi^+_{,i}}\right) = -\psi_{ii} = -\Box \psi \; ; \quad \frac{\partial \mathcal{L}}{\partial \psi^+} = -m^2 \psi,$$

it is easily seen that the Euler-Lagrange equation written for ψ^+

$$\frac{\partial \mathcal{L}}{\partial \psi^+} - \frac{\partial}{\partial x_i}\left(\frac{\partial \mathcal{L}}{\partial \psi^+_{,i}}\right) = 0$$

leads to (3.1.7). Similarly, it can be shown that the equation (3.1.8) is obtained by means of

$$\frac{\partial \mathcal{L}}{\partial \psi} - \frac{\partial}{\partial x_i}\left(\frac{\partial \mathcal{L}}{\partial \psi_{,i}}\right) = 0.$$

In view of (1.1.7) and (3.1.9), the momentum densities are

$$\pi_i = \frac{\partial \mathcal{L}}{\partial \psi^+_{,i}} = -\psi_{,i} \; ; \quad \pi_i^+ = \frac{\partial \mathcal{L}}{\partial \psi_{,i}} = -\psi_{,i}^+. \tag{3.1.11}$$

In view of the considerations developed in the last paragraph of Chapter I, charged mesons are described by complex scalar fields, because in this case the current density four vector is non-zero ($s_l \neq 0$; (see (1.8.8)), while neutral mesons are quanta of the real scalar fields, with $s_l = 0$.

To write the Lagrangian density of a real scalar field, we first note that here the field functions ψ and ψ^+ cannot be considered as being independent, which means that the number of field variables is halved. Then we set everywhere $\psi^+ = \psi$, and the Lagrangian density writes

$$\mathcal{L} = -\frac{1}{2}\psi_{,i}\psi_{,i} - \frac{1}{2}m^2\psi^2. \tag{3.1.12}$$

Using the general framework developed in Chapter I, we are now able to write the fundamental quantities associated with the complex

51

The quantization of the scalar field

scalar field (energy, momentum, and angular momentum). The expressions of these quantities for the real scalar field will then follow without any difficulty.

Momentum four-vector. First of all, we notice that the tensor $T_{ik}^{(c)}$ is already symmetric (there is no need to be symmetrized), while the scalar field quanta are spinless. According to (1.5.4) we then have:

$$T_{ik}^{(c)} = T_{ik}^{(m)} = T_{ik} = \pi_k^+ \psi_{,i} + \psi_{,i}^+ \pi_k - \mathcal{L}\delta_{ik}, \qquad (3.1.13)$$

or, in view of (3.1.9) and (3.1.11),

$$T_{ik} = -\psi_{,i}^+ \psi_{,k} - \psi_{,k}^+ \psi_{,i} + (\psi_{,l}^+ \psi_{,l} + m^2 \psi^+ \psi)\,\delta_{ik}. \qquad (3.1.14)$$

The momentum four-vector, in SLV unit system, is (see (1.5.7))

$$P_i = i \int T_{ik}\, dS_k, \qquad (3.1.15)$$

where dS_k is and element of three-dimensional hypersurface, bounding the four-dimensional domain Ω. Here we consider the momentum at a given time $t = const.$, in which case the only nonzero surface element is

$$dS_4 = dx_1\, dx_2\, dx_3 = d\vec{x}. \qquad (3.1.16)$$

Then

$$P_i = i \int T_{i4}\, d\vec{x}, \qquad (3.1.17)$$

where the integration is performed over the usual three-dimensional space. The spatial components of the momentum four-vector are then

$$P_\alpha = i \int T_{\alpha 4}\, d\vec{x} = -\int (\psi_{,\alpha}^+ \psi_{,4} + \psi_{,4}^+ \psi_{,\alpha})\, d\vec{x}, \qquad (3.1.18)$$

while the temporal component is

$$P_4 = i\,E = i\,H = i \int T_{44}\, d\vec{x}, \qquad (3.1.9)$$

in which $H = E$ is the total energy of the field. In view of (3.1.14), we still have

$$T_{44} = -\psi_{,4}^+ \psi_{,4} - \psi_{,4}^+ \psi_{,4} + \psi_{,\alpha}^+ \psi_{,\alpha} + \psi_{,4}^+ \psi_{,4} + m^2 \psi^+ \psi$$

$$= \psi_{,\alpha}^+ \psi_{,\alpha} - \psi_{,4}^+ \psi_{,4} + m^2 \psi^+ \psi,$$

and (3.1.19) yields

$$H = \int (\psi^+_{,\alpha}\psi_{,\alpha} - \psi^+_{,4}\psi_{,4} + m^2\psi^+\psi)\, d\bar{x}. \tag{3.1.20}$$

Charge. According to (1.8.8), the current density is

$$s_l = i\,(\pi^+_l\psi - \psi^+\pi_l) = i\,(\psi^+\psi_{,l} - \psi^+_{,l}\psi),$$

where (3.1.11) has been considered. If we want (1.8.8) to represent the electric current density, then we have to include the elementary electric charge $e = 1.6 \times 10^{-19}C$ in s_l, that is

$$s_l = i\,e\,(\psi^+\psi_{,l} - \psi^+_{,l}\psi). \tag{3.1.21}$$

The *total charge* of the field, in view of (1.8.10), is

$$Q' = \int s_l\, dS_l.$$

Assuming $t = const.$ and choosing $s_1 = s_x$, $s_2 = s_y$, $s_3 = s_z$, $s_4 = i\rho_e$, where ρ_e is the electric charge density, we obtain the total charge of the mesonic field as

$$Q = \int \rho_e\, d\bar{x} = \frac{1}{i}\int s_4\, d\bar{x} = e\int (\psi^+\psi_{,4} - \psi^+_{,4}\psi)\, d\bar{x}. \tag{3.1.22}$$

3.2. Momentum representation

To establish the fundamental properties of the complex scalar field (energy, charge), in the previous section we worked in the *coordinate representation* (or *coordinate configuration*). This means that the field variables were considered functions of the independent variables x_k, while the integrations have been performed over either a hypersurface, or a domain of the space defined by x_k.

In most cases, our experimental opportunities do not allow us to determine the coordinates x_k, but rather the conjugate momenta p_k. It is then more convenient to express the aforementioned fundamental quantities using the *momentum space* (*momentum representation*). The transition from one representation to the other is performed by means of the four-dimensional Fourier transform

$$\psi(x) = \frac{1}{(2\pi)^2 i}\int \psi(p)\, e^{ipx}\, dp, \tag{3.2.1}$$

The quantization of the scalar field

where a factor of $(2\pi)^{-1/2}$ for each dimension has been counted. In (3.2.1) x and p are four-vectors, that is

$$px = \vec{p} \cdot \vec{x} + p_4 x_4; \qquad (3.2.2)$$

$$dp = dp_1 \, dp_2 \, dp_3 \, dp_4, \qquad (3.2.3)$$

while i in (3.2.1) has been inserted to compensate i of $dp_4 = i \, dE$.

To obtain the inverse transform of (3.2.1), we multiply it by $[1/(2\pi)^2 i]e^{-ip'x}$ and integrate over the whole configuration space. We have:

$$\frac{1}{(2\pi)^2 i} \int e^{-ip'x} \, \psi(x) \, dx = \frac{1}{i} \int \psi(p) \left[\frac{1}{(2\pi)^4 i} \int e^{i(p-p')x} \, dx \right] dp.$$

But, in view of (A.44)

$$\frac{1}{(2\pi)^2 i} \int e^{-ip'x} \, \psi(x) \, dx = \frac{1}{i} \int \psi(p)\delta(p - p') \, d\vec{p} \, i \, dp_0. \qquad (3.2.4)$$

Using (A.4), we have

$$\psi(p) = \frac{1}{(2\pi)^2 i} \int e^{-ipx} \, \psi(x) \, dx, \qquad (3.2.5)$$

where p' has been replaced by p.

Using the momentum representation, let us consider that the state function $\psi(x)$ is a solution of the Klein-Gordon-Schrödinger equation. Imposing condition that $\psi(x)$ given by (3.2.1) must be a solution of the equation (3.1.7), we get

$$(\Box - m^2) \, \psi(x) = \frac{1}{(2\pi)^2 i} \int \psi(p)(\Box - m^2)e^{ipx} \, dp = 0, \qquad (3.2.6)$$

where the fact that x and p are independent variables and, therefore, the D'Alembertian operator does not act on $\psi(p)$, has been taken into account. We have

$$\Box \, e^{ipx} = \frac{\partial^2}{\partial x_l \partial x_l} \, e^{ip_k x_k} = - \, p_l p_l \, e^{ip_k x_k} = - \, p^2 \, e^{ipx}, \qquad (3.2.7)$$

and (3.2.6) yields

$$\int \psi(p) \, (p^2 + m^2) \, e^{ipx} \, dp = 0. \qquad (3.2.8)$$

Since the functions e^{ipx} are linearly independent, this relation is satisfied if and only if

$$\psi(p)\,(p^2 + m^2) = 0. \tag{3.2.9}$$

This equation displays two possibilities:

$$\psi(p) = 0 \;;\quad p^2 + m^2 \neq 0;$$

$$\psi(p) \neq 0 \;;\quad p^2 + m^2 = 0.$$

In other words, the state function $\psi(p)$ is zero only for situations in which energy-momentum relation (3.1.4) is not satisfied, and vice-versa. As known, energy-momentum relation can be written as

$$\vec{p}^2 - E^2 + m^2 = 0, \tag{3.2.10}$$

or

$$E = \pm\sqrt{\vec{p}^2 + m^2}. \tag{3.2.11}$$

As will be shown, this equation represents a family of generalized two-sheet hyperboloids. Therefore, $\psi(x)$ differs from zero on a measureless domain in momentum space, i.e. on a hypersurface in this space.

To fulfill all these conditions and, in addition, have the integral in (3.2.1) be different from zero, we choose $\psi(p)$ as

$$\psi(p) = \sqrt{2\pi}\,\delta(p^2 + m^2)\,\varphi(p). \tag{3.2.12}$$

The factor $\sqrt{2\pi}$ appearing in (3.2.12) will be further justified. Expression (3.2.1) then becomes

$$\psi(x) = \frac{1}{(2\pi)^{3/2}i} \int \varphi(p)\,\delta(p^2 + m^2)\,e^{ipx}\,dp. \tag{3.2.13}$$

Denoting by E_p the absolute value of the square root in (3.2.11)

$$E = \pm E_p = \pm\sqrt{\vec{p}^2 + m^2}, \tag{3.2.14}$$

we can write

$$p^2 + m^2 = E^2 - E_p^2 = (E - E_p)(E + E_p) \tag{3.2.15}$$

and (3.2.13) becomes

$$\psi(x) = \frac{1}{(2\pi)^{3/2}i} \int \varphi(p)\,\delta[(E - E_p)(E + E_p)]\,e^{ipx}\,dp. \tag{3.2.16}$$

The quantization of the scalar field

According to (A.9), we have

$$\delta[(E - E_p)(E + E_p)] = \frac{\delta(E - E_p) + \delta(E + E_p)}{2E_p}. \qquad (3.2.17)$$

If we now expand the vector notations, using the spatial and temporal components

$$dp = d\vec{p} \cdot i \, dE \; ; \quad \varphi(p) = \varphi(\vec{p}, E) \; ; \quad ipx = i(\vec{p} \cdot \vec{x} - Et), \qquad (3.2.18)$$

and substitute (3.2.17) and (3.2.18) into (3.2.16), we obtain

$$\psi(x) = \frac{1}{(2\pi)^{3/2}} \int \frac{1}{2E_p} \varphi(\vec{p}, E) \, \delta(E - E_p) \, e^{i(\vec{p} \cdot \vec{x} - Et)} \, d\vec{p} \, dE$$

$$+ \frac{1}{(2\pi)^{3/2}} \int \frac{1}{2E_p} \varphi(\vec{p}, E) \, \delta(E + E_p) \, e^{i(\vec{p} \cdot \vec{x} - Et)} \, d\vec{p} \, dE, \qquad (3.2.19)$$

or, integrating with respect to E,

$$\psi(x) = \frac{1}{(2\pi)^{3/2}} \int \frac{1}{2E_p} \varphi(\vec{p}, E_p) \, e^{i(\vec{p} \cdot \vec{x} - E_p t)} \, d\vec{p}$$

$$+ \frac{1}{(2\pi)^{3/2}} \int \frac{1}{2E_p} \varphi(\vec{p}, - E_p) \, e^{i(\vec{p} \cdot \vec{x} + E_p t)} \, d\vec{p}. \qquad (3.2.20)$$

Here we see the justification for the factor $\sqrt{2\pi}$ in (3.2.13): since our integral is triple and the Fourier transform introduces $(2\pi)^{-1/2}$ for each dimension, one needs an extra $\sqrt{2\pi}$ to balance the terms.

Take now $-\vec{p}$ instead of $\vec{p}$ in the second integral of (3.2.0), and denote

$$\varphi(\vec{p}, E_p) = \bar{\varphi}_+(\vec{p}) \; ; \quad \varphi(\vec{p}, -E_p) = \bar{\varphi}_-(\vec{p}). \qquad (3.2.21)$$

Then (3.2.20) yields

$$\psi(x) = \frac{1}{(2\pi)^{3/2}} \int \frac{1}{2E_p} \bar{\varphi}_+(\vec{p}) e^{i(\vec{p} \cdot \vec{x} - E_p t)} \, d\vec{p}$$

$$+ \frac{1}{(2\pi)^{3/2}} \int \frac{1}{2E_p} \bar{\varphi}_-(\vec{p}) \, e^{-i(\vec{p} \cdot \vec{x} - E_p t)} \, d\vec{p}$$

$$= \frac{1}{(2\pi)^{3/2}} \int \frac{1}{2E_p} \bar{\varphi}_+(\vec{p}) \, e^{ipx} \, d\vec{p} + \frac{1}{(2\pi)^{3/2}} \int \frac{1}{2E_p} \bar{\varphi}_-(\vec{p}) \, e^{-ipx} \, d\vec{p}.$$

$$(3.2.22)$$

Here we have used $\bar{\varphi}$ instead of φ, and kept the notation φ for other purposes.

Denoting

$$\psi_\pm(x) = \frac{1}{(2\pi)^{3/2}} \int \frac{1}{2E_p} \bar{\varphi}_\pm(\vec{p}) \, e^{\pm ipx} \, d\vec{p}, \tag{3.2.23}$$

we have

$$\psi(x) = \psi_+(x) + \psi_-(x). \tag{3.2.24}$$

Functions $\psi_+(x)$ and $\psi_-(x)$ are called *positive and negative frequency parts of* $\psi(x)$, respectively.

Taking the Hermitian conjugate of $\psi(x)$ given by (3.2.22), we have:

$$\psi(x)^+ = \frac{1}{(2\pi)^{3/2}} \int \frac{1}{2E_p} [\bar{\varphi}_+(\vec{p})]^+ \, e^{-ipx} \, d\vec{p}$$

$$+ \frac{1}{(2\pi)^{3/2}} \int \frac{1}{2E_p} [\bar{\varphi}_-(\vec{p})]^+ \, e^{ipx} \, d\vec{p}. \tag{3.2.25}$$

On the other hand, according to (3.2.1), the Fourier transform of $\psi^+(x)$ is

$$\psi^+(x) = \frac{1}{(2\pi)^2 i} \int \psi^+(p) \, e^{ipx} \, dp, \tag{3.2.26}$$

where $\psi^+(p) \neq [\psi(p)]^+$. Following the same steps as before, one finds

$$\psi^+(x) = \frac{1}{(2\pi)^{3/2}} \int \frac{1}{2E_p} \bar{\varphi}_+^+(\vec{p}) \, e^{ipx} \, d\vec{p}$$

$$+ \frac{1}{(2\pi)^{3/2}} \int \frac{1}{2E_p} \bar{\varphi}_-^+(\vec{p}) \, e^{-ipx} \, d\vec{p}$$

$$= \psi_+^+(x) + \psi_-^+(x). \tag{3.2.27}$$

Equating (3.2.25) with (3.2.27), we finally obtain

$$[\bar{\varphi}_+(\vec{p})]^+ = \bar{\varphi}_-^+(\vec{p}) \; ; \quad [\bar{\varphi}_-(\vec{p})]^+ = \bar{\varphi}_+^+(\vec{p}),$$

or, written in a condensed form

$$[\bar{\varphi}_\pm(\vec{p})]^+ = \bar{\varphi}_\mp^+(\vec{p}). \tag{3.2.28}$$

3.3. Momentum , energy and charge of the complex scalar field in momentum representation

a) **Momentum.** As seen in (3.1.18), to calculate the spatial components P_α of the momentum one must know $\psi_{,\alpha}$ and $\psi_{,4}$. In view of (3.2.23), we have

$$\psi_{\pm,\alpha}(x) = \frac{1}{(2\pi)^{3/2}} \int \frac{\pm i p_\alpha}{2E_p} \, \bar\varphi_\pm(\vec p) \, e^{\pm i p x} \, d\vec p; \qquad (3.3.1)$$

$$\psi_{\pm,4}(x) = \frac{1}{(2\pi)^{3/2}} \int \frac{\mp E_p}{2E_p} \, \bar\varphi_\pm(\vec p) \, e^{\pm i p x} \, d\vec p; \qquad (3.3.2)$$

while the Hermitian conjugates are

$$\psi_{\pm,\alpha}^+(x) = \frac{1}{(2\pi)^{3/2}} \int \frac{\pm i p_\alpha}{2E_p} \, \bar\varphi_\pm^+(\vec p) \, e^{\pm i p x} \, d\vec p \; ; \qquad (3.3.3)$$

$$\psi_{\pm,4}^+(x) = \frac{1}{(2\pi)^{3/2}} \int \frac{\mp E_p}{2E_p} \, \bar\varphi_\pm^+(\vec p) \, e^{\pm i p x} \, d\vec p \; . \qquad (3.3.4)$$

By means of (3.2.24) and (3.3.1)-(3.3.4), relation (3.1.18) becomes

$$P_\alpha = -i \int [(\psi_{+,4}^+ + \psi_{-,4}^+)(\psi_{+,\alpha} + \psi_{-,\alpha})$$

$$+ (\psi_{+,\alpha}^+ + \psi_{-,\alpha}^+)(\psi_{+,4} + \psi_{-,4})] \, d\vec x$$

$$= P_{\alpha 1+} + P_{\alpha 1-} + P_{\alpha 2+} + P_{\alpha 2-}, \qquad (3.3.5)$$

where the following notations have been used

$$P_{\alpha 1\pm} = -i \int (\psi_{\pm,4}^+ \psi_{\pm,\alpha} + \psi_{\pm,\alpha}^+ \psi_{\pm,4}) \, d\vec x \; ; \qquad (3.3.6)$$

$$P_{\alpha 2\pm} = -i \int (\psi_{\pm,4}^+ \psi_{\mp,\alpha} + \psi_{\pm,\alpha}^+ \psi_{\mp,4}) \, d\vec x. \qquad (3.3.7)$$

Let us first calculate $P_{\alpha 1+}$. We can write:

$$P_{\alpha 1+} = -\frac{i}{(2\pi)^3} \int \left[-\frac{1}{2} \, \bar\varphi_+^+(\vec p) \frac{i p_\alpha'}{2E_{p'}} \, \bar\varphi_+(\vec p') \, e^{i(p+p')x} \right.$$

$$\left. + \frac{i p_\alpha}{2E_p} \, \bar\varphi_+^+(\vec p) \left(-\frac{1}{2} \right) \bar\varphi_+(\vec p') \, e^{i(p+p')x} \right] d\vec x \, d\vec p \, d\vec p'.$$

Momentum, energy and charge

Since

$$i(p + p')x = i(\vec{p} + \vec{p}') \cdot \vec{x} - i(E_p + E_{p'})t, \tag{3.3.8}$$

we have

$$P_{\alpha 1+} = -\frac{1}{2} \int \left\{ \frac{1}{(2\pi)^3} \int e^{i(\vec{p}+\vec{p}')\cdot\vec{x}} \, d\vec{x} \left[\frac{p'_\alpha}{2E_{p'}} \, \bar{\varphi}^+_+(\vec{p}) \, \bar{\varphi}_+(\vec{p}') \right. \right.$$

$$\left. \left. + \frac{p_\alpha}{2E_p} \, \bar{\varphi}^+_+(\vec{p}) \, \bar{\varphi}_+(\vec{p}') \right] e^{-i(E_p+E_{p'})t} \right\} \, d\vec{p} \, d\vec{p}'. \tag{3.3.9}$$

But

$$\frac{1}{(2\pi)^3} \int e^{i(\vec{p}+\vec{p}')\cdot\vec{x}} \, d\vec{x} = \delta(\vec{p} + \vec{p}'), \tag{3.3.10}$$

and performing integration with respect to $\vec{p}'$ in (3.3.9), we have

$$P_{\alpha 1+} = -\frac{1}{2} \int \left[\frac{-p_\alpha}{2E_p} \, \bar{\varphi}^+_+(\vec{p}) \, \bar{\varphi}_+(-\vec{p}) \right.$$

$$\left. + \frac{p_\alpha}{2E_p} \, \bar{\varphi}^+_+(\vec{p}) \, \bar{\varphi}_+(-\vec{p}) \right] e^{-2iE_p t} \, d\vec{p} = 0. \tag{3.3.11}$$

We note that following the integration with respect to $\vec{p}'$, everywhere $\vec{p}' \to -\vec{p}$, according to the well-known relation

$$\int f(\vec{p}') \, \delta(\vec{p} + \vec{p}') \, d\vec{p}' = f(-\vec{p}), \tag{3.3.12}$$

while $E_{p'} \to E_p$, which is obvious, because $E_{p'}$ and E_p are absolute values (see (3.2.14)). In a similar way one can deduce

$$P_{\alpha 1-} = 0. \tag{3.3.13}$$

Let us now derive $P_{\alpha 2+}$. We have:

$$P_{\alpha 2+} = -\frac{i}{(2\pi)^3} \int \left[-\frac{1}{2} \, \bar{\varphi}^+_+(\vec{p}) \frac{-ip'_\alpha}{2E_{p'}} \, \bar{\varphi}_-(\vec{p}') \, e^{i(p-p')x} \right.$$

$$\left. + \frac{ip_\alpha}{2E_p} \, \bar{\varphi}^+_+(\vec{p}) \frac{1}{2} \, \bar{\varphi}_-(\vec{p}') \, e^{i(p-p')x} \right] d\vec{x} \, d\vec{p} \, d\vec{p}'.$$

Observing that

$$i(p - p')x = i(\vec{p} - \vec{p}') \cdot \vec{x} - i(E_p - E_{p'})t, \tag{3.3.14}$$

59

we have

$$P_{\alpha 2+} = \frac{1}{2} \int \left\{ \frac{1}{(2\pi)^3} \int e^{i(\vec{p}-\vec{p}')\cdot\vec{x}} \, d\vec{x} \left[\frac{p'_\alpha}{2E_{p'}} \, \bar{\varphi}_+^+(\vec{p}) \, \bar{\varphi}_-(\vec{p}') \right. \right.$$

$$\left. \left. + \frac{p_\alpha}{2E_p} \, \bar{\varphi}_+^+(\vec{p}) \, \bar{\varphi}_-(\vec{p}') \right] e^{-i(E_p - E_{p'})t} \right\} \, d\vec{p} \, d\vec{p}'. \qquad (3.3.15)$$

Taking into account that

$$\frac{1}{(2\pi)^3} \int e^{i(\vec{p}-\vec{p}')\cdot\vec{x}} \, d\vec{x} = \delta(\vec{p}-\vec{p}'), \qquad (3.3.16)$$

and performing the integration with respect to $\vec{p}'$ in (3.3.15), we get

$$P_{\alpha 2+} = \int \frac{p_\alpha}{2E_p} \, \bar{\varphi}_+^+(\vec{p}) \, \bar{\varphi}_-(\vec{p}) \, d\vec{p}. \qquad (3.3.17)$$

Similarly, one obtains

$$P_{\alpha 2-} = \int \frac{p_\alpha}{2E_p} \, \bar{\varphi}_-^+(\vec{p}) \, \bar{\varphi}_+(\vec{p}) \, d\vec{p}. \qquad (3.3.18)$$

Substituting (3.3.17) and (3.3.18) into (3.3.4), and taking into account (3.3.11) and (3.3.12), we finally obtain the spatial components P_α of the momentum of the complex scalar field in the momentum representation:

$$P_\alpha = \int \frac{p_\alpha}{2E_p} \left[\bar{\varphi}_+^+(\vec{p}) \, \bar{\varphi}_-(\vec{p}) + \bar{\varphi}_-^+(\vec{p}) \, \bar{\varphi}_+(\vec{p}) \right] \, d\vec{p}. \qquad (3.3.19)$$

b) **Hamiltonian.** By virtue of (3.1.20), we can write

$$H = H_{1+} + H_{1-} + H_{2+} + H_{2-}, \qquad (3.3.20)$$

where we denoted

$$H_{1\pm} = \int (\psi_{\pm,\alpha}^+ \psi_{\pm,\alpha} - \psi_{\pm,4}^+ \psi_{\pm,4} + m^2 \psi_\pm^+ \psi_\pm) \, d\vec{x} \; ; \qquad (3.3.21)$$

$$H_{2\pm} = \int (\psi_{\pm,\alpha}^+ \psi_{\mp,\alpha} - \psi_{\pm,4}^+ \psi_{\mp,4} + m^2 \psi_\pm^+ \psi_\mp) \, d\vec{x} \; . \qquad (3.3.22)$$

We shall first calculate H_{1+}. Using (3.2.23), (3.3.1)-(3.3.4), we have

$$H_{1+} = \frac{1}{(2\pi)^3} \int \left[\frac{ip_\alpha}{2E_p} \frac{ip'_\alpha}{2E_{p'}} \, \bar{\varphi}_+^+(\vec{p}) \, \bar{\varphi}_+(\vec{p}') \right.$$

$$-\left(-\frac{E_p}{2E_p}\right)\left(-\frac{E_{p'}}{2E_{p'}}\right)\bar{\varphi}_+^\dagger(\vec{p})\,\bar{\varphi}_+(\vec{p}')$$

$$+m^2\,\frac{1}{2E_p}\frac{1}{2E_{p'}}\,\bar{\varphi}_+^\dagger(\vec{p})\,\bar{\varphi}_+(\vec{p}')\Bigg]\,e^{i(p+p')x}\,d\vec{x}\,d\vec{p}\,d\vec{p}',$$

or, in view of (3.3.8) and (3.3.10)

$$H_{1+}=\int\frac{1}{4E_pE_{p'}}(-p_\alpha p'_\alpha-E_pE_{p'}+m^2)\,\bar{\varphi}_+^\dagger(\vec{p})\,\bar{\varphi}_+(\vec{p}')$$

$$\times\,\delta(\vec{p}+\vec{p}')\,e^{-i(E_p+E_{p'})t}\,d\vec{p}\,d\vec{p}'.$$

Integrating with respect to $\vec{p}'$, one observes that, according to energy-momentum relation (3.1.4) the expression between parentheses becomes

$$p_\alpha p_\alpha-E_pE_p+m^2=p_\alpha p_\alpha+p_4p_4+m^2=p^2+m^2=0,$$

which leads to

$$H_{1+}=0. \tag{3.3.23}$$

Similarly,

$$H_{1-}=0. \tag{3.3.24}$$

Let us now calculate H_{2+}. We have:

$$H_{2+}=\frac{1}{(2\pi)^3}\int\frac{1}{4E_pE_{p'}}\,\bar{\varphi}_+^\dagger(\vec{p})\,\bar{\varphi}_-(\vec{p}')$$

$$\times\left[(ip_\alpha)(-ip'_\alpha)-(-E_p)(E_{p'})+m^2\right]e^{i(p-p')x}\,d\vec{x}\,d\vec{p}\,d\vec{p}'.$$

Using (3.3.14) and (3.3.16), we can write

$$H_{2+}=\int\frac{1}{4E_pE_{p'}}\bar{\varphi}_+^\dagger(\vec{p})\,\bar{\varphi}_-(\vec{p}')(p_\alpha p'_\alpha+E_pE_{p'}+m^2)$$

$$\times\,\delta(\vec{p}-\vec{p}')\,e^{-(E_p-E_{p'})t}\,d\vec{p}\,d\vec{p}',$$

Integrating with respect to $\vec{p}'$, everywhere $\vec{p}'\to\vec{p}$, $E_{p'}\to E_p$, and the expression between parentheses writes

$$p_\alpha p_\alpha+E_p^2+m^2=2E_p^2,$$

where (3.2.14) has been taken into account. Then we are left with

$$H_{2+}=\int\frac{E_p}{2E_p}\bar{\varphi}_+^\dagger(\vec{p})\,\bar{\varphi}_-(\vec{p})\,d\vec{p}. \tag{3.3.25}$$

61

The quantization of the scalar field

Using the same procedure, we also have

$$H_{2-} = \int \frac{E_p}{2E_p} \bar{\varphi}_-^+(\vec{p})\bar{\varphi}_+(\vec{p}) \, d\vec{p}. \tag{3.3.26}$$

Introducing (3.3.23)-(3.3.26) into (3.3.20), we finally obtain the momentum representation of the Hamiltonian (i.e. energy) of the complex scalar field:

$$H = \int \frac{E_p}{2E_p} [\bar{\varphi}_+^+(\vec{p})\,\bar{\varphi}_-(\vec{p}) + \bar{\varphi}_-^+(\vec{p})\bar{\varphi}_+(\vec{p})] \, d\vec{p}. \tag{3.3.27}$$

c) **Charge.** In view of (3.1.22) and (3.2.24), we have

$$Q = Q_{1+} + Q_{1-} + Q_{2+} + Q_{2-}, \tag{3.3.28}$$

where we denoted

$$Q_{1\pm} = e \int (\psi_\pm^+ \psi_{\pm,4} - \psi_{\pm,4}^+ \psi_\pm) \, d\vec{x} \; ; \tag{3.3.29}$$

$$Q_{2\pm} = e \int (\psi_\pm^+ \psi_{\mp,4} - \psi_{\mp,4}^+ \psi_\pm) \, d\vec{x} \; ; \tag{3.3.30}$$

Using (3.2.23), (3.3.2) and (3.3.4), we have for Q_{1+}

$$Q_{1+} = \frac{1}{(2\pi)^3} \int \frac{e}{4E_p E_{p'}} [(-E_p - (-E_{p'})]$$

$$\times \bar{\varphi}_+^+(\vec{p})\,\bar{\varphi}_-(\vec{p}')\, e^{i(\vec{p}+\vec{p}')\cdot\vec{x}}\, e^{-i(E_p+E_{p'})t} \, d\vec{x}\, d\vec{p}\, d\vec{p}'. \tag{3.3.31}$$

Integrating with respect to $\vec{x}$ and $\vec{p}'$, successively, one observes that the expression between square brackets is zero, so that

$$Q_{1+} = 0. \tag{3.3.32}$$

In a similar way we find

$$Q_{1-} = 0. \tag{3.3.33}$$

Expanding Q_{2+}, we have

$$Q_{2+} = \frac{1}{(2\pi)^3} \int \frac{e}{4E_p E_{p'}} [E_{p'}\, \bar{\varphi}_+^+(\vec{p})\,\bar{\varphi}_-(\vec{p}') - E_p\, \bar{\varphi}_-^+(\vec{p})\,\bar{\varphi}_+(\vec{p}')]$$

$$\times e^{i(\vec{p}-\vec{p}')\cdot\vec{x}}\, e^{-i(E_p-E_{p'})t} \, d\vec{x}\, d\vec{p}\, d\vec{p}'.$$

Using (3.3.16), then integrating with respect to $\vec{p}'$ and $\vec{p}$, since

$$\int f(\vec{p}') \, \delta(\vec{p}' - \vec{p}) \, d\vec{p}' = f(\vec{p}),$$

one obtains for Q_{2+}:

$$Q_{2+} = \int \frac{e}{4E_p} [\, \bar{\varphi}_+^+(\vec{p}) \, \bar{\varphi}_-(\vec{p}) - \bar{\varphi}_-^+(\vec{p}) \, \bar{\varphi}_+(\vec{p})] \, d\vec{p}. \qquad (3.3.34)$$

Similarly,

$$Q_{2-} = \int \frac{e}{4E_p} [\, \bar{\varphi}_+^+(\vec{p}) \, \bar{\varphi}_-(\vec{p}) - \bar{\varphi}_-^+(\vec{p}) \, \bar{\varphi}_+(\vec{p})] \, d\vec{p} = Q_{2+}. \qquad (3.3.35)$$

Using (3.3.32)-(3.3.35), we finally obtain from (3.3.28) the total charge of the complex scalar field quanta

$$Q = \int \frac{e}{2E_p} [\, \bar{\varphi}_+^+(\vec{p}) \, \bar{\varphi}_-(\vec{p}) - \bar{\varphi}_-^+(\vec{p}) \, \bar{\varphi}_+(\vec{p})] \, d\vec{p}. \qquad (3.3.36)$$

Conclusion. Comparing the expressions (3.3.19), (3.3.27) and (3.3.36) which represent the spatial components P_α of the momentum, the total energy. and the total charge of the complex scalar field, respectively, we realize that they are similar. The factors that multiply the quantity in square brackets have the same denominator $2E_p$. This fact suggests the idea of renormalization of the state function. Choosing

$$\bar{\varphi}_\pm(\vec{p}) = \sqrt{2E_p} \; \varphi_\pm(\vec{p}), \qquad (3.3.37)$$

we arrive at

$$\psi_\pm(x) = \frac{1}{(2\pi)^{3/2}} \int \frac{1}{\sqrt{2E_p}} \, \varphi_\pm(\vec{p}) \, e^{\pm ipx} \, d\vec{p}. \qquad (3.3.38)$$

Substituting (3.3.37) in (3.3.19), (3.3.27) and (3.3.36), we obtain the momentum, energy, and charge of the complex scalar field as follows:

$$P_\alpha = \int p_\alpha [\, \varphi_+^+(\vec{p}) \, \varphi_-(\vec{p}) + \varphi_-^+(\vec{p}) \, \varphi_+(\vec{p})] \, d\vec{p} \; ; \qquad (3.3.39)$$

$$H = \int E_p [\, \varphi_+^+(\vec{p}) \, \varphi_-(\vec{p}) + \varphi_-^+(\vec{p}) \, \varphi_+(\vec{p})] \, d\vec{p} \; ; \qquad (3.3.40)$$

$$Q = \int e [\, \varphi_+^+(\vec{p}) \, \varphi_-(\vec{p}) - \varphi_-^+(\vec{p}) \, \varphi_+(\vec{p})] \, d\vec{p} \; ; \qquad (3.3.41)$$

The quantization of the scalar field

The last three relations show that p_α and H are always positive, while Q can be positive, negative, or zero. The meaning of the operators $\varphi_\pm(\vec{p})$, $\varphi_\pm^\dagger(\vec{p})$, and their products will be revealed in the next section.

We do mention, however, that our notations (3.3.37) do not affect the validity of relations (3.2.28), that is, we can write

$$[\varphi_\pm(\vec{p})]^+ = \varphi_\mp^\dagger(\vec{p}). \tag{3.3.42}$$

Observation. In the case of the *real scalar field*, $\psi^+(x) = \psi(x)$, and then we have

$$\varphi_\pm^\dagger(\vec{p}) = \varphi_\pm(\vec{p}) = [\varphi_\mp(\vec{p})]^+. \tag{3.3.43}$$

Since the Lagrangian density of the real scalar field is chosen as (see (3.1.12))

$$\mathcal{L} = -\frac{1}{2}\,\psi_{,i}\psi_{,i} - \frac{1}{2}m^2\psi^2,$$

the momentum, the energy and the charge of the real scalar field are:

$$P_\alpha = \frac{1}{2}\int p_\alpha[\varphi_+(\vec{p})\varphi_-(\vec{p}) + \varphi_-(\vec{p})\varphi_+(\vec{p})]\,d\vec{p}\;; \tag{3.3.44}$$

$$H = \frac{1}{2}\int E_p\,[\varphi_+(\vec{p})\varphi_-(\vec{p}) + \varphi_-(\vec{p})\varphi_+(\vec{p})]\,d\vec{p}\;; \tag{3.3.45}$$

$$Q = 0, \tag{3.3.46}$$

the last equality being obvious.

3.4. Commutators of the free scalar field

As we have shown in Chapter II, there are several methods of field quantization. Apart from the already mentioned possibilities, field quantization can also be performed using the Fourier series expansion, and Fourier integrals of the field variables. In this section, we shall apply both these methods, namely by using the results of Section 2.5 in conjunction with (a) the Fourier expansions, and (b) the Fourier integrals.

a) To quantize a field one first needs to know its Hamiltonian density. According to (3.3.40), this is

$$H = \sum_{\vec{p}'} E_{\vec{p}'}(\varphi_{+\vec{p}'}^+\varphi_{-\vec{p}'} + \varphi_{-\vec{p}'}^+\varphi_{+\vec{p}'}), \tag{3.4.1}$$

where we used $\vec{p}\,'$ as a lower index, because momentum is a discrete variable in the Fourier series expansion.

Let us choose c_i as $\varphi_{+\vec{p}}$. In this case f_i must be replaced by $E_{\vec{p}}\,\varphi_{-\vec{p}}^+$, $\partial H / \partial f_i$ by $\varphi_{+\vec{p}}$, ω by $E_{\vec{p}}$, and the quantization condition (2.5.9) writes

$$E_{\vec{p}}\,\varphi_{+\vec{p}} = -\,[H, \varphi_{+\vec{p}}], \qquad (3.4.2)$$

or, by means of (3.4.1),

$$E_{\vec{p}}\,\varphi_{+\vec{p}} = \sum_{\vec{p}\,'} E_{\vec{p}\,'} (\varphi_{+\vec{p}}\varphi_{+\vec{p}\,'}^+ \varphi_{-\vec{p}\,'} + \varphi_{+\vec{p}}\varphi_{-\vec{p}\,'}^+ \varphi_{+\vec{p}\,'}$$

$$-\varphi_{+\vec{p}\,'}^+ \varphi_{-\vec{p}\,'} \varphi_{+\vec{p}} - \varphi_{-\vec{p}\,'}^+ \varphi_{+\vec{p}\,'} \varphi_{+\vec{p}}). \qquad (3.4.3)$$

Keeping in mind that the quanta of a scalar field obey the Bose-Einstein statistics, we rewrite (3.4.3) as a convenient combination of commutators, as follows:

$$\varphi_{+\vec{p}}\varphi_{+\vec{p}\,'}^+ \varphi_{-\vec{p}\,'} + \varphi_{+\vec{p}}\varphi_{-\vec{p}\,'}^+ \varphi_{+\vec{p}\,'} - \varphi_{+\vec{p}\,'}^+ \varphi_{-\vec{p}\,'} \varphi_{+\vec{p}} - \varphi_{-\vec{p}\,'}^+ \varphi_{+\vec{p}\,'} \varphi_{+\vec{p}}$$

$$= [\varphi_{+\vec{p}}, \varphi_{+\vec{p}\,'}^+]\varphi_{-\vec{p}\,'} + \varphi_{+\vec{p}\,'}^+ \varphi_{+\vec{p}}\varphi_{-\vec{p}\,'} + [\varphi_{+\vec{p}}, \varphi_{-\vec{p}\,'}^+]\varphi_{+\vec{p}\,'}$$

$$+\varphi_{-\vec{p}\,'}^+ \varphi_{+\vec{p}}\varphi_{+\vec{p}\,'} - \varphi_{+\vec{p}\,'}^+ \varphi_{-\vec{p}\,'} \varphi_{+\vec{p}} - \varphi_{-\vec{p}\,'}^+ \varphi_{+\vec{p}\,'} \varphi_{+\vec{p}}$$

$$= [\varphi_{+\vec{p}}, \varphi_{+\vec{p}\,'}^+]\varphi_{-\vec{p}\,'} + \varphi_{+\vec{p}\,'}^+ [\varphi_{+\vec{p}}, \varphi_{-\vec{p}\,'}]$$

$$+[\varphi_{+\vec{p}}, \varphi_{-\vec{p}\,'}^+]\varphi_{+\vec{p}\,'} + \varphi_{-\vec{p}\,'}^+ [\varphi_{+\vec{p}}, \varphi_{+\vec{p}\,'}].$$

Relation (3.4.3) then becomes

$$\sum_{\vec{p}\,'} E_{\vec{p}\,'} \varphi_{+\vec{p}\,'} \delta_{\vec{p},\vec{p}\,'} = \sum_{\vec{p}\,'} E_{\vec{p}\,'} \{[\varphi_{+\vec{p}}, \varphi_{+\vec{p}\,'}^+]\varphi_{-\vec{p}\,'} + \varphi_{+\vec{p}\,'}^+ [\varphi_{+\vec{p}}, \varphi_{-\vec{p}\,'}]$$

$$+[\varphi_{+\vec{p}}, \varphi_{-\vec{p}\,'}^+]\varphi_{+\vec{p}\,'} + \varphi_{-\vec{p}\,'}^+ [\varphi_{+\vec{p}}, \varphi_{+\vec{p}\,'}]\},$$

which is true only if

$$\begin{cases} [\varphi_{+\vec{p}}, \varphi_{-\vec{p}\,'}^+] = \delta_{\vec{p},\vec{p}\,'}\ ; \\ [\varphi_{+\vec{p}}, \varphi_{+\vec{p}\,'}^+] = [\varphi_{+\vec{p}}, \varphi_{-\vec{p}\,'}] = [\varphi_{+\vec{p}}, \varphi_{+\vec{p}\,'}] = 0. \end{cases} \qquad (3.4.4)$$

Performing similar calculations with $\varphi_{-\vec{p}}$ or $\varphi_{-\vec{p}}^+$ replacing c_i, and $-E_{\vec{p}}$ instead of ω, we obtain

$$\begin{cases} [\varphi_{+\vec{p}}, \varphi_{-\vec{p}\,'}^+] = \delta_{\vec{p},\vec{p}\,'}\ ; \\ [\varphi_{+\vec{p}}^+, \varphi_{+\vec{p}\,'}^+] = [\varphi_{+\vec{p}}^+, \varphi_{-\vec{p}\,'}^+] = [\varphi_{+\vec{p}}^+, \varphi_{+\vec{p}\,'}] = 0. \end{cases} \qquad (3.4.5)$$

The quantization of the scalar field

The remaining commutation relations are obtained by taking $\varphi_{-\vec{p}}$ and $\varphi_{-\vec{p}}^+$ instead of c_i, and $-E_{\vec{p}}$ instead of ω. The full set of commutation relations then is:

$$\begin{cases} [\varphi_{+\vec{p}}, \varphi_{-\vec{p}'}^+] = \delta_{\vec{p},\vec{p}'} \ ; \quad [\varphi_{+\vec{p}}^+, \varphi_{-\vec{p}'}] = \delta_{\vec{p},\vec{p}'} \ ; \\ [\varphi_{\pm\vec{p}}, \varphi_{\pm\vec{p}'}] = [\varphi_{\pm\vec{p}}, \varphi_{\pm\vec{p}'}^+] = [\varphi_{\pm\vec{p}}^+, \varphi_{\pm\vec{p}'}^+] = 0 \ ; \\ [\varphi_{+\vec{p}}, \varphi_{-\vec{p}'}] = [\varphi_{+\vec{p}}^+, \varphi_{-\vec{p}'}^+] = 0 \ . \end{cases} \tag{3.4.6}$$

The commutation relations (3.4.6), together with the results of Section 2.3, show that the complex scalar field is characterized by two types of particles of momentum $\vec{p}$:

$$N_{+\vec{p}} = \varphi_{-\vec{p}} \varphi_{+\vec{p}}^+ , \quad \text{and} \quad N_{-\vec{p}} = \varphi_{-\vec{p}}^+ \varphi_{+\vec{p}} \ . \tag{3.4.7}$$

Therefore, $\varphi_{-\vec{p}}$ and $\varphi_{+\vec{p}}^+$ are the creation and annihilation operators, respectively, of particles of the first kind, while $\varphi_{-\vec{p}}^+$ and $\varphi_{+\vec{p}}$ are the creation and annihilation operators, respectively, of particles of the second kind.

Since in (3.3.39), (3.3.40) and (3.3.41) the field was not quantized yet, and the order of functions was arbitrary, we have to arrange the factors such that each of the resulting terms will represent a number of particles. (As the reader will see in the next section, with the help of the *normal product* concept, this procedure will get a deeper signification). Using (3.4.7), we then have:

$$\vec{P} = \sum_{\vec{p}} \vec{p} (N_{+\vec{p}} + N_{-\vec{p}}) \ ;$$

$$H = \sum_{\vec{p}} E_p (N_{+\vec{p}} + N_{-\vec{p}}) \ ; \tag{3.4.8}$$

$$Q = e \sum_{\vec{p}} (N_{+\vec{p}} - N_{-\vec{p}}) \ .$$

It then follows that $N_{+\vec{p}}$ is the number of particles carrying a positive charge, while $N_{-\vec{p}}$ denotes the field quanta with negative charge.

In the case of a real scalar field, the number of commutation relations (3.4.6) decreases, and the remaining ones are

$$[\varphi_{+\vec{p}}, \varphi_{-\vec{p}'}] = \delta_{\vec{p},\vec{p}'} \ ; \quad [\varphi_{\pm\vec{p}}, \varphi_{\pm\vec{p}'}] = 0. \tag{3.4.9}$$

The number of particles whose creation operator is $\varphi_{-\vec{p}}$ and the annihilation operator is $\varphi_{+\vec{p}}$, then writes

$$N_{\vec{p}} = \varphi_{-\vec{p}} \varphi_{+\vec{p}}. \tag{3.4.10}$$

66

Commutators of the free scalar field

We obtain

$$\vec{P} = \sum_{\vec{p}} \vec{p}\, N_{\vec{p}} \;; \quad H = \sum_{\vec{p}} E_{\vec{p}}\, N_{\vec{p}} \;; \quad Q = 0, \tag{3.4.11}$$

meaning that $N_{\vec{p}}$ is the number of neutral field quanta.

Observation. The use of the quantization method developed in Sections 2.3 and 3.3 could have straightforwardly led to (3.4.6) and (3.4.9) for the complex and real scalar field, respectively.

b) We shall redo the quantization of the scalar field, this time using the commutation relation (2.5.15), and the Fourier integral expansion of the field variables. In view of (3.1.11), we have

$$U_{(k)} = \psi \;; \quad \pi_0^+ = \frac{\partial \mathcal{L}}{\partial \psi_{,0}} = \psi_{,0}^+,$$

and (2.5.15) yields

$$[\psi_{,0}^+(\vec{x},t),\ \psi(\vec{x}',t)] = -\,i\,\delta(\vec{x}-\vec{x}'). \tag{3.4.12}$$

It is worthwhile to emphasize, once more, that the momenta $\pi_{0(i)}^+$ are defined for those points where the field variables are $U_{(k)}$ *at the same time* t. Using (3.2.24), we can write

$$[\psi_{,0}^+(\vec{x},t),\ \psi(\vec{x}',t)] = [\psi_{+,0}^+(\vec{x},t),\ \psi_+(\vec{x}',t)] + [\psi_{+,0}^+(\vec{x},t),\ \psi_-(\vec{x}',t)]$$

$$+[\psi_{-,0}^+(\vec{x},t),\ \psi_+(\vec{x}',t)] + [\psi_{-,0}^+(\vec{x},t),\ \psi_-(\vec{x}',t)]. \tag{3.4.13}$$

To calculate each commutator on the r.h.s. of (3.4.13), one first observes that

$$\psi_{\pm,0}^+(x) = \frac{1}{(2\pi)^{3/2}} \int \frac{\mp i E_p}{\sqrt{2E_p}}\, \varphi_{\pm}^+(\vec{p})\, e^{\pm i p x}\, d\vec{p} \;; \tag{3.4.14}$$

$$[\psi_{+,0}^+(\vec{x},t),\psi_+(\vec{x}',t)] = \frac{1}{(2\pi)^3} \int \frac{-i E_p}{\sqrt{4 E_p E_{p'}}}\, [\varphi_+^+(\vec{p}),\varphi_+(\vec{p}')]$$

$$\times e^{\vec{p}\cdot\vec{x}+\vec{p}'\cdot\vec{x}'}\, e^{-i(E_p+E_{p'})t}\, d\vec{p}\, d\vec{p}'. \tag{3.4.15}$$

Examination of (3.4.15) shows that, regardless of the choice of $\vec{p}$ and $\vec{p}'$, the exponent $[-i(E_p + E_{p'})]$ is always different from zero. One also notes that $\delta(\vec{x}-\vec{x}')$ does not explicitly contains time, therefore none of the commutators in (3.4.12) can explicitly depend on time.

The quantization of the scalar field

This means that the commutator (3.4.15) must vanish, and consequently,

$$[\varphi_+^+(\vec{p}), \varphi_+(\vec{p}')] = 0. \tag{3.4.16}$$

Similarly, it can be shown that

$$[\varphi_-^\pm(\vec{p}), \varphi_-(\vec{p}')] = 0. \tag{3.4.17}$$

The last two relations can be compressed as

$$[\varphi_\pm(\vec{p}), \varphi_\pm^\pm(\vec{p}')] = 0. \tag{3.4.18}$$

Let us now calculate the second commutator on the r.h.s. of (3.4.13). Using (3.3.38) and (3.4.14), we obtain

$$[\psi_{+,0}^+(\vec{x}, t), \psi_-(\vec{x}', t)] = \frac{1}{(2\pi)^3} \int \frac{-iE_p}{\sqrt{4E_p E_{p'}}} [\varphi_+^+(\vec{p}), \varphi_-(\vec{p}')]$$

$$\times e^{i(\vec{p}\cdot\vec{x} - \vec{p}'\cdot\vec{x}')} e^{-i(E_p - E_{p'})t} \, d\vec{p} \, d\vec{p}'. \tag{3.4.19}$$

In view of the commutation relations (3.4.12), we choose

$$[\varphi_+^+(\vec{p}), \varphi_-(\vec{p}')] = \delta(\vec{p} - \vec{p}'). \tag{3.4.20}$$

Substituting (3.4.20) in (3.4.19) and performing the integration with respect to $\vec{p}'$, we obtain

$$[\psi_{+,0}^+(\vec{x}, t), \psi_-(\vec{x}', t)] = -\frac{i}{2}\frac{1}{(2\pi)^3} \int e^{i\vec{p}\cdot(\vec{x} - \vec{x}')} \, d\vec{p} = -\frac{i}{2}\delta(\vec{x} - \vec{x}'). \tag{3.4.21}$$

In a similar way, the third commutator in (3.4.13) gives

$$[\psi_{-,0}^+(\vec{x}, t), \psi_+(\vec{x}', t)] = -\frac{i}{2}\delta(\vec{x} - \vec{x}'), \tag{3.4.22}$$

where we have used

$$[\varphi_-^\pm(\vec{p}), \varphi_+(\vec{p}')] = -\delta(\vec{p} - \vec{p}'). \tag{3.4.23}$$

Taking into account (3.4.21) and (3.4.22), as well as the fact that the first and the fourth commutators in (3.4.13) vanish, one observes that the commutation relation (3.4.12) is proved.

The aforementioned considerations attest the validity of (3.4.20) and (3.4.23). Since the Dirac delta function is even, the last relation can also be written as

$$[\varphi_+(\vec{p}), \varphi_-^\pm(\vec{p}')] = \delta(\vec{p} - \vec{p}'). \tag{3.4.24}$$

68

Commutators of the free scalar field

In the case of the real scalar field, the commutation relations (3.4.20) and (3.4.24) become a single relation

$$[\varphi_+(\vec{p}), \varphi_-(\vec{p}')] = \delta(\vec{p} - \vec{p}'), \tag{3.4.25}$$

the rest of commutators being zero

$$[\varphi_\pm(\vec{p}), \varphi_\pm(\vec{p}')] = 0.$$

A similar calculation leads to:

$$[\psi_{,0}(\vec{x}, t), \psi(\vec{x}', t)] = [\psi_{,0}^+(\vec{x}, t), \psi^+(\vec{x}', t)] = 0. \tag{3.4.26}$$

Let us now compare the commutation relations (3.4.24) with the commutations relations

$$[c_i, c_k^+] = \delta_{ik},$$

already established in the previous section for the creation and annihilation operators of bosons, where $c_i^+ c_i = n_i$ (no summation) signifies the number of particles. We observe that the quantities

$$N_+(\vec{p}) = \varphi_-(\vec{p})\,\varphi_+^+(\vec{p}) ; \tag{3.4.27}$$

$$N_-(\vec{p}) = \varphi_-^+(\vec{p})\,\varphi_+(\vec{p}) , \tag{3.4.28}$$

represent particle number densities, the operators $\varphi_-(\vec{p})$, $\varphi_+^+(\vec{p})$ being creation operators, and $\varphi_+(\vec{p})$, $\varphi_+^+(\vec{p})$- annihilation operators. Combining the results leading to (3.3.39), (3.3.40), (3.3.41), (3.4.27), and (3.4.28), we obtain:

$$P_\alpha = \int p_\alpha [N_+(\vec{p}) + N_-(\vec{p})]\, d\vec{p}, \tag{3.4.29}$$

$$H = \int E_p [N_+(\vec{p}) + N_-(\vec{p})]\, d\vec{p}, \tag{3.4.30}$$

$$Q = e \int [N_+(\vec{p}) - N_-(\vec{p})]\, d\vec{p}, \tag{3.4.31}$$

Here we omitted the zero-state quantities

$$P_{\alpha 0} = \int \delta(0)\, p_\alpha\, d\vec{p} ; \quad E_0 = \int \delta(0)\, E_p\, d\vec{p} ; \quad Q_o = \int \delta(0)\, d\vec{p}$$

$$\tag{3.4.31'},$$

which are infinite. If the scalar field is real, the two types of particles (3.4.27) and (3.4.28) coincide, so that

$$N(\vec{p}) = \varphi_-(\vec{p}) \; \varphi_+(\vec{p}), \tag{3.4.32}$$

while (3.3.44), (3.3.45), and (3.3.46) become

$$P_\alpha = \int p_a N(\vec{p}) \; d\vec{p} \; ; \tag{3.4.33}$$

$$H = \int E_p N(\vec{p}) \; d\vec{p} \; ; \tag{3.4.34}$$

$$Q = 0. \tag{3.4.35}$$

Relations (3.4.27) and (3.4.28) show that the complex scalar field is characterized by two types of particles. The operators $\varphi_-(\vec{p})$ and $\varphi_+^+(\vec{p})$ are operators of creation and annihilation of one type of particles, with charge $+e$, while $\varphi_-^+(\vec{p})$ and $\varphi_+(\vec{p})$ are operators of creation and annihilation of another type of particles, with charge $-e$. In view of (3.4.29), (3.4.30), and (3.4.31), the two types of particles have the same momentum and the same energy, but different values for the charge. If the scalar field is real, then $\varphi_-(\vec{p})$ is the emission operator of a particle with momentum $\vec{p}$ and charge zero, while $\varphi_+(\vec{p})$ is the absorption operator of the same type of particle. Consequently, we have mesons π^+ and π^-, described by the complex scalar field, and mesons π^0, as quanta of the real scalar field.

As a final remark, we note that various methods of quantization lead - obviously - to the same result.

c) We are now prepared to calculate the commutator

$$[\psi(x), \psi^+(x')].$$

Recalling that x and x' are four-vectors, and using (3.2.24) and (3.2.27), we have:

$$[\psi(x), \psi^+(x')] = [\psi_+(x), \psi_+^+(x')] + [\psi_-(x), \psi_-^+(x')]$$

$$+ [\psi_+(x), \psi_-^+(x')] + [\psi_-(x), \psi_+^+(x')]. \tag{3.4.36}$$

Using the method already described, the first two commutators on the r.h.s. of (3.4.36) give

$$[\psi_\pm(x), \psi_\pm^+(x')] = \frac{1}{(2\pi)^3} \int \frac{1}{\sqrt{4E_p E_{p'}}} [\varphi_\pm(x), \varphi_\pm^+(x')]$$

$$\times e^{\pm(px+p'x')} \, d\vec{p} \, d\vec{p}' = 0, \qquad (3.4.37)$$

where the commutation relations (3.4.18) have been taken into account. Using (3.4.29) and (3.4.24), we have

$$[\psi_\pm(x), \psi_\mp^+(x')] = \frac{1}{(2\pi)^3} \int \frac{1}{\sqrt{4E_p E_{p'}}} [\varphi_\pm(x), \varphi_\mp^+(x')]$$

$$\times e^{\pm(px-p'x')} \, d\vec{p} \, d\vec{p}'$$

$$= \pm \frac{1}{(2\pi)^3} \int \frac{1}{\sqrt{4E_p E_{p'}}} \delta(\vec{p} - \vec{p}') \, e^{\pm i(px-p'x')} \, d\vec{p} \, d\vec{p}'. \qquad (3.4.38)$$

Performing now the integration with respect to $\vec{p}'$ in (3.4.38), we arrive at

$$[\psi_\pm(x), \psi_\mp^+(x')] = \pm \frac{1}{(2\pi)^3} \int \frac{1}{2E_p} e^{\pm ip(x-x')} \, d\vec{p} = i \, D_{m\pm}(x - x'),$$
$$(3.4.39)$$

where we denoted (see (A.57))

$$D_{m\pm}(x - x') = \pm \frac{1}{(2\pi)^3 i} \int \frac{1}{2E_p} e^{\pm ip(x-x')} \, d\vec{p}. \qquad (3.4.40)$$

If we also denote (see (A.58.1))

$$D_m(x - x') = D_{m+}(x - x') + D_{m-}(x - x') \qquad (3.4.41)$$

and combine (3.4.37), (3.4.39), and (3.4.41), then (3.4.36) yields

$$[\psi(x), \psi^+(x')] = i \, D_m(x - x'). \qquad (3.4.42)$$

Using a similar procedure, it can be shown that

$$[\psi(x), \psi(x')] = [\psi^+(x), \psi^+(x')] = 0. \qquad (3.4.43)$$

3.5. Products of operators

a) **Normal product.** As we have seen in Chapter II regarding the creation and annihilation operators, the disappearance of one quantum state is always accompanied by the appearance of another quantum state. In quantum field theory language this represents a successive application of two operators, one which annihilates the initial quantum state, and one which creates the final quantum state. In this respect, an important role is played by the *order* in which the operators act. This order has to be logical, i.e. a product of operators must be written in such a way that the annihilation operator acts *first*, and the creation operator acts *second*.

Consider, for example, the product of operators $\psi(x)$ and $\psi^+(y)$. In view of (3.2.24), we have:

$$\psi(x)\psi^+(y) = \psi_+(x)\psi_+^+(y)+\psi_-(x)\psi_-^+(y)+\psi_+(x)\psi_-^+(y)+\psi_-(x)\psi_+^+(y).$$
$$(3.5.1)$$

If we apply (3.5.1) to a state function, one observes that both factors in the first two terms on the r.h.s. are annihilation and creation operators, respectively, meaning that from our point of view the order of operators is unimportant. In the last term, the annihilation operator is on the right, i.e. the order is natural. But in the last term remaining, the third, the operators are written in a non-logical order. To avoid this difficulty, one defines the *normal product* of a product of operators, as being the arrangement of the operators in which the annihilation operators are always written on the right, as opposed to the creation operators. For example, in our case the normal product of operators $\psi(x)$ and $\psi^+(y)$ is

$$N[\psi(x)\psi^+(y)] = \psi_+(x)\psi_+^+(y) + \psi_-(x)\psi_-^+(y)$$

$$+\psi_-^+(x)\psi_+(y) + \psi_-(x)\psi_+^+(y).\qquad(3.5.2)$$

Subtracting (3.5.2) from (3.5.1), and making use of (3.4.39), we obtain

$$\psi(x)\psi^+(y)-N[\psi(x)\psi^+(y)] = [\psi_+(x), \psi_-^+(y)] = i\,D_{m+}(x-y). \quad(3.5.3)$$

Since the order of factors appearing in classical quantities is not relevant, we shall agree to choose the order indicated by the normal product, whenever the quantum equivalent of classical quantities is taken. This constitutes one of the principles of quantum field theory, the *correspondence principle: the operators associated with physical quantities are obtained by substituting ordinary products by normal products.*

As an example, the Lagrangian density of the complex scalar field writes

$$\mathcal{L} = -\,N[\psi_{,i}^+\psi_{,i}] - m^2\,N[\psi^+\psi].\qquad(3.5.4)$$

This way we also justify the convenient arrangement of operators in (3.4.8), (3.4.11), (3.4.29), (3.4.30), and (3.4.31).

b) **time-ordered product**. Going now back to the product $\psi(x)\psi^+(y)$, we require that the order of operators is chronologically correct. If for example, $x_0 > y_0$, where x_0 and y_0 represent the time corresponding to events x and y, respectively, then the operator $\psi^+(y)$ acts first, and the operator $\psi(x)$ acts next. In this respect, we define the *time-ordered product* of an arbitrary number of operators as follows: the product of any pair of operators, corresponding to the time moments t_1 and t_2, with $t_1 > t_2$, must be written in such a way that the operator corresponding to t_2 acts before the operator associated to t_1. We then define the *time-ordered product* or *chronological product* as

$$T[\psi(x)\psi^+(y)] = \begin{cases} \psi(x)\,\psi^+(y) & \text{if} \quad x_0 > y_0\ ; \\ \psi^+(y)\,\psi(x) & \text{if} \quad y_0 > x_0. \end{cases} \qquad (3.5.5)$$

If, by definition

$$\theta(x) = \begin{cases} 1, & x > 0\ ; \\ 0, & x < 0, \end{cases} \qquad (3.5.6)$$

then the time-ordered product can also be written as

$$T[\psi(x)\psi^+(y)] = \theta(x_0 - y_0)\psi(x)\psi^+(y) + \theta(y_0 - x_0)\psi^+(y)\psi(x). \quad (3.5.7)$$

Next, we analyze the connection between the two types of products, normal and time-ordered. By means of (3.5.3) and (3.5.7), we can write

$$T[\psi(x)\psi^+(y)] - N[\psi(x)\psi^+(y)] = [\theta(x_0 - y_0) - 1]\psi(x)\psi^+(y)$$
$$+\theta(y_0 - x_0)\psi^+(y)\psi(x) + i\,D_{m+}(x - y).$$

Since

$$[\theta(x_0 - y_0) - 1] = -\,\theta(y_0 - x_0), \qquad (3.5.8)$$

and using (3.4.41), (3.4.42), we have:

$$T[\psi(x)\psi^+(y)] - N[\psi(x)\psi^+(y)]$$
$$= -\,\theta(y_0 - x_0)[\psi(x), \psi^+(y)] + i\,D_{m+}(x - y)$$
$$= i\left[-\,\theta(y_0 - x_0)\,D_{m+}(x - y) - \theta(y_0 - x_0)\,D_{m-}(x - y) + D_{m+}(x - y)\right].$$

But

$$1 - \theta(y_0 - x_0) = \theta(x_0 - y_0) \qquad (3.5.9)$$

and we finally obtain

$$T[\psi(x)\psi^+(y)] - N[\psi(x)\psi^+(y)] = D_m^c(x - y), \qquad (3.5.10)$$

The quantization of the scalar field

where

$$D_m^c(x-y) = i\,[\theta(x_0-y_0)\,D_{m+}(x-y)-\theta(y_0-x_0)D_{m-}(x-y)]. \quad (3.5.11)$$

In this expression, $D_m^c(x - y)$ is a singular function, called *causal function* (see the Appendix).

If we set $y = 0$, $x_0 = t$ in (3.5.11), and take into consideration (3.4.40), we arrive at

$$D_m^c(x) = i\,[\theta(t)\,D_{m+}(x - y) - \theta(-t)\,D_{m-}(x - y)]$$

$$= \begin{cases} \frac{1}{(2\pi)^3}\int \frac{1}{2E_p}\,e^{i(\vec{p}\cdot\vec{x}-E_p t)}\,d\vec{p}\,, & (t>0); \\ \frac{1}{(2\pi)^3}\int \frac{1}{2E_p}\,e^{-i(\vec{p}\cdot\vec{x}-E_p t)}\,d\vec{p}\,, & (t<0). \end{cases} \quad (3.5.12)$$

If in the second integral in (3.5.12) we replece $\vec{p}$ by $-\vec{p}$, we can write (3.5.12) in the following condensed form

$$D_m^c(x) = \frac{1}{(2\pi)^3}\int \frac{1}{2E_p}\,e^{i(\vec{p}\cdot\vec{x}-E_p|t|)}\,d\vec{p}. \quad (3.5.13)$$

3.6. Vacuum states. The Fock representation

A *vacuum state*, denoted by $|0>$, is a state vector with the property that, when an annihilation operator acts upon it, the result is zero. In other words, the vacuum state is a state with no particles. For example, if the annihilation operators (already familiar) $\varphi_+(\vec{p})$ and $\varphi_+^+(\vec{p})$ act on the vector ket $|0>$, then

$$\varphi_+(\vec{p})|0>=0\,; \quad \varphi_+^+(\vec{p})|0>=0. \quad (3.6.1)$$

Taking the Hermitian conjugates of (3.6.1), and using (3.3.42), we have

$$<0|\varphi_-^+(\vec{p}) = 0\,; \quad <0|\varphi_-(\vec{p}) = 0. \quad (3.6.2)$$

Therefore, the result of the creation operators $\varphi_-^+(\vec{p})$ and $\varphi_-(\vec{p})$ acting on the vector bra $<0|$ is also zero. On the other hand, the action of a creation operator on $|0>$, or of an annihilation operator on $<0|$, gives a result which is different from zero. For example, in the case of a real scalar field we have

$$|p>= \varphi_-(\vec{p})|0>\neq 0, \quad (3.6.3)$$

74

because otherwise the commutation rules for the creation and annihilation operators could not be satisfied.

Consider the operator relation

$$|p_2, p_1 > = \varphi_-(\vec{p}_2)\,\varphi_-(\vec{p}_1)|0 > \neq 0. \qquad (3.6.4)$$

The application of the creation operator $\varphi_-(\vec{p}_1)$ on the vector $|0 >$ leads to a particle with momentum $\vec{p}_1$. Next, by applying $\varphi_-(\vec{p}_2)$, we get an additional particle, of momentum $\vec{p}_2$. If this procedure is repeated, one obtains a set of functions, one for each of the possible states of an arbitrary number of particles. The resulting eigenvectors can be orthonormalized, and any given state of this set of particles can be expressed as a series expansion in terms of these vectors. Such a representation of the state functions is called *the Fock representation*.

In the following, let us define the expectation value of the observable quantities (charge, energy, momentum) as being zero in the vacuum state. If, for example, we denote by $\hat{F}$ the operator corresponding to an observable F, then the vacuum expectation value of $\hat{F}$ is

$$< 0\,|\hat{F}|\,0 > = 0,$$

which is usually denoted by

$$< \hat{F} >_0 = 0. \qquad (3.6.5)$$

To exemplify the usefulness of this convention, let us write down the Hamiltonian density of the complex scalar field. In view of (3.3.40), we have:

$$H = \int E_p[\varphi_+^+(\vec{p})\varphi_-(\vec{p}) + \varphi_-^+(\vec{p})\varphi_+(\vec{p})]\,d\vec{p}. \qquad (3.6.6)$$

Since the first term of (3.6.6) is not written under the sign of normal product, we take advantage of the commutation relation (3.4.23), as seen in Section 3.4. Omitting the zero state energy, one obtains:

$$H = \int E_p[\varphi_-(\vec{p})\varphi_+^+(\vec{p}) + \varphi_-^+(\vec{p})\varphi_+(\vec{p})]\,d\vec{p}. \qquad (3.6.7)$$

But according to (3.6.6) we would have $< 0|H|0 > \neq 0$, or even infinite, which is contrary to our aforementioned convention. But if we write the Hamiltonian density under the sign of normal product, we have

$$< 0\,|H|\,0 > = < H >_0 = 0, \qquad (3.6.8)$$

in agreement with all our conventions.

The quantization of the scalar field

As we have already established, all non-quantized field quantities are assumed in QFT under the sign of normal product, meaning that the annihilation operators are always placed on the right relative to the creation operators. According to these conventions, *the vacuum expectation value of a normal product is always zero.* Returning to our example concerning the product of the operators $\psi(x)$ and $\psi^+(y)$, we therefore have:

$$< N[\psi(x)\,\psi^+(y)] >_0 = 0. \qquad (3.6.9)$$

In view of (3.5.3), we then have:

$$< \psi(x)\,\psi^+(y) >_0 = i\,D_{m+}(x - y), \qquad (3.6.10)$$

because

$$< 0|D_{m+}(x - y)|0 > = D_{m+}(x - y) < 0|0 > = D_{m+}(x - y)$$

where we used the fact that the vacuum state is orthonormal :
$< 0|0 > = 1$. Also , according to (3.5.10) the vacuum expectation value of the time-ordered product is

$$< T[\psi(x)\,\psi^+(y)] >_0 = D_m^c(x - y). \qquad (3.6.11)$$

The vacuum expectation value of the product $\psi(x)\,\psi^+(y)$ is usually denoted by

$$< \psi(x)\,\psi^+(y) >_0 = \underbrace{\psi(x)\,\psi^+(y)} \qquad (3.6.12)$$

and is called *vacuum contraction* (or, simply, *contraction*) of the operators $\psi(x)$ and $\psi^+(y)$. On the other hand, the vacuum expectation value of the time-ordered product of the same two operators is designated as

$$< T[< \psi(x)\,\psi^+(y)] >_0 = \overbrace{\psi(x)\,\psi^+(y)}, \qquad (3.6.13)$$

and is named *time-ordered contraction*, or *T-contraction*, of the two operators. By means of (3.6.12), (3.6.13), Eqs. (3.5.3) and (3.5.10) yield

$$\psi(x)\,\psi^+(y) = N[\psi(x)\,\psi^+(y)] + \underbrace{\psi(x)\,\psi^+(y)} \ ; \qquad (3.6.14)$$

$$T[\psi(x)\,\psi^+(y)] = N[\psi(x)\,\psi^+(y)] + \overbrace{\psi(x)\,\psi^+(y)} \ . \qquad (3.6.15)$$

A contraction under the sign of the normal product can be written outside (usually in front of) this product, as in the following example:

$$N[\psi_1(x_1)...\underbrace{\psi_j(x_j)...\psi_k(x_k)}...\psi_n(x_n)]$$

$$= \underbrace{\psi_j(x_j)\psi_k(x_k)} \ N[\psi_1(x_1)...\psi_{j-1}(x_{j-1})$$

$$\times \psi_{j+1}(x_{j+1})....\psi_{k-1}(x_{k-1})\psi_{k+1}(x_{k+1})...\psi_n(x_n)]. \tag{3.6.16}$$

3.7. Wick's theorems

In Section 3.5 we defined the notions of normal product and time-ordered product in the case of two operators. If the product contains more than two factors, say n, there exist three theorems, known as *Wick's theorems*, which we are going to prove in this section.

There are two points we need to establish in the beginning. First, we shall abandon the use of Greek letters and use capital Latin letters instead. Second, our demonstration uses operators that satisfy commutation rules, i.e. *bosonic operators*. However, Wick's theorems are also valid in case of fermionic operators, except for some sign changes.

Theorem I. *A product of operators equals the sum of all normal products of these operators, with all possible contractions, including the ordinary normal product (i.e. without contractions), that is*

$$A_1 A_2 A_3...A_n = N[A_1 A_2 A_3...A_n] + N[\underbrace{A_1 A_2} A_3...A_n]$$

$$+N[\underbrace{A_1 A_2 A_3}...A_n] + ... + N[\underbrace{A_1 A_2 A_3...A_n}] + N[\underbrace{A_1 A_2} \underbrace{A_3 A_4}...A_n]$$

$$+N[\underbrace{A_1 A_2} \underbrace{A_3 A_4}...A_n] + ... + N[\underbrace{A_1 A_2} \underbrace{A_3 A_4} A_5...A_n] + ... \tag{3.7.1}$$

The proof of this theorem is based on *the method of complete induction*. As a preliminary step, let us prove the following *lemma*:

$$N[A_1 A_2 A_3...A_n]B = N[A_1 A_2 A_3...A_n B] + \sum_{i=1}^{n} N[A_1... \underbrace{A_i...A_n B}].$$

$$\tag{3.7.2}$$

We distinguish the following two cases:

1) B **is an annihilation operator.** In this case, irrespective of the order of operators $A_1, ..., A_n$, the operator B can be introduced under the sign of the normal product in (3.7.2). We have

$$\underbrace{A_i B} =< A_i B >_0 = 0, \text{ because } B|0>= 0,$$

and the lemma is proved.

2) B **is a creation operator**. Here we also have two possibilities, depending on whether $A_1, ..., A_n$ are operators of creation or annihilation:

a) If $A_1, ..., A_n$ are operators of creation, then B can be placed anywhere, therefore it can be introduced under the sign of the normal product. We have $\underbrace{A_i B} = <\ A_i B\ >_0 = 0$, because $< 0|A_i = 0$, and lemma (3.7.2) is again satisfied.

b) If $A_1, ..., A_n$ are operators of annihilation, we have:

$$N[A_1 A_2 ... A_n B] = B A_1 A_2 ... A_n\ ;$$

$$N[A_1 A_2 ... A_n]B = A_1 A_2 ... A_n B\ ,$$

and subtracting member by member

$$N[A_1 A_2 ... A_n]B - N[A_1 A_2 ... A_n B] = A_1 A_2 ... A_n B - B A_1 A_2 ... A_n.$$

But, according to (3.6.14),

$$A_n B = N[A_n B] + \underbrace{A_n B} = B A_n + \underbrace{A_n B},$$

and then

$$N[A_1 A_2 ... A_n]B - N[A_1 A_2 ... A_n B]$$
$$= A_1 ... A_{n-1} B A_n + A_1 ... \underbrace{A_n B} - B A_1 ... A_n\ .$$

Reiterating this reasoning, we have:

$$N[A_1 A_2 ... A_n]B - N[A_1 A_2 ... A_n B] = A_1 ... A_{n-2} B A_{n-1} A_n$$

$$+ A_1 ... \underbrace{A_{n-1} B}\ A_n + A_1 ... \underbrace{A_n B} - B A_1 ... A_n = ...$$

$$= B A_1 ... A_n + A_1 \underbrace{B} A_2 ... A_n + ...$$

$$+ A_1 ... \underbrace{A_{n-1} B}\ A_n + A_1 ... \underbrace{A_n B} - B A_1 ... A_n$$

$$= \sum_{i=1}^{n} A_1 ... \underbrace{A_i B}\ A_{i+1} ... A_n.$$

But

$$A_1 ... \underbrace{A_i B}\ A_{i+1} ... A_n = N[A_1 ... \underbrace{A_i ... A_n B}],$$

which completes the proof of lemma (3.7.2).

Let us now apply the complete induction method. The relation (3.7.1) is, obviously, valid for $n = 2$, since according to (3.6.14)

$$A_1 A_2 = N[A_1 A_2] + \underbrace{A_1 A_2} = N[A_1 A_2] + N[\underbrace{A_1 A_2}].$$

We may then presume that (3.7.1) is valid for n operators, and proceed to show that it is also valid for $n + 1$ operators. To this end, we multiply (3.7.1) on the right by the operator A_{n+1} and obtain

$$A_1 A_2 ... A_n A_{n+1} = N[A_1 ... A_n] A_{n+1} + N[\underbrace{A_1 A_2} ... A_n] A_{n+1}$$

$$+ N[\underbrace{A_1 A_2 A_3} ... A_n] A_{n+1} + ... + N[\underbrace{A_1 ... A_n}] A_{n+1} + ...$$

On the other hand, if we take $B = A_{n+1}$ in (3.7.2), the result is

$$N[A_1 A_2 ... A_n] A_{n+1} = N[A_1 A_2 ... A_n A_{n+1}] + \sum_{i=1}^{n} N[A_1 ... \underbrace{A_i ... A_n A_{n+1}}],$$

so that

$$A_1 A_2 ... A_{n+1} = N[A_1 A_2 ... A_{n+1}] + \sum_{i=1}^{n} N[A_1 ... \underbrace{A_i ... A_n A_{n+1}}]$$

$$+ N[\underbrace{A_1 A_2} A_3 ... A_n] A_{n+1} + ... + N[\underbrace{A_1 ... A_n}] A_{n+1} + ... \qquad (3.7.3)$$

Applying again lemma (3.7.2) to the third term of the r.h.s. of (3.7.3), we still have

$$A_1 A_2 ... A_{n+1} = N[A_1 A_2 ... A_{n+1}] + \sum_{i=1}^{n} N[A_1 ... \underbrace{A_i ... A_n A_{n+1}}]$$

$$+ N[\underbrace{A_1 A_2} A_3 ... A_{n+1}] + \sum_{i=3}^{n} N[\underbrace{A_1 A_2} A_3 ... \underbrace{A_i ... A_n A_{n+1}}]$$

$$+ N[\underbrace{A_1 A_2 A_3} ... A_n] A_{n+1} + ...$$

Repeating the procedure, we obtain precisely the relation (3.7.1), but this time written for $n + 1$ operators. The proof of the theorem is therefore completed.

Theorem II. *The time-ordered product of n operators equals the sum of all normal products of the operators, with all possible T-contractions, including the ordinary normal product (i.e. without contractions), that is*

$$T[A_1...A_n] = N[A_1...A_n] + N[\overset{\frown}{A_1 A_2}...A_n] + N[\overset{\frown}{A_1 A_2 A_3}...A_n] + ...$$

$$+N[\overset{\frown}{A_1...A_n}] + N[\overset{\frown}{A_1 A_2}\,\overset{\frown}{A_3 A_4}...A_n] + ... + N[\overset{\frown}{A_1 A_2}\,\overset{\frown}{A_3...A_n}] + ... \quad (3.7.4)$$

The proof of this theorem is based on the previous theorem. According to (3.6.14) and (3.6.15), two operators A_1 and A_2 obey the relations

$$A_1 A_2 = N[A_1 A_2] + \underset{\smile}{A_1 A_2} \; ; \qquad\qquad (3.7.5)$$

$$T[A_1 A_2] = N[A_1 A_2] + \overset{\frown}{A_1 A_2} \; . \qquad\qquad (3.7.6)$$

Next, we define the action of the operator T on the sum $B_1 + B_2$, where B_1 and B_2 are products of operators, as

$$T[B_1 + B_2] = T[B_1] + T[B_2]. \qquad\qquad (3.7.7)$$

Relation (3.7.5) then yields

$$T[A_1 A_2] = T\big[N[A_1 A_2]\big] + T\big[\underset{\smile}{A_1 A_2}\big] \qquad\qquad (3.7.8)$$

and, by means of (3.7.6)

$$T\big[N[A_1 A_2]\big] = N[A_1 A_2] \; ; \qquad T\big[\underset{\smile}{A_1 A_2}\big] = \overset{\frown}{A_1 A_2} \; . \qquad (3.7.9)$$

If we now apply the operator T to (3.7.1), we get:

$$T[A_1...A_2] = T\big[N[A_1...A_n]\big]$$

$$+T\big[N[\underset{\smile}{A_1 A_2}...A_n]\big]$$

$$+T\big[N[\underset{\smile}{A_1 A_2 A_3}...A_n]\big] + ...$$

$$+T\big[N[\underset{\smile}{A_1...A_n}]\big] + ... + T\big[N[\underset{\smile}{A_1 A_2}\,\underset{\smile}{A_3 A_4}...A_n]\big] + ...$$

$$+T\big[N[\underset{\smile}{A_1 A_2}...\underset{\smile}{A_{n-1} A_n}]\big] + ... + T\big[N[\underset{\smile}{A_1 A_2}\,\underset{\smile}{A_3 A_4}\,\underset{\smile}{A_5...A_n}]\big] + ...$$

$$= N[A_1...A_n] + T\left[\underbrace{A_1 A_2}\, N[A_3...A_n]\right] + T\left[\underbrace{A_1 A_3}\, N[A_2 A_4...A_n]\right] + ...$$

$$+ T\left[\underbrace{A_1 A_n}\, N[A_2...A_{n-1}]\right] + ... + T\left[\underbrace{A_1 A_2}\,\underbrace{A_3 A_4}\, N[A_5...A_n]\right] + ...$$

$$T\left[\underbrace{A_1 A_2}\,\underbrace{A_{n-1} A_n}\, N[A_3...A_{n-2}]\right] + ...$$

$$+ T\left[\underbrace{A_1 A_2}\,\underbrace{A_3 A_4}\,\underbrace{A_5 A_n}\, N[A_6...A_{n-1}]\right] + ...$$

$$= N[A_1...A_n] + \overbrace{A_1 A_2}\, N[A_3...A_n] + \overbrace{A_1 A_3}\, N[A_2 A_4...A_n] + ...$$

$$+ \overbrace{A_1 A_n}\, N[A_2...A_{n-1}] + ... + \overbrace{A_1 A_2}\,\overbrace{A_3 A_4}\, N[A_5...A_n] + ...$$

$$+ \overbrace{A_1 A_2}\,\overbrace{A_{n-1} A_n}\, N[A_3...A_{n-2}] + ...$$

$$+ \overbrace{A_1 A_2}\,\overbrace{A_3 A_4}\,\overbrace{A_5 A_n}\, N[A_6...A_{n-1}] + ...$$

Reintroducing the T-contractions under the sign of the time-ordered product, we get (3.7.4). The theorem is therefore proved.

Theorem III. *The time-ordered product of a product of normal products of operators equals the sum of all normal products of the operators, with all possible T-contractions, including the ordinary normal product (i.e. without contractions), except for the terms containing contractions under the sign of the same normal product,* that is

$$T\left[N[A_1...A_{n_1}]N[B_1...B_{n_2}]...N[X_1...X_{n_k}]\right]$$

$$= N[A_1...A_{n_1} B_1...B_{n_2}...X_{n_k}]$$

$$+ N[\overbrace{A_1...A_{n_1} B_1}\,...B_{n_2}...X_{n_k}] + N[\overbrace{A_1...A_{n_1} B_1 B_2}\,...B_{n_2}...X_{n_k}]$$

$$+ N[\overbrace{A_1...A_{n_1} B_1...B_{n_2}...X_{n_k}}] + ... + N[\overbrace{A_1...A_{n_1} B_1}\,...\,\overbrace{B_{n_2}...X_{n_k}}] + ...$$
$$(3.7.10)$$

Since the proof of this theorem is tedious, we shall limit ourselves to its justification in two simple cases.

a) Under the sign of time-ordered product there is a single normal product of two operators. In view of (3.7.5)-(3.7.8), one gets

$$T\left[N[A_1 A_2]\right] = T[A_1 A_2 - \underbrace{A_1 A_2}] = T[A_1 A_2] - \overbrace{A_1 A_2}$$

$$= N[A_1 A_2] + \overbrace{A_1 A_2} - \overbrace{A_1 A_2} = N[A_1 A_2]. \qquad (3.7.11)$$

The quantization of the scalar field

b) There are two normal products, each of them containing two operators. In view of (3.7.5)-(3.7.8) and Wick's second theorem (3.7.4), we can write:

$$T\big[N[A_1 A_2]N[B_1 B_2]\big] = T\big[(A_1 A_2 - \underbrace{A_1 A_2})(B_1 B_2 - \underbrace{B_1 B_2})\big]$$

$$= T[A_1 A_2 B_1 B_2] - T[A_1 A_2]\,\overbrace{B_1 B_2} - \overbrace{A_1 A_2}\,T[B_1 B_2] + \overbrace{A_1 A_2}\,\overbrace{B_1 B_2}$$

$$= N[A_1 A_2 B_1 B_2] + N[\overbrace{A_1 A_2}\,B_1 B_2] + N[\overbrace{A_1 A_2 B_1}\,B_2] + N[\overbrace{A_1 A_2 B_1 B_2}]$$

$$+ N[A_1\,\overbrace{A_2 B_1}\,B_2] + N[A_1\,\overbrace{A_2 B_1 B_2}] + N[A_1 A_2\,\overbrace{B_1 B_2}]$$

$$+ N[\overbrace{A_1 A_2}\,\overbrace{B_1 B_2}] + N[\{A_1\,\overbrace{A_2 B_1}\}B_2] + N[A_1\,\overbrace{\overbrace{A_2 B_1}\,B_2}]$$

$$- N[A_1 A_2]\,\overbrace{B_1 B_2} - \overbrace{A_1 A_2}\,\overbrace{B_1 B_2} - \overbrace{A_1 A_2}\,N[B_1 B_2]$$

$$- \overbrace{A_1 A_2}\,\overbrace{B_1 B_2} + \overbrace{A_1 A_2}\,\overbrace{B_1 B_2}. \tag{3.7.12}$$

Since the T-contractions can be either introduced into or extracted from the sign of the normal product, some terms in (3.7.12) disappear, and we finally are left with

$$T\big[N[A_1 A_2]N[B_1 B_2]\big] = N[A_1 A_2 B_1 B_2] + N[\overbrace{A_1 A_2 B_1}\,B_2]$$

$$+ N[\overbrace{A_1 A_2 B_1 B_2}] + N[A_1\,\overbrace{A_2 B_1}\,B_2] + N[A_1\,\overbrace{A_2 B_1 B_2}]$$

$$+ N[\overbrace{A_1\{A_2 B_1}\,B_2\}] + N[A_1\,\overbrace{\overbrace{A_2 B_1}\,B_2}], \tag{3.7.13}$$

and the validity of (3.7.10) is again satisfied. Using the same procedure, a long but elementary calculation proves Wick's third theorem (3.7.10).

82

CHAPTER IV

THE QUANTIZATION OF THE ELECTROMAGNETIC FIELD

> *Whatever Nature has in store for mankind, unpleasant as it may be, men must accept, for ignorance is never better than knowledge.*
>
> Enrico Fermi

4.1. The Lagrangian formalism

As we mentioned in the Preface, one can say that QFT was "born" in 1927, when P.A.M. Dirac succeeded in quantizing the electromagnetic field.

As well-known, the fundamental equations of the electromagnetic field (i.e. Maxwell's equations) are obtained starting from of the Lagrangian density

$$\mathcal{L} = -\frac{1}{4}F_{ik}F_{ik}, \qquad (4.1.1)$$

where

$$F_{ik} = \frac{\partial A_k}{\partial x_i} - \frac{\partial A_i}{\partial x_k} = A_{k,i} - A_{i,k} \qquad (4.1.2)$$

is the electromagnetic field tensor, and A_i is the potential four-vector. It is also known that A_i is not uniquely determined by (4.1.2). Indeed, if we replace A_i by

$$A_i' = A_i + \frac{\partial f}{\partial x_i}, \qquad (4.1.3)$$

we obtain the same F_{ik}, i.e. the same field. The transformation (4.1.3) is known as a *gauge condition*. To uniquely determine the four-potential A_i, one introduces a supplementary condition, known as *the Lorentz gauge condition*

$$\frac{\partial A_i}{\partial x_i} = A_{i,i} = 0. \qquad (4.1.4)$$

The quantization of the electromagnetic field

Using (4.1.4), Maxwell's equations yield

$$\Box A_i = A_{i,kk} = 0 \qquad (4.1.5)$$

which is the wave equation for A_i.

A straight quantum transposition of this formalism meets with some difficulties, and here are three of them:

a) The role of the field variables $U_{(r)}$ is played in our case by the four-potential components A_i, which satisfy the wave equations (4.1.5). Unfortunately, these equations do not result *directly* by means of the Lagrangian density (4.1.1), but only through Maxwell's equations and Lorentz' condition (4.1.4).

b) The quantization condition

$$[\pi_{0(i)}^+(\vec{x},t),\; U_{(k)}(\vec{x}',t)] = \frac{1}{i}\delta_{(i)(k)}\delta(\vec{x}-\vec{x}') \qquad (4.1.6)$$

cannot be satisfied. Indeed, taking into account the fact that the electromagnetic field is real, and using (1.1.16) and (4.1.1), we have:

$$\pi_{lr} = \frac{\partial \mathcal{L}}{\partial A_{r,l}} = -\frac{1}{2}F_{ik}\frac{\partial}{\partial A_{r,l}}(A_{k,i} - A_{i,k})$$

$$= -\frac{1}{2}F_{ik}(\delta_{kr}\delta_{il} - \delta_{ir}\delta_{kl}) = -\frac{1}{2}(F_{lr} - F_{rl}) = F_{rl}, \qquad (4.1.7)$$

where we used the property $F_{ik} = -F_{ki}$. However, in this case the quantization condition (4.1.6) is not satisfied. Taking, for example, $l = r = 4$ in (4.1.7), we get $\pi_{44} = F_{44} = 0,$ and therefore

$$[\pi_{44}(\vec{x},t),\; A_4(\vec{x}',t)] = 0,$$

which is in contradiction with (4.1.6).

c) Consider the Dirac equation (see 5.1.9), written in a condensed form

$$\hat{D}\,|\psi>= 0, \qquad (4.1.8)$$

as well as the equation

$$\hat{A}_{i,i}\,|\psi>= 0, \qquad (4.1.9)$$

written by virtue of the fact that in QFT $\hat{A}_i$ is an operator. It can be shown that $\hat{D}$ and $\hat{A}_{i,i}$ do not commute, that is

$$[\hat{D}, \hat{A}_{i,i}] \neq 0. \qquad (4.1.10)$$

This would mean that these two operators cannot have common proper states. In other words, the electromagnetic and the electron-positron fields couldn't coexist. Such a conclusion would be absurd, because it would contradict reality.

Analyzing the above contradictions, we realize that they emerge from the fact that we considered some macroscopic laws as being valid in microcosmos. For example, it follows from (b) that we are not allowed to apply the Lorentz gauge condition in its classical form (4.1.4) but, at the same time, we cannot discard it, as being essential in macroscopic field theory. As we shall see, condition (4.1.4) will be replaced by a *weaker* condition, with the property that it leads to (4.1.4) for expectation values, i.e. for macrocosmos. We therefore have to modify the classical (macroscopic) formalism in such a way that the classical laws implying non-quantum quantities remain unchanged. In the following we propose ourselves to find a microscopic theory, leading to the classical formalism for expectation values of the operators, and able to eliminate the foregoing contradictions.

First, we have to choose a suitable Lagrangian density, capable of leading *straight* to equation (4.1.5). Such a Lagrangian density was given by *Dirac* and *Fock* in 1932 as

$$\mathcal{L} = -\frac{1}{4}F_{ik}F_{ik} - \frac{1}{2}A_{i,i}A_{k,k}. \qquad (4.1.11)$$

This Lagrangian density can be written in a more appropriate form observing that

$$\mathcal{L} = -\frac{1}{4}(A_{k,i} - A_{i,k})(A_{k,i} - A_{i,k}) - \frac{1}{2}A_{i,i}A_{k,k}$$

$$= -\frac{1}{2}(A_{k,i}A_{k,i} - A_{i,k}A_{k,i} + A_{i,i}A_{k,k})$$

$$= -\frac{1}{2}A_{k,i}A_{k,i} + \frac{1}{2}\frac{\partial}{\partial x_i}(A_{i,k}A_k - A_iA_{k,k}) - \frac{1}{2}A_kA_{i,ki} + \frac{1}{2}A_iA_{k,ki}. \qquad (4.1.12)$$

A convenient change of summation indices shows that the sum of the last two terms is zero. Recalling that the field equations do not change if one drops a four-divergence, we finally obtain the Lagrangian density

$$\mathcal{L} = -\frac{1}{2}A_{k,i}A_{k,i}. \qquad (4.1.13)$$

Comparing (4.1.13) with the Lagrangian density of the real scalar field

$$\mathcal{L} = -\frac{1}{2}\psi_{,i}\psi_{,i} - \frac{1}{2}m^2\psi^2$$

The quantization of the electromagnetic field

and keeping in mind that the photon mass is zero, a remarkable analogy is observed.

According to (1.1.16), the momentum densities are

$$\pi_{kr} = \frac{\partial \mathcal{L}}{\partial A_{r,k}} = - A_{r,k}.\tag{4.1.14}$$

Substituting now (4.1.13) and (4.1.14) into (1.1.17), we obtain directly equation (4.1.5). This way the difficulties mentioned at points (a) and (b) are solved (since $\pi_{44} \neq 0$), while point (c) will be addressed in Section 4.6.

Following (1.5.6), the canonic energy tensor writes

$$T_{ik}^{(c)} = \pi_{kr} A_{r,i} - \mathcal{L}\delta_{ik} = \frac{1}{2} A_{l,r} A_{l,r} \delta_{ik} - A_{r,i} A_{r,k}.\tag{4.1.15}$$

The space components of the momentum four-vector, according to (3.1.17), are

$$P_\alpha = i \int T_{\alpha 4}^{(c)} \, d\vec{x} = - i \int A_{r,\alpha} A_{r,4} \, d\vec{x},\tag{4.1.16}$$

while the Hamiltonian is

$$H = \int T_{44}^{(c)} \, d\vec{x} = \int \left[\frac{1}{2} A_{r,l} A_{r,l} - A_{r,4} A_{r,4} \right] d\vec{x}$$

$$= \frac{1}{2} \int (A_{r,\alpha} A_{r,\alpha} - A_{r,4} A_{r,4}) \, d\vec{x}.\tag{4.1.17}$$

Since the electromagnetic field is real, the photon charge is zero. But, unlike the scalar field, the spin of the electromagnetic field quanta differs from zero. Using (1.3.6), (1.6.14), and (4.1.14), we obtain

$$S_{ikl} = i(\pi_{li} A_k - \pi_{lk} A_i) = i(A_i A_{k,l} - A_k A_{i,l}).\tag{4.1.19}$$

4.2. Momentum representation

By analogy with the theory developed in Chapter III, we write the potential four-vector as

$$A_j(x) = A_{j+}(x) + A_{j-}(x)\tag{4.2.1}$$

where $A_{j+}(x)$ and $A_{j-}(x)$ are the positive and negative frequency parts of $A_j(x)$, respectively. By means of (3.3.38), the Fourier transform of $A_{j\pm}(x)$ is

$$A_{j\pm}(x) = \frac{1}{(2\pi)^{3/2}} \int \frac{1}{\sqrt{2\omega}} A_{j\pm}(\vec{k}) e^{\pm ikx} \, d\vec{k}, \qquad (4.2.2)$$

where k is the momentum four-vector of the photon

$$k_1 = k_x, \ k_2 = k_y, \ k_3 = k_z, \ k_4 = i k_0 = i\omega. \qquad (4.2.3)$$

Since photons do not have mass, energy-momentum relation reduces to

$$k^2 = \vec{k}^2 - \omega^2 = 0, \qquad (4.2.4)$$

which yields

$$|\vec{k}| = \omega. \qquad (4.2.5)$$

By means of (4.2.1) and (4.2.2), we can write

$$A_j(x) = \frac{1}{(2\pi)^{3/2}} \int \frac{1}{\sqrt{2\omega}} \left[A_{j+}(\vec{k}) e^{ikx} + A_{j-}(\vec{k}) e^{-ikx} \right] d\vec{k}. \qquad (4.2.6)$$

The series expansion of $A_{j+}(x)$ is

$$A_j^+(x) = \frac{1}{(2\pi)^{3/2}} \int \frac{1}{\sqrt{2\omega}} \left[A_{j+}^+(\vec{k}) e^{ikx} + A_{j-}^+(\vec{k}) e^{-ikx} \right] d\vec{k}, \qquad (4.2.7)$$

while its complex conjugate writes

$$A_j^*(x) = \frac{1}{(2\pi)^{3/2}} \int \frac{1}{\sqrt{2\omega}} \left[A_{j+}^*(\vec{k}) e^{-ikx} + A_{j-}^*(\vec{k}) e^{ikx} \right] d\vec{k}. \qquad (4.2.8)$$

Let us now take the components of the four-potential A_j

$$A_1 = A_x, \ A_2 = A_y, \ A_3 = A_z, A_4 = iA_0 = iV, \qquad (4.2.9)$$

and impose condition that these components are observable quantities. Then

$$A_j^+(x) = A_j(x) \,; \quad A_{j\pm}^+(x) = A_{j\pm}(x) \,, \quad \text{but } \left[A_{j\pm}(x)\right]^+ = A_{j\mp}(x). \qquad (4.2.10)$$

Comparing (4.2.6) with (4.2.7), we also have

$$A_{j\pm}^+(\vec{k}) = A_{j\pm}(\vec{k}) \,, \quad \text{but } \left[A_{j\pm}(\vec{k})\right]^+ = A_{j\mp}(\vec{k}). \qquad (4.2.11)$$

The quantization of the electromagnetic field

On the other hand, according to (4.2.9) the first three components of $A_j(x)$ are real, that is

$$A_\alpha^*(x) = A_\alpha(x) \; ; \; (\alpha = 1, 2, 3), \tag{4.2.12}$$

and then, comparing (4.2.6) with (4.2.8), we have

$$A_{\alpha\pm}^*(\vec{k}) = A_{\alpha\mp}(\vec{k}). \tag{4.2.13}$$

The fourth component being imaginary, we can write

$$A_4^*(x) = -A_4(x), \tag{4.2.14}$$

and comparing again (4.2.6) with (4.2.8), we obtain

$$A_{4\pm}^*(\vec{k}) = -A_{4\mp}(\vec{k}). \tag{4.2.15}$$

This last relation shows that the expectation value of A_4 is also imaginary, which contradicts our initial hypothesis. We shall circumvent this difficulty later in this Chapter.

The quantization of the electromagnetic field is intimately connected with *photon polarization*. In this respect, we define a set of unit polarization vectors, associated with the direction of motion of the photon. At any point, we can define a system of four orthogonal unit vectors $e_j^1, e_j^2, e_j^3, e_j^4$ that satisfy the orthogonality condition

$$e_j^l e_j^{l'} = \delta^{ll'}. \tag{4.2.16}$$

We choose e_j^3 oriented along the direction of motion of the photon (i.e. in the $\vec{k}$ direction), that is

$$\vec{e}_j^{\,3} = \frac{\vec{k}}{|\vec{k}|} = \frac{\vec{k}}{\omega}, \tag{4.2.17}$$

while e_j^4 is chosen along the direction of time. In this case, e_j^1 and e_j^2 ore orthogonal on the direction of motion of the photon. This choice allows us to call the photons directed along e_j^1 and e_j^2 - *transversal photons*, those associated with e_j^3 - *longitudinal photons*, and the photons associated with e_j^4 - *temporal photons*. The transversal photons are also called *real photons*, while the longitudinal and temporal photons are called *virtual photons* or *pseudo-photons*. Even though macroscopic measurements detect only real photons, virtual photons are indispensable in microscopic theory. For example, the electrostatic interaction is explained through constant emission and absorption of virtual photons, while the nuclear interaction consists in the continuous emission and absorption of virtual mesons.

In the particular case when the unit vectors e_j^l are oriented along the coordinate axes, that is $\bar{e}_j^l = \delta_j^l$, we have

$$\bar{e}_j^l \, \bar{e}_{j'}^l = \delta_{jj'}, \tag{4.2.18}$$

and, since the scalar product is invariant with respect to coordinate transformations, we can also write the completeness relation

$$e_j^l \, e_{j'}^l = \delta_{jj'}. \tag{4.2.19}$$

It is instructive to arrive at the last relation in a different manner. Let E_{lj} be the matrix with e_j^l as elements. If $\tilde{E}$ is the transpose of matrix E, then (4.2.19) becomes

$$E \, \tilde{E} = I \quad \text{(unit matrix)} .$$

On the other hand, since the matrix E is nonsingular, it belongs to a group, which means that

$$\tilde{E} \, E = E \, \tilde{E} = I,$$

that is

$$\tilde{E}_{jl} \, E_{lj'} = \delta_{jj'},$$

and (4.2.19) follows immediately.

Using the unit polarization vectors e_j^l, any four-vector $A_j(\vec{k})$ can be written as

$$A_j(\vec{k}) = e_j^l \, a_l(\vec{k}). \tag{4.2.20}$$

Obviously, we also have

$$A_{j\pm}(\vec{k}) = e_j^l \, a_{l\pm}(\vec{k}). \tag{4.2.21}$$

4.3. Momentum, energy and spin of the electromagnetic field in momentum representation

a) **Momentum.** Using (4.1.16), we can write

$$P_\alpha = -i \int A_{r,\alpha} A_{r,4} \, d\vec{x} = P_{\alpha 1+} + P_{\alpha 1-} + P_{\alpha 2+} + P_{\alpha 2-}, \tag{4.3.1}$$

The quantization of the electromagnetic field

where

$$P_{\alpha 1\pm} = -i \int A_{r\pm,\alpha} A_{r\pm,4} \, d\vec{x} \; ; \tag{4.3.2}$$

$$P_{\alpha 2\pm} = -i \int A_{r\pm,\alpha} A_{r\mp,4} \, d\vec{x} \; . \tag{4.3.3}$$

In view of (4.2.2), we have

$$A_{r\pm,\alpha}(x) = \frac{1}{(2\pi)^{3/2}} \int \frac{\pm ik_\alpha}{\sqrt{2\omega}} A_{r\pm}(\vec{k}) \, e^{\pm ikx} \, d\vec{k} \; ; \tag{4.3.4}$$

$$A_{r\pm,4}(x) = \frac{1}{(2\pi)^{3/2}} \int \frac{\mp \omega}{\sqrt{2\omega}} A_{r\pm}(\vec{k}) \, e^{\pm ikx} \, d\vec{k} \; ; \tag{4.3.5}$$

Let us first calculate $P_{\alpha 1+}$. Using (4.3.2), (4.3.4), and (4.3.5), we can write:

$$P_{\alpha 1+} = -i \int A_{r+,\alpha} A_{r+,4} \, d\vec{x}$$

$$= -\frac{i}{(2\pi)^3} \int \frac{ik_\alpha}{\sqrt{2\omega}} \frac{(-\omega')}{\sqrt{2\omega'}} A_{r+}(\vec{k}) A_{r+}(\vec{k}') \, e^{i(k+k')x} \, d\vec{x} \, d\vec{k} \, d\vec{k}'. \tag{4.3.6}$$

Since

$$i(k+k')x = i(\vec{k}+\vec{k}')\cdot\vec{x} - i\,(\omega+\omega')t, \tag{4.3.7}$$

and

$$\frac{1}{(2\pi)^3} \int e^{i(\vec{k}+\vec{k}')\cdot\vec{x}} \, d\vec{x} = \delta(\vec{k}+\vec{k}'), \tag{4.3.8}$$

we have

$$P_{\alpha 1+} = -\int \frac{k_\alpha \omega'}{\sqrt{4\omega\omega'}} A_{r+}(\vec{k}) A_{r+}(\vec{k}') \, \delta(\vec{k}+\vec{k}') \, e^{-i(\omega+\omega')t} \, d\vec{k} \, d\vec{k}', \tag{4.3.9}$$

and, integrating over $\vec{k}'$

$$P_{\alpha 1+} = -\int \frac{k_\alpha}{2} A_{r+}(\vec{k}) A_{r+}(-\vec{k}) \, e^{-2i\omega t} \, d\vec{k}. \tag{4.3.10}$$

Replacing $\vec{k}$ by $-\vec{k}$, we finally obtain

$$P_{\alpha 1+} = \int \frac{k_\alpha}{2} A_{r+}(-\vec{k}) A_+(\vec{k}) \, e^{-2i\omega t} \, d\vec{k} = -P_{\alpha 1+},$$

that is

$$P_{\alpha 1+} = 0. \tag{4.3.11}$$

Momentum, energy and spin

Similarly, we find
$$P_{\alpha 1-} = 0. \qquad (4.3.12)$$

Let us now calculate $P_{\alpha 2+}$. We have:

$$P_{\alpha 2+} = -i \int A_{r+,\alpha} A_{r-,4}\, d\vec{x} = -\frac{i}{(2\pi)^3} \int \frac{ik_\alpha}{\sqrt{2\omega}} \frac{\omega'}{\sqrt{2\omega'}} A_{r+}(\vec{k}) A_{r-}(\vec{k}')$$

$$\times e^{i(\vec{k}-\vec{k}')\cdot\vec{x}} e^{-i(\omega-\omega')t}\, d\vec{x}\, d\vec{k}\, d\vec{k}'.$$

Integrating with respect to $\vec{x}$, then $\vec{k}'$, we are left with

$$P_{\alpha 2+} = \int \frac{k_\alpha}{2} A_{r+}(\vec{k}) A_{r-}(\vec{k})\, d\vec{k}. \qquad (4.3.13)$$

Following the same procedure, we find

$$P_{\alpha 2-} = \int \frac{k_\alpha}{2} A_{r-}(\vec{k}) A_{r+}(\vec{k})\, d\vec{k}. \qquad (4.3.14)$$

Substituting (4.3.11)-(4.3.14) in (4.3.1), we obtain

$$P_\alpha = \frac{1}{2} \int k_\alpha \left[A_{r+}(\vec{k}) A_{r-}(\vec{k}) + A_{r-}(\vec{k}) A_{r+}(\vec{k}) \right] d\vec{k}. \qquad (4.3.15)$$

We mention that in (4.3.15) the operators $A_{r+}(\vec{k})$ and $A_{r-}(\vec{k})$ have to be taken under the sign of normal product, but we did not pay attention to this aspect, for the moment.

In view of (4.2.21), the product $A_{r\pm}(\vec{k}) A_{r\mp}(\vec{k})$ can also be written as

$$A_{r\pm}(\vec{k}) A_{r\mp}(\vec{k}) = e_r^l e_r^{l'} a_{l\pm}(\vec{k}) a_{l'\mp}(\vec{k})$$

$$= \delta^{ll'} a_{l\pm}(\vec{k}) a_{l'\mp}(\vec{k}) = a_{l\pm}(\vec{k}) a_{l\mp}(\vec{k}), \qquad (4.3.16)$$

and (4.3.15) becomes

$$P_\alpha = \int k_\alpha \left[a_{r+}(\vec{k}) a_{r-}(\vec{k}) + a_{r-}(\vec{k}) a_{r+}(\vec{k}) \right] d\vec{k}. \qquad (4.3.17)$$

b) **Hamiltonian.** By means of (4.1.17), we can write:

$$H = \frac{1}{2} \int (A_{r,\alpha} A_{r,\alpha} - A_{r,4} A_{r,4})\, d\vec{x} = H_{1+} + H_{1-} + H_{2+} + H_{2-}, \qquad (4.3.18)$$

91

where

$$H_{1\pm} = \frac{1}{2} \int (A_{r\pm,\alpha} A_{r\pm,\alpha} - A_{r\pm,4} A_{r\pm,4})\, d\vec{x} = H_{1\pm}^{(1)} - H_{1\pm}^{(2)} \; ; \quad (4.3.19)$$

$$H_{2\pm} = \frac{1}{2} \int (A_{r\pm,\alpha} A_{r\mp,\alpha} - A_{r\pm,4} A_{r\mp,4})\, d\vec{x} = H_{2\pm}^{(1)} - H_{2\pm}^{(2)} \; ; \quad (4.3.20)$$

We have:

$$H_{1+}^{(1)} = \frac{1}{2} \int A_{r+,\alpha} A_{r+,\alpha}\, d\vec{x}$$

$$= \frac{1}{2} \frac{1}{(2\pi)^3} \int \frac{-k_\alpha k_\alpha'}{\sqrt{4\omega\omega'}} A_{r+}(\vec{k}) A_{r+}(\vec{k}')\, e^{i(k+k')x}\, d\vec{x}\, d\vec{k}\, d\vec{k}'. \quad (4.3.21)$$

Using (4.3.7) and (4.3.8), and integrating with respect to $\vec{x}$, then $\vec{x}'$, we also have:

$$H_{1+}^{(1)} = \frac{1}{2} \int \frac{-k_\alpha k_\alpha'}{\sqrt{4\omega\omega'}} A_{r+}(\vec{k}) A_{r+}(\vec{k}') \delta(\vec{k} + \vec{k}')\, e^{-i(\omega+\omega')t}\, d\vec{k}\, d\vec{k}'$$

$$= \frac{1}{2} \int \frac{k_\alpha k_\alpha}{2\omega} A_{r+}(\vec{k}) A_{r+}(-\vec{k}) e^{-2i\omega t}\, d\vec{k}. \quad (4.3.22)$$

Following the same procedure, we also obtain

$$H_{1+}^{(2)} = \frac{1}{2} \int A_{r+,4} A_{r+,4}\, d\vec{x}$$

$$= \frac{1}{2} \frac{1}{(2\pi)^3} \int \frac{\omega\omega'}{\sqrt{4\omega\omega'}} A_{r+}(\vec{k}) A_{r+}(\vec{k}')\, e^{i(k+k')x}\, d\vec{x}\, d\vec{k}\, d\vec{k}'$$

$$= \frac{1}{2} \int \frac{\omega\omega'}{\sqrt{4\omega\omega'}} A_{r+}(\vec{k}) A_{r+}(\vec{k}')\, \delta(\vec{k} + \vec{k}')\, e^{-i(\omega+\omega')t}\, d\vec{k}\, d\vec{k}'$$

$$= \frac{1}{2} \int \frac{\omega^2}{2\omega} A_{r+}(\vec{k}) A_{r+}(-\vec{k}) e^{-2i\omega t}\, d\vec{k}. \quad (4.3.23)$$

Substituting (4.3.22) and (4.3.23) in (4.3.19), we get

$$H_{1+} = \frac{1}{2} \int \frac{k_\alpha k_\alpha - \omega^2}{2\omega} A_{r+}(\vec{k}) A_{r+}(-\vec{k}) e^{-2i\omega t}\, d\vec{k} = 0, \quad (4.3.24)$$

since, according to the energy-momentum relation, $k_\alpha k_\alpha - \omega^2 = k^2 = 0$. A similar calculation gives

$$H_{1-} = 0. \quad (4.3.25)$$

Momentum, energy and spin

Next, we also have:

$$H_{2+}^{(1)} = \frac{1}{2} \int \frac{k_\alpha k_\alpha}{2\omega} A_{r+}(\vec{k}) A_{r-}(\vec{k}) \, d\vec{k} \; ;$$

$$H_{2+}^{(2)} = \frac{1}{2} \int \frac{-\omega^2}{2\omega} A_{r+}(\vec{k}) A_{r-}(\vec{k}) \, d\vec{k} \; .$$

Since $k_\alpha k_\alpha + \omega^2 = |\vec{k}|^2 + \omega^2 = 2\omega^2$, we can write

$$H_{2+} = H_{2+}^{(1)} - H_{2+}^{(2)} = \frac{1}{2} \int \omega A_{r+}(\vec{k}) A_{r-}(\vec{k}) \, d\vec{k}. \qquad (4.3.26)$$

In the same way, we find

$$H_{2-} = \frac{1}{2} \int \omega A_{r-}(\vec{k}) A_{r+}(\vec{k}) \, d\vec{k}. \qquad (4.3.27)$$

Using (4.3.24)-(4.3.27) and (4.3.18), we are left with the momentum representation of the Hamiltonian of the electromagnetic field

$$H = \frac{1}{2} \int \omega [A_{r+}(\vec{k}) A_{r-}(\vec{k}) + A_{r-}(\vec{k}) A_{r+}(\vec{k})] \, d\vec{k}, \qquad (4.3.28)$$

or, if we consider (4.3.16),

$$H = \frac{1}{2} \int \omega [a_{r+}(\vec{k}) a_{r-}(\vec{k}) + a_{r-}(\vec{k}) a_{r+}(\vec{k})] \, d\vec{k}, \qquad (4.3.29)$$

c) **Photon spin.** As we already know (see (1.6.7) and (1.6.11)), the proper angular momentum of the field is given by

$$I_{ik} = \int S_{ikl} \, dS_l, \qquad (4.3.30)$$

where S_{ikl} is the proper angular momentum density, or *spin density*. Suppose, once more, that the time t is fixed. In this case, the only non-zero hypersurface element is $dS_4 \equiv d\vec{x}$. The spatial components of the spin momentum then are

$$I_{\alpha\beta} = \int S_{\alpha\beta4} \, d\vec{x}. \qquad (4.3.31)$$

Taking into account (4.1.19), we have

$$S_{\alpha\beta4} = i(A_\alpha A_{\beta,4} - A_\beta A_{\alpha,4}). \qquad (4.3.32)$$

93

The quantization of the electromagnetic field

Let $\vec{S}$ be the axial vector attached to the antisymmetric, second order tensor $I_{\alpha\beta}$. Its components are

$$S_1 = \int S_{234} \, d\vec{x} \; ; \quad S_2 = \int S_{314} \, d\vec{x} \; ; \quad S_3 = \int S_{124} \, d\vec{x}. \quad (4.3.33)$$

Using these notations, (4.3.31) can be written as

$$I_{\alpha\beta} = I'_{\alpha\beta} - I''_{\alpha\beta}, \quad (4.3.34)$$

with

$$I'_{\alpha\beta} = i \int A_\alpha A_{\beta,4} \, d\vec{x} \; ; \quad (4.3.35)$$

$$I''_{\alpha\beta} = i \int A_\beta A_{\alpha,4} \, d\vec{x} \; ; \quad (4.3.36)$$

Using the already known procedure, we decompose $I'_{\alpha\beta}$ as

$$I'_{\alpha\beta} = I'_{\alpha\beta1+} + I'_{\alpha\beta1-} + I'_{\alpha\beta2+} + I'_{\alpha\beta2-}, \quad (4.3.37)$$

where

$$I'_{\alpha\beta1\pm} = i \int A_{\alpha\pm} A_{\beta\pm,4} \, d\vec{x} \; ; \quad (4.3.38)$$

$$I'_{\alpha\beta2\pm} = i \int A_{\alpha\pm} A_{\beta\mp,4} \, d\vec{x} \; . \quad (4.3.39)$$

Then, we have:

$$I'_{\alpha\beta1+} = i \int A_{\alpha+} A_{\beta+,4} \, d\vec{x}$$

$$= \frac{i}{(2\pi)^3} \int \frac{1}{\sqrt{2\omega}} \frac{-\omega'}{\sqrt{2\omega'}} A_{\alpha+}(\vec{k}) A_{\beta+}(\vec{k}') \quad (4.3.40)$$

$$\times e^{i(\vec{k}+\vec{k}')\cdot\vec{x}} e^{-i(\omega+\omega')t} \, d\vec{x} \, d\vec{k} \, d\vec{k}'$$

$$= -i \int \frac{\omega'}{\sqrt{4\omega\omega'}} A_{\alpha+}(\vec{k}) A_{\beta+}(\vec{k}') \, \delta(\vec{k}+\vec{k}') \, e^{-i(\omega+\omega')t} \, d\vec{k} \, d\vec{k}'$$

$$= -\frac{i}{2} \int A_{\alpha+}(\vec{k}) A_{\beta+}(-\vec{k}) \, e^{-2i\omega t} \, d\vec{k} \; .$$

Similarly, we find:

$$I'_{\alpha\beta1-} = \frac{i}{2} \int A_{\alpha-}(\vec{k}) A_{\beta-}(-\vec{k}) \, e^{2i\omega t} \, d\vec{k} \; ; \quad (4.3.41)$$

94

Momentum, energy and spin

$$I'_{\alpha\beta2+} = \frac{i}{2} \int A_{\alpha+}(\vec{k})A_{\beta-}(\vec{k}) \, d\vec{k} \; ; \qquad (4.3.42)$$

$$I'_{\alpha\beta2-} = -\frac{i}{2} \int A_{\alpha-}(\vec{k})A_{\beta+}(\vec{k}) \, d\vec{k} \; ; \qquad (4.3.43)$$

Substituting (4.3.40)-(4.3.43) in (4.3.37), we obtain:

$$I'_{\alpha\beta} = \frac{i}{2} \int \left[- A_{\alpha+}(\vec{k})A_{\beta+}(-\vec{k})e^{-2i\omega t} + A_{\alpha-}(\vec{k})A_{\beta-}(-\vec{k})e^{2i\omega t} \right.$$

$$\left. + A_{\alpha+}(\vec{k})A_{\beta-}(\vec{k}) + A_{\alpha-}(\vec{k})A_{\beta+}(\vec{k}) \right] \, d\vec{k}. \qquad (4.3.44)$$

As one observes, $I'_{\alpha\beta}$ becomes $I''_{\alpha\beta}$ following index interchange $\alpha \leftrightarrow \beta$, that is

$$I''_{\alpha\beta} = \frac{i}{2} \int \left[- A_{\beta+}(\vec{k})A_{\alpha+}(-\vec{k})e^{-2i\omega t} + A_{\beta-}(\vec{k})A_{\alpha-}(-\vec{k})e^{2i\omega t} \right.$$

$$\left. + A_{\beta+}(\vec{k})A_{\alpha-}(\vec{k}) + A_{\beta-}(\vec{k})A_{\alpha+}(\vec{k}) \right] \, d\vec{k}. \qquad (4.3.45)$$

Using $-\vec{k}$ instead of $\vec{k}$ in the first two terms of (3.4.45), then substituting (4.5.44) and (4.3.45) in (4.3.34), we obtain

$$I_{\alpha\beta} = \frac{i}{2} \int \left[A_{\alpha+}(\vec{k})A_{\beta-}(\vec{k}) + A_{\beta-}(\vec{k})A_{\alpha+}(\vec{k}) \right.$$

$$\left. - A_{\alpha-}(\vec{k})A_{\beta+}(\vec{k}) - A_{\beta+}(\vec{k})A_{\alpha-}(\vec{k}) \right] \, d\vec{k}, \qquad (4.3.46)$$

where, anticipating, we have changed the order of the operators $A_{\beta\pm}(-\vec{k})$ and $A_{\alpha\pm}(\vec{k})$ (see (4.4.1)).

The third component $S_3 = S_z$ of the spin angular momentum is, according to (4.3.33),

$$S_z = I_{12} = \frac{i}{2} \int \left[A_{1+}(\vec{k})A_{2-}(\vec{k}) + A_{2-}(\vec{k})A_{1+}(\vec{k}) \right.$$

$$\left. - A_{1-}(\vec{k})A_{2+}(\vec{k}) - A_{2+}(\vec{k})A_{1-}(\vec{k}) \right] \, d\vec{k}. \qquad (4.3.47)$$

If the propagation takes place along the z-axis, which means $e^l_j = \delta^l_j$ (see (4.2.17)), we can write

$$A_{1\pm}A_{2\mp} = e^l_1 a_{l\pm} e^{l'}_2 a_{l'\mp} = \delta^l_1 \delta^{l'}_2 a_{l\pm} a_{l'\mp} = a_{1\pm}a_{2\mp},$$

as well as

$$A_{2\pm}A_{1\mp} = a_{2\pm}a_{1\mp}.$$

95

The quantization of the electromagnetic field

Therefore,

$$S_z = \frac{i}{2} \int \left[a_{1+}(\vec{k})a_{2-}(\vec{k}) + a_{2-}(\vec{k})a_{1+}(\vec{k}) \right.$$

$$\left. - a_{1-}(\vec{k})a_{2+}(\vec{k}) - a_{2+}(\vec{k})a_{1-}(\vec{k}) \right] d\vec{k}. \qquad (4.3.48)$$

To obtain the eigenvalues of the spin component S_z, this expression has to be put in diagonal form. To this end, we define a new set of operators $b_{1\pm}$ and $b_{2\pm}$, related to $a_{1\pm}$ and $a_{2\pm}$ by

$$a_{1\pm} = \frac{1}{\sqrt{2}}(b_{1\pm} + b_{2\pm}) ; \quad a_{3\pm} = b_{3\pm} ;$$

$$a_{2\pm} = \frac{\pm i}{\sqrt{2}}(b_{1\pm} - b_{2\pm}) ; \quad a_{4\pm} = b_{4\pm}. \qquad (4.3.49)$$

As can be noted, the field quantities P_α and H do not change under the transformation (4.3.49). Indeed, we have:

$$a_{r+}a_{r-} = \frac{1}{2}[(b_{1+} + b_{2+})(b_{1-} + b_{2-})]$$

$$+ \frac{1}{2}[(b_{1+} - b_{2+})(b_{1-} - b_{2-})] + b_{3+}b_{3-} + b_{4+}b_{4-} = b_{r+}b_{r-} ;$$

$$a_{r-}a_{r+} = b_{r-}b_{r+}. \qquad (4.3.50)$$

Using (4.3.49), relation (4.3.88) becomes

$$S_z = \frac{1}{4} \int \left[(b_{1+} + b_{2+})(b_{1-} - b_{2-}) + (b_{1-} - b_{2-})(b_{1+} + b_{2+}) \right.$$

$$\left. + (b_{1-} + b_{2-})(b_{1+} - b_{2+}) + (b_{1+} - b_{2+})(b_{1-} + b_{2-}) \right] d\vec{k},$$

or, after some simplifications

$$S_z = \frac{1}{2} \int \left[b_{1+}(\vec{k})b_{1-}(\vec{k}) + b_{1-}(\vec{k})b_{1+}(\vec{k}) \right.$$

$$\left. - b_{2+}(\vec{k})b_{2-}(\vec{k}) - b_{2-}(\vec{k})b_{2+}(\vec{k}) \right] d\vec{k}. \qquad (4.3.51)$$

As we shall see in the next section, $b_{1-}b_{1+}$ represents the number of photons with momentum $\vec{k}$, oriented along z-axis, with spin $+1$, while $b_{2-}b_{2+}$ gives the number of photons with momentum $\vec{k}$ and spin -1. Therefore, b_{1-} and b_{1+} are operators of creation and annihilation, respectively, of the first category of particles, while b_{2-} and b_{2+} play the same role for the second type of particles.

96

4.4. Commutators of the free electromagnetic field

Proceeding in the same manner as we did in the previous Chapter, let us first focus our attention on the quantization of the electromagnetic field. By means of (4.1.6) and (4.1.14), we obtain the following quantization condition

$$[A_{i,0}(\vec{x}, t), A_j(\vec{x}', t)] = -i\, \delta_{ij}\, \delta(\vec{x} - \vec{x}').$$

A similar calculation (see (3.4.12), (3.4.26)) gives

$$[A_{i\pm}(\vec{k}), A_{j\pm}(\vec{k}')] = 0 ; \tag{4.4.1}$$

$$[A_{i+}(\vec{k}), A_{j-}(\vec{k}')] = \delta_{ij}\, \delta(\vec{k} - \vec{k}'). \tag{4.4.2}$$

Using (4.2.18) and (4.2.21), it is not difficult to show that the commutation relations (4.4.1) and (4.4.2) are also valid for the operators $a_{i\pm}(\vec{k})$, that is

$$[a_{i\pm}(\vec{k}), a_{j\pm}(\vec{k}')] = 0 ; \tag{4.4.3}$$

$$[a_{i+}(\vec{k}), a_{j-}(\vec{k}')] = \delta_{ij}\, \delta(\vec{k} - \vec{k}'). \tag{4.4.4}$$

Furthermore, if $a_{i-}(\vec{k})$ and $a_{i+}(\vec{k})$ are the emission and absorption operators, respectively, for a photon with momentum $\vec{k}$ and polarization e_j^l, then

$$N_i(\vec{k}) = a_{i-}(\vec{k})\, a_{i+}(\vec{k}) \tag{4.4.5}$$

represents the number density of photons with momentum $\vec{k}$. In view of (4.4.3)-(4.4.5), the relations (4.3.17) and (4.3.29) become

$$P_\alpha = \int k_\alpha \sum_r N_r(\vec{k})\, d\vec{k} ; \qquad H = \int \omega \sum_r N_r(\vec{k})\, d\vec{k}, \tag{4.4.6}$$

where the vacuum values

$$P_{\alpha 0} = \frac{1}{2}\delta(0) \int k_\alpha\, d\vec{k}, \qquad E_0 = \frac{1}{2}\delta(0) \int \omega\, d\vec{k}$$

have been omitted.

To give a similar expression for the S_z component of the spin, it is necessary to show that the operators $b_{i\pm}(\vec{k})$ defined by (4.3.49) satisfy the same commutation relations as $a_{i\pm}(\vec{k})$. It is easily seen that

$$b_{1\pm} = \frac{1}{\sqrt{2}}(a_{1\pm} \mp ia_{2\pm}) ; \qquad b_{2\pm} = \frac{1}{\sqrt{2}}(a_{1\pm} \pm ia_{2\pm}). \tag{4.4.7}$$

The quantization of the electromagnetic field

For example,

$$[[b_{1+}, b_{1-}] = \frac{1}{2}[(a_{1+} - ia_{2+}), (a_{1-} + ia_{2-})]$$

$$= \frac{1}{2}\{[a_{1+}, a_{1-}] + i[a_{1+}, a_{2-}] - i[a_{2+}, a_{1-}] + [a_{2+}, a_{2-}]\},$$

and similarly for $[b_{2+}, b_{2-}]$. By virtue of (4.4.4), one then obtains

$$[b_{1+}(\vec{k}), b_{1-}(\vec{k}')] = [b_{2+}(\vec{k}), b_{2-}(\vec{k}')] = \delta(\vec{k} - \vec{k}'). \qquad (4.4.8)$$

Denoting

$$N_1'(\vec{k}) = b_{1-}(\vec{k}) \, b_{1+}(\vec{k}) \; ; \qquad N_2'(\vec{k}) = b_{2-}(\vec{k}) \, b_{2+}(\vec{k}) \qquad (4.4.9)$$

and omitting the vacuum terms, the z-component of the spin momentum (4.3.51) writes

$$S_z = \int [N_1'(\vec{k}) - N_2'(\vec{k})] \, d\vec{k}. \qquad (4.4.10)$$

Here $N_1'(\vec{k})$ stands for the number density of photons with momentum $\vec{k}$ and spin $+1$, while $N_2'(\vec{k})$ represents the number density of photons with momentum $\vec{k}$ and spin -1, for the wave propagating in z-direction.

The commutators of the field observables defined at different moments of time take the form (3.4.37) or (3.4.38). For instance,

$$[A_{1-}(x), A_{j+}(x')] = \frac{1}{(2\pi)^3} \int \frac{1}{\sqrt{4\omega\omega'}} [A_{i-}(\vec{k}), A_{j+}(\vec{k}')]$$

$$\times e^{-i(kx - k'x')} \, d\vec{k} \, d\vec{k}',$$

where (4.2.2) has also been taken into account. Using the commutation relation (4.4.2), we have

$$[A_{i-}(x), A_{j+}(x')] = -\frac{1}{(2\pi)^3} \int \frac{1}{\sqrt{4\omega\omega'}} \delta(\vec{k} - \vec{k}') \, \delta_{ij} \, e^{-i(kx - k'x')} \, d\vec{k} \, d\vec{k}',$$

or, if the integration with respect to $\vec{k}'$ is performed

$$[A_{i-}(x), A_{j+}(x')] = -\frac{1}{(2\pi)^3} \int \frac{1}{2\omega} \delta_{ij} \, e^{-ik(x - x')} \, d\vec{k}$$

$$= i \, \delta_{ij} \, D_{0-}(x - x'). \qquad (4.4.11)$$

Here the singular function $D_{0-}(x - x')$ is analogous to $D_{m-}(x - x')$ defined for the scalar field (see (3.4.40)), with the mention that the index 0 indicates that the the photon rest mass is zero (see also A.92). We also have

$$[A_{i+}(x), A_{j-}(x')] = -[A_{j-}(x'), A_{i+}(x)] = -i \, \delta_{ji} \, D_{0-}(x' - x),$$

and since

$$D_{0-}(x' - x) = -D_{0+}(x - x'),$$

it follows that

$$[A_{i+}(x), A_{j-}(x')] = i \, \delta_{ij} \, D_{0+}(x - x'). \qquad (4.4.12)$$

Relations (4.4.10) and (4.4.11) can be written in a condensed form as

$$[A_{i\pm}(x), A_{j\mp}(x')] = i \, \delta_{ij} \, D_{0\pm}(x - x'), \qquad (4.4.13)$$

the rest of the commutators being zero.

By means of (4.4.13), we can now define the commutator $[A_i(x), A_j(x')]$. Using (4.2.1), we have

$$[A_i(x), A_j(x')] = [A_{i+}(x), A_{j-}(x')] + [A_{i-}(x), A_{j+}(x')]$$

$$= i \, \delta_{ij} \, [D_{0+}(x - x') + D_{0-}(x - x')] = i \, \delta_{ij} \, D_0(x - x'), \qquad (4.4.14)$$

where we denoted

$$D_0(x - x') = D_{0+}(x - x') + D_{0-}(x - x'). \qquad (4.4.15)$$

The notions of *vacuum expectation value* and *vacuum contraction* are preserved in the case of the electromagnetic field. Therefore, the vacuum expectation value of the product of operators $A_i(x)$ and $A_j(y)$ is defined by

$$< A_i(x) \, A_j(y) >_0 = \underbrace{A_i(x) \, A_j(y)} = i \, \delta_{ij} \, D_{0+}(x - y), \qquad (4.4.16)$$

while the vacuum expectation value of the time-ordered product of these operators writes

$$< T[A_i(x) \, A_j(y)] >_0 = \overbrace{A_i(x) \, A_j(y)} = \delta_{ij} \, D_0^c(x - y). \qquad (4.4.17)$$

4.5. The indefinite metric formalism

As we mentioned in Section 2, there are some difficulties in quantizing the electromagnetic field. These problems are connected with certain commutation relations of the four-vector operator A_j, whose components are given by (4.2.9). If A_j is supposed to be Hermitian, i.e. $A_j^+ = A_j$, possessing real eigenvalues, then according to

$$A_4 = i\, A_0 \; ; \qquad A_{4\pm} = i\, A_{0\pm} \qquad (4.5.1)$$

one obtains

$$A_4^+ = -\, iA_0^+ \; ; \qquad A_{4\pm}^+ = -\, iA_{0\pm}^+, \qquad (4.5.2)$$

and one has

$$A_0^+ = -\, A_0 \; ; \qquad A_{0\pm} = -\, A_{0\pm}, \qquad (4.5.3)$$

meaning that $A_{0\pm}$ is antihermitian, and has imaginary eigenvalues. At the same time, in view of (4.5.2), we can write

$$[A_{0+}^+, \, A_{0-}] = [A_{4+}^+, \, A_{4-}], \qquad (4.5.4)$$

meaning that both sets of operators $A_{0\pm}$ and $A_{4\pm}$ satisfy the same commutation rule. However, both the eigenvalues and the expectation values of the operators $A_{0\pm}$ must be real. On the other hand, even if the operators $A_{4\pm}$ are Hermitian, they should have both imaginary eigenvalues and imaginary expectation values. We also remember that, if $A_{0\pm}$ is antihermitian, then its expectation value is imaginary. Indeed, we have:

$$< \psi|A_{0\pm}|\psi >^* = < \psi|A_{0\pm}^+|\psi >= -\; < \psi|A_{0\pm}|\psi >,$$

or

$$< A_{0\pm} >^* = -\; < A_{0\pm} > . \qquad (4.5.5)$$

These difficulties have been eliminated in by *K.Bleuler* (Helv. Phys.Acta, v.23, p.567586, 1950) and *S.Gupta* (Proc.Phys.Soc, v.A63, nr.267, p.681691, 1950), who developed the *indefinite metric formalism*. Here is the essence of this formalism.

Let η be an operator that converts the indefinite metric from Minkowski to Hilbert space, and let $|i>$ be a base in this space. By definition, the norm of $|i>$ is

$$< i|\eta|i >= \pm 1.$$

The indefinite metric formalism

If

$$< i|\eta|k >= 0 , \; i \neq k ,$$

the two last relations can be combined to give

$$< i|\eta|k >= \pm\delta_{ik}, \qquad (4.5.6)$$

which is known as *the orthonormality condition*.

The expectation value of an operator L is defined by

$$\overline{L} =< \psi|\eta L|\psi >, \qquad (4.5.7)$$

and its complex conjugate is

$$\overline{L}^* =< \psi|\eta L|\psi >^* =< \psi|L^+\eta^+|\psi > . \qquad (4.5.8)$$

In order that the above defined expectation value be real, we must have $\overline{L}^* = \overline{L}$, that is, according to (4.5.7) and (4.5.8),

$$\eta L = l^+\eta^+, \qquad (4.5.9)$$

while the expectation value is imaginary if $\overline{L}^* = -\overline{L}$, that is

$$\eta L = -L^+\eta^+ . \qquad (4.5.10)$$

Note that the eigenvalues of η are $+1$ and -1, which are real. This means that η can be represented as a unitary Hermitian operator $(\eta^+ = \eta)$, and then (4.5.9) writes

$$\eta L = L^+\eta, \qquad (4.5.11)$$

while (4.5.10) becomes

$$\eta L = - L^+\eta. \qquad (4.5.12)$$

Multiplying (4.3.11), on the right, by η^{-1}, we have

$$L^+ = \eta L \eta^{-1}. \qquad (4.5.13)$$

The operator η is called *metric operator*. Let us analyze two cases:

a) *The operator L is Hermitian:* $L^+ = L$. In this case, according to
(4.5.11), the expectation value of L is real if

$$[\eta, L] = 0, \qquad (4.5.14)$$

while, in view of (4.5.12), the expectation value of L is imaginary if

$$\{\eta, L\} = 0. \tag{4.5.15}$$

b) *The operator L is antihermitian:* $L^+ = -L$. According to (4.5.11), the condition that the expectation value of L be real is then

$$\{\eta, L\} = 0, \tag{4.5.16}$$

while following (4.5.12) the condition that the expectation value of L be imaginary is

$$[\eta, L] = 0. \tag{4.5.17}$$

Let us now apply the indefinite metric formalism to the operators A_4 and A_0. Clearly:

a) $A_4 = iV$ is Hermitian, but its expectation value has to be imaginary, that is

$$\{\eta, A_4\} = 0 ; \qquad \{\eta, A_{4\pm}\} = 0 . \tag{4.5.18}$$

b) $A_0 = V$ is antihermitian, but its expectation value has to be real, i.e.

$$\{\eta, A_0\} = 0 ; \qquad \{\eta, A_{0\pm}\} = 0 . \tag{4.5.19}$$

Formulas (4.5.18) and (4.5.19) show that A_4 and A_0 satisfy *the same* anti-commutation relation. The contradiction is, therefore, eliminated.

Obviously, the operator η commutes with the spatial components A_α of the four-vector operator A_j, which are Hermitian and have real expectation values, that is

$$[\eta, A_\alpha] = 0 ; \qquad [\eta, A_{\alpha\pm}] = 0 . \tag{4.5.20}$$

4.6. The Lorentz-Fermi condition

a) As we mentioned in §1 of this chapter, we cannot abandon the Lorentz condition in a macroscopic theory, which means that the expectation value of the operator $A_{i,i}$ must be zero. Using definition (4.5.7), the Lorentz condition (4.1.9) writes

$$\overline{A}_{i,i} =< \psi|\eta A_{i,i}|\psi >= 0, \tag{4.6.1}$$

or, in view of (4.2.1),

$$< \psi|\eta A_{i+,i}|\psi > + < \psi|\eta A_{i-,i}|\psi >= 0. \tag{4.6.2}$$

The Lorentz-Fermi condition

Let us now show that the first term of this relation is the complex conjugate of the second. Using (4.5.18) and (4.5.20), as well as the hermiticity of η, we have

$$< \psi|\eta A_{i+,i}|\psi >=< \psi|\eta A_{\alpha+,\alpha}|\psi > + < \psi|\eta A_{4+,4}|\psi >$$

$$=< \psi|A_{\alpha+,\alpha}\eta|\psi > - < \psi|A_{4+,4}\eta|\psi >$$

$$=< \psi|\eta(A_{\alpha+,\alpha})^{+}|\psi >^{*} - < \psi|\eta(A_{4+,4})^{+}|\psi >^{*} .$$

Using (4.2.10), we can write

$$(A_{\alpha+})^{+} = A_{\alpha-} ; \qquad A_{4+}^{+} = A_{4-}.$$

Since x_α are real, and x_4 is imaginary, once the derivatives are performed we have

$$(A_{\alpha+,\alpha})^{+} = A_{\alpha-,\alpha} ; \qquad (A_{4+,4})^{+} = - A_{4-,4},$$

so that

$$< \psi|\eta A_{i+,i}|\psi >=< \psi|\eta(A_{\alpha-,\alpha} + A_{4-,4})|\psi >^{*}=< \psi|\eta A_{i-,i}|\psi >^{*},$$

which proves our statement.

Consequently, the condition (4.6 1) is satisfied if $< \psi|\eta A_{i+,i}|\psi >= 0$. Therefore, we shall require that

$$A_{i+,i}|\psi >= 0, \qquad (4.6.3)$$

which is *the Fermi-Lorentz condition*, in a form given by Bleuler and Gupta. (Note that (4.6.1) can also be satisfied if $A_{i-,i}|\psi >= 0$.) According to this condition, the only possible states are those satisfying (4.6.3). One of the main consequences of this constraint is the *elimination of virtual photons* out of the expectation value of an observable. Multiplying (4.6.3), on the right, by η, and taking into account (4.5.18), (4.5.20), and (4.6.2), the Fermi-Lorentz condition (4.6.3) writes

$$< \psi|\eta A_{i-,i} = 0, \qquad (4.6.4)$$

where the presence of η allows us to express the scalar product in this new formalism.

Let us express (4.6.3) and (4.6.4) in the momentum representation. In view of (4.2.21), relation (4.6.3) yields

$$ik_j \, e_j^l \, a_{l+}(\vec{k})|\psi >= 0. \qquad (4.6.5)$$

The quantization of the electromagnetic field

If propagation takes place along the z-axis, the only non-zero components of the four-vector k_j are k_3 and k_4, and (4.6.5) can also be written as

$$(k_3 e_3^l a_{l+} + k_4 e_4^l a_{l+})|\psi> = 0.$$

Since $e_3^l = \delta_3^l$; $e_4^l = \delta_4^l$, we finally obtain

$$(k_3 a_{3+} + k_4 a_{4+})|\psi> = 0. \tag{4.6.6}$$

Following a similar reasoning, the adjoint (4.6.4) of the Fermi-Lorentz condition results in

$$< \psi|\eta(k_3 a_{3-} + k_4 a_{4-}) = 0. \tag{4.6.7}$$

Since

$$k_3 = |\vec{k}| = \omega ; \qquad k_4 = i k_0 = i\omega, \qquad \text{with } \vec{k} \equiv \vec{k}(0,0,k_3),$$

we have

$$(a_{3+} + i\, a_{4+})|\psi> = 0 ; \tag{4.6.8}$$

$$< \psi|\eta(a_{3-} + i\, a_{4-}) = 0. \tag{4.6.9}$$

It then follows that the operator form of the Lorentz gauge condition (4.1.9) is not valid in QFT, being replaced by the weaker condition (4.6.3), which leads to the Lorentz condition (4.1.9) for expectation values.

b) Let us now show that the expectation value of the Lagrangian density (4.1.11) leads to the classical Lagrangian density (4.1.1). To this end, we must prove that the expectation value of the last term in (4.1.1) vanishes. We have:

$$\overline{A_{i,i} A_{j,j}} = < \psi|\eta A_{i,i} A_{j,j}|\psi> = < \psi|\eta A_{i+,i} A_{j-,j}|\psi>, \tag{4.6.10}$$

where we have omitted the terms which, by virtue of (4.6.3) and (4.6.4), are zero. Subtract now from the r.h.s. of (4.6.10) the term $< \psi|\eta A_{j-,j} A_{i+,i}|\psi>$ which, according to (4.6.3) (or (4.6.4)) is also zero. We then have

$$\overline{A_{i,i} A_{j,j}} = < \psi|\eta[A_{i+,i}, A_{j-,j}]|\psi>, \tag{4.6.11}$$

We note that

$$[A_{i+,i}(x), A_{j-,j}(x')] = \frac{1}{(2\pi)^3}\int \frac{(ik_i)(-ik_j')}{\sqrt{4\omega\omega'}}[A_{i+}(\vec{k}), A_{j-}(\vec{k}')]$$

104

$$\times e^{i(kx - k'x')} \, d\vec{k} \, d\vec{k}',$$

or, taking into account the commutation relation (4.2.14), and the energy-momentum relation $k_i k_i = \sum_{i=1}^{4} k_i^2 = k^2 = 0$, and integrating with respect to $\vec{k}'$,

$$[A_{i+,i}(x), A_{j-,j}(x')] = \frac{1}{(2\pi)^3} \int \frac{k_i k_i}{2\omega} \, d\vec{k} = 0. \qquad (4.6.12)$$

This result shows that $\overline{A_{i,i} A_{j,j}} = 0$, and therefore

$$\overline{L} = -\frac{1}{4} F_{ik} F_{ik}. \qquad (4.6.13)$$

c) As an additional application, let us calculate the expectation value of the number of virtual photons corresponding to allowed states, that is to those states that satisfy conditions (4.6.8) and (4.6.9). We have

$$< N_3 + N_4 > = < \psi | \eta(a_{3-} a_{3+} + a_{4-} a_{4+} | \psi >$$

$$= < \psi | \eta \, a_{3-} a_{3+} | \psi > + < \psi | \eta \, a_{4-} a_{4+} | \psi > .$$

However, from (4.6.8) we have

$$a_{4+} | \psi > = i \, a_{3+} | \psi >,$$

so that

$$< N_3 + N_4 > = < \psi | \eta \, a_{3-} a_{3+} | \psi > + i < \psi | \eta \, a_{4-} a_{3+} | \psi >$$

$$= < \psi | \eta(a_{3-} + i a_{4-}) a_{3+} | \psi >, = 0, \qquad (4.6.14)$$

where (4.6.9) has also been used. Consequently, *the expectation value of the number of virtual photons is zero.* Therefore, as expected, the virtual photons do not play a role in the expectation values of physical quantities (energy, momentum, etc.)

d) Lastly, we shall prove that the norm of the state function ψ does not depend on the number of virtual photons. Since the number of transverse photons does not enter the following analysis, let us denote them by n_1 and n_2, which will be kept fixed.

Suppose that $|n_3 n_4 >$ is the orthonormal set of eigenvectors belonging to the states with n_3 longitudinal photons, and n_4 temporal photons. The state function ψ can then be expressed in terms of this set as

$$|\psi > = \sum_{n_3, n_4} C_{n_3 n_4} |n_3 n_4 > . \qquad (4.6.15)$$

The quantization of the electromagnetic field

We shall retain only that part of the sum for which $n_3 + n_4 = n$, that is

$$\sum_{n_4} C_{n_3 n_4} |n_3 n_4 >= \sum_{n_4} C_{n-n_4,n_4} |n - n_4, n_4 >= C_{n,0} |n >, \quad (4.6.16)$$

where $C_{n,0}|n >$ is used as a shorthand. Then, we have

$$|\psi >= \sum_{n} C_{n,0} |n >= \sum_{n} \sum_{n_4=0} C_{n-n_4,n_4} |n - n_4, n_4 > . \quad (4.6.17)$$

On the other hand, recalling that a_{3+} and a_{4+} are annihilation operators, and using (4.6.8) and (4.6.15), we have

$$(a_{3+} + i a_{4+})|\psi >= \sum_{n_3,n_4} C_{n_3 n_4} (a_{3+} + i a_{4+})|n_3 n_4 >$$

$$= \sum_{n_3,n_4} C_{n_3 n_4} \sqrt{n_3}|n_3 - 1, n_4 > \quad (4.6.18)$$

$$+ i \sum_{n_3,n_4} C_{n_3 n_4} \sqrt{n_4}|n_3, n_4 - 1 >= 0,$$

where (2.3.13) has also been taken into account. Let us now set $n_3 - 1 = n_3'$ in the first sum, $n_4 - 1 = n_4'$ in the second, and drop the "prime". Then,

$$\sum_{n_3,n_4} (\sqrt{n_3 + 1}\, C_{n_3+1,n_4} + i\sqrt{n_4 + 1}\, C_{n_3,n_4+1})|n_3 n_4 >= 0, \quad (4.6.19)$$

where the $C's$ are zero for any negative index.

Since the states are linearly independent, in order for (4.6.19) to be true, each coefficient has to be zero, that is

$$\sqrt{n_3 + 1}\, C_{n_3+1,n_4} + i\sqrt{n_4 + 1}\, C_{n_3,n_4+1} = 0. \quad (4.6.20)$$

Let us rewrite now relation (4.6.20) for $n_3 = n - n_4$, $n_4 \to n_4 - 1$. The result is

$$\sqrt{n - n_4 + 1}\, C_{n-n_4+1,n_4-1} + i\sqrt{n_4}\, C_{n-n_4,n_4} = 0,$$

or, equivalently,

$$\sqrt{n_4}\, C_{n-n_4,n_4} = i\sqrt{n - n_4 + 1}\, C_{n-n_4+1,n_4-1}. \quad (4.6.21)$$

106

If n_4 takes the values $n_4 - 2$, $n_4 - 3, ..., 3, 2, 1$ seriatim, we have successively:

$$\sqrt{n_4 - 1}\, C_{n-n_4+1,n_4-1} = i\sqrt{n - n_4 + 2}\, C_{n-n_4+2,n_4-2} \; ;$$

$$\sqrt{n_4 - 2}\, C_{n-n_4+2,n_4-2} = i\sqrt{n - n_4 + 3}\, C_{n-n_4+3,n_4-3} \; ;$$

$$\cdots \cdots \cdots \cdots$$

$$\sqrt{3}\, C_{n-3,3} = i\sqrt{n - 2}\, C_{n-2,2} \; ;$$

$$\sqrt{2}\, C_{n-2,2} = i\sqrt{n - 1}\, C_{n-1,1} \; ;$$

$$\sqrt{1}\, C_{n-1,1} = i\sqrt{n}\, C_{n,0} \; .$$

Multiplying these relations member by member, and taking into account (4.6.21), after simplifications we are left with

$$C_{n-n_4,n_4} = (i)^{n_4} \sqrt{C_n^{n_4}}\, c_{n,0}. \tag{4.6.22}$$

Since

$$C_n^k = \frac{n!}{k!(n-k)!},$$

relation (4.6.16) yields

$$C_{n,0}|n>= \sum_{n_4}(i)^{n_4} \sqrt{C_n^{n_4}}\, c_{n,0}|n - n_4, n_4 >,$$

or, simplifying by $C_{n,0}$

$$|n>= \sum_{n_4}(i)^{n_4} \sqrt{C_n^{n_4}}\, |n - n_4, n_4 > . \tag{4.6.23}$$

Similarly, for the vector bra $< n|$ one obtains:

$$< n| = \sum_{n_4} < n - n_4, n_4|(-i)^{n_4} \sqrt{C_n^{n_4}}. \tag{4.6.24}$$

If n takes the values $0, 1, 2, ...$ in (4.6.23), we have:

$$n = 0 \quad (n_4 = 0) : \quad |0>= |0,0 > \; ;$$

$$n = 1 \quad (n_4 = 0, 1) \; ; \; |1>= |1,0 > +i|0, 1 > \; ;$$

$$n = 2 \quad (n_4 = 0, 1, 2) \; ; \quad |2>= |2,0 > +i\sqrt{2}|1, 1 > -|0, 2 > \; ;$$

The quantization of the electromagnetic field

$$n = 3 \quad (n_4 = 0, 1, 2) \quad :$$

$$|3> = |3, 0> + i\sqrt{3}|2, 1> - \sqrt{3}|1, 2> - i|0, 3>, \; etc.$$

For deriving the norm of ψ, we first note that the number of sign changes introduced by the metric operator η equals the number of operators a_4, which is also the number of particles in the corresponding state. We can write

$$|\psi> = |0> + |1> + |2> + ... \tag{4.6.25}$$

The norm of ψ is then

$$<\psi|\eta|\psi> = <0|\eta|0> + <1|\eta|1> + ...+ <n|\eta|n> +... \tag{4.6.26}$$

In view of (4.6.23) and (4.6.24), a generic term in (4.6.26) can be written as

$$<n|\eta|n> = \sum_{n_4} <n - n_4, n_4|(-i)^{n_4}\sqrt{C_n^{n_4}}\eta(i)^{n_4}\sqrt{C_n^{n_4}}|n - n_4, n_4>.$$

$$\tag{4.6.27}$$

Taking into account the sign changes introduced by the operator η, we may write $\eta = (-1)^{n_4}$. Therefore,

$$<n|\eta|n> = \sum_{n_4}(-1)^{n_4} C_n^{n_4}. \tag{4.6.28}$$

To find the result of the summation in (4.6.28), let us take, for example,

$$(a - b)^n = \sum_{k=o} C_n^k a^{n-k}(-b)^k,$$

and let $a = b = 1$, and $k = n_4$. Then,

$$0 = (1 - 1)^n = \sum_{n_4} C_n^{n_4}(1)^{n-n_4}(-1)^{n_4} = \sum_{n_4}(-1)^{n_4}C_n^{n_4},$$

which yields

$$<n|\eta|n> = 0 ; \quad (n = 1, 2, 3, ...),$$

or, equivalently,

$$<\psi|\eta|\psi> = \sum_{n} <n|\eta|n> = C_{0,0} <0|\eta|0> = <0|0> . \tag{4.6.29}$$

In conclusion, it has been proven that *the norm of the state function ψ does not depend on the number of virtual photons.*

Observation. Virtual photons do not contribute to the expectation values of physical observables, but play an essential role in the study of field interactions.

CHAPTER V

THE QUANTIZATION OF THE SPINORIAL FIELD

> *Well, in the first place, it leads to great anxiety as to whether it's going to be correct or not...*
>
> P.A.M. Dirac
> (At age 60, when asked about his feelings on discovering the Dirac equation)

5.1. The Dirac equation and the algebra of gamma matrices

It is well-known that the Dirac equation is obtained by linearizing the energy-momentum relation

$$p^2 + m^2 = 0, \tag{5.1.1}$$

where p is the momentum four-vector, and m is the electron rest mass. In the case of a free electron, the Dirac equation writes

$$(\vec{\alpha} \cdot \hat{\vec{p}} + \beta m - \hat{E})|\psi> = 0, \tag{5.1.2}$$

where $\alpha_1, \alpha_2, \alpha_3$ and β are the Dirac matrices:

$$\alpha_1 = \begin{pmatrix} 0 & 0 & 0 & 1 \\ 0 & 0 & 1 & 0 \\ 0 & 1 & 0 & 0 \\ 1 & 0 & 0 & 0 \end{pmatrix} \; ; \quad \alpha_2 = \begin{pmatrix} 0 & 0 & 0 & -i \\ 0 & 0 & i & 0 \\ 0 & -i & 0 & 0 \\ i & 0 & 0 & 0 \end{pmatrix} \; ;$$

$$\alpha_3 = \begin{pmatrix} 0 & 0 & 1 & 0 \\ 0 & 0 & 0 & -1 \\ 1 & 0 & 0 & 0 \\ 0 & -1 & 0 & 0 \end{pmatrix} \; ; \quad \beta = \begin{pmatrix} 1 & 0 & 0 & 0 \\ 0 & 1 & 0 & 0 \\ 0 & 0 & -1 & 0 \\ 0 & 0 & 0 & -1 \end{pmatrix} , \tag{5.1.3}$$

while $\hat{E}$ is the energy operator. For the sake of simplicity, we again drop the notation $\hat{}$, which designates operators, and we shall reserve this sign for other purposes.

109

The quantization of the spinorial field

As can be seen, the Dirac equation (5.1.2) can also be written as

$$[\vec{\alpha} \cdot \vec{p} + (iI)(iE) + \beta m]|\psi> = 0, \qquad (5.1.4)$$

where I is the 4×4 unit matrix.

Analyzing the equation (5.1.4), we note that, by analogy with the momentum four-vector p_k, one can define a four-vector α_k, with the components $\vec{\alpha}, iI$. Equation (5.1.4) then becomes

$$(\alpha_k p_k + \beta m)|\psi> = 0 , \qquad \text{with } \alpha_4 = iI. \qquad (5.1.5)$$

Multiplying this equation, on the left, by $-i\beta$, and denoting

$$-i\beta\alpha_k = \gamma_k, \qquad (5.1.6)$$

we obtain

$$(\gamma_k p_k - i\, m)|\psi> = 0. \qquad (5.1.7)$$

From now on, we shall use the notation $\hat{}$ to denote a four-vector contracted with the matrix γ_k. For instance,

$$\gamma_k p_k = \hat{p}, \qquad (5.1.8)$$

and the Dirac equation writes

$$(\hat{p} - i\, m)|\psi> = 0. \qquad (5.1.9)$$

Since

$$p_k = \frac{1}{i} \frac{\partial}{\partial x_k} = \frac{1}{i} \partial_k,$$

this convention leads to

$$\hat{p} = \frac{1}{i}\hat{\partial}, \qquad \text{with} \qquad \hat{\partial} = \gamma_k \partial_k, \qquad (5.1.10)$$

and the Dirac equation takes the simplified form

$$(\hat{\partial} + m)|\psi> = 0. \qquad (5.1.11)$$

Let us now establish two essential properties of the matrices γ_k.

1. *The matrices γ_k obey the same anti-commutation rule as Dirac's matrices $\vec{\alpha}$ and β, that is*

$$\{\gamma_i, \gamma_k\} = \gamma_i \gamma_k + \gamma_k \gamma_i = 2\,\delta_{ik}\, I. \qquad (5.1.12)$$

The Dirac equation and the algebra of gamma matrices

In view of (5.1.6), the last equation can also be expressed as

$$- (\beta \alpha_i \beta \alpha_k + \beta \alpha_k \beta \alpha_i) = 2 \, \delta_{ik} \, I. \tag{5.1.13}$$

There are three possible situations:

a) $i, k = 1, 2, 3$. In this case α_i, α_k anticommute with β, and, since $\beta^2 = I$, one obtains

$$\alpha_i \alpha_k + \alpha_k \alpha_i = 2 \, \delta_{ik} \, I \; ; \; (i, k = 1, 2, 3). \tag{5.1.14}$$

b) $i = 1, 2, 3 \; ; \; k = 4$. This choice, in view of (5.1.5) and (5.1.13), yields

$$- (\beta \alpha_i \beta \alpha_4 + \beta \alpha_4 \beta \alpha_i) = - i(\beta \alpha_i \beta + \beta \beta \alpha_i) = - i\beta(\alpha_i \beta + \beta \alpha_i) = 0,$$

and (5.1.13) is again satisfied.

c) $i = k = 4.$. This time, we have:

$$- (\beta \alpha_4 \beta \alpha_4 + \beta \alpha_4 \beta \alpha_4) = (\beta I \beta I + \beta I \beta I) = 2,$$

which also shows that (5.1.13) is true. To conclude, the anti-commutation relation (5.1.12) is also true.

2. *The matrices γ_k are Hermitian*. Since Dirac's matrices $\vec{\alpha}$ are Hermitian, we distinguish two cases:

a) $k = 1, 2, 3$. Then (5.1.6) and (5.1.14) yield

$$\gamma_k = - i\beta \alpha_k \; ; \quad \gamma_k^+ = i\alpha_k^+ \beta^+ = i\alpha_k \beta = - i\beta \alpha_k = \gamma_k \; ; \quad (k = 1, 2, 3). \tag{5.1.15}$$

b) $k = 4$. We can write

$$\gamma_4 = - i\beta \alpha_4 = - (i\beta)(iI) = \beta \quad \text{(Hermitian)} \, . \tag{5.1.16}$$

Therefore the matrices γ_k are, indeed, Hermitian.

We mention that many authors use the notation α_4 instead of β, but we shall continue using the notation $\alpha_4 = iI$.

Let us now obtain the *adjoint Dirac equation*. Ignoring the ket vector symbol, equation (5.1.11) writes

$$(\hat{\partial} + m)\psi = 0. \tag{5.1.17}$$

Separating the spatial and temporal parts, we have

$$\gamma_\alpha (\partial_\alpha \psi) + \frac{1}{i} \gamma_4 \left(\frac{\partial}{\partial t} \psi \right) + m \, \psi = 0,$$

The quantization of the spinorial field

as well as its adjoint

$$(\partial_\alpha \psi^+)\gamma_\alpha - \frac{1}{i}\left(\frac{\partial}{\partial t}\psi^+\right)\gamma_4 + m\,\psi^+ = 0, \qquad (5.1.18)$$

where the hermiticity of γ_k has been used. Multiplying now this equation on the right by γ_4, we have

$$(\partial_\alpha \psi^+)\gamma_\alpha\gamma_4 - \frac{1}{i}\left(\frac{\partial}{\partial t}\psi^+\right)\gamma_4\gamma_4 + m\,\psi^+\gamma_4 = 0.$$

Using the property (5.1.12), and denoting

$$\overline{\psi} = \psi^+\gamma_4, \qquad (5.1.19)$$

the adjoint Dirac equation writes

$$(\partial_k\overline{\psi})\gamma_k - m\overline{\psi} = 0,$$

or, in a shorter form,

$$\overline{\psi}(\overleftarrow{\partial} - m) = 0, \qquad (5.1.20)$$

where, by convention, the arrow over $\overleftarrow{\partial}$ shows that the operator acts upon quantities on its left.

Next, we shall establish some *algebraic properties* of the matrices γ_k. Starting from the four matrices γ_k, we can construct 16 linearly independent matrices, as follows:

$C_4^0 \Rightarrow$ one (unit) matrix I ;
$C_4^1 \Rightarrow$ four matrices γ_k ;
$C_4^2 \Rightarrow$ six matrices $\gamma_{kl} = \gamma_k\gamma_l, \quad k < l$;   (5.1.21)
$C_4^3 \Rightarrow$ four matrices $\gamma_{klm} = \gamma_k\gamma_l\gamma_m, \quad k < l < m$;
$C_4^4 \Rightarrow$ one matrix $\gamma_5 = \gamma_k\gamma_l\gamma_m\gamma_p = \gamma_1\gamma_2\gamma_3\gamma_4, \quad k < l < m < p$.

The linear independence of these 16 matrices will be proved below. Using the anti-commutation rule (5.1.12), we have

$$\gamma_{kl}^2 = \gamma_k\gamma_l\gamma_k\gamma_l = -\gamma_k\gamma_l\gamma_l\gamma_k = -\gamma_k\gamma_k = -I , \quad \text{(no summation)} ; \qquad (5.1.22)$$
$$\gamma_{klm}^2 = \gamma_k\gamma_l\gamma_m\gamma_k\gamma_l\gamma_m = \gamma_k\gamma_l\gamma_l\gamma_k = -I , \quad \text{(no summation)} ; \qquad (5.1.23)$$
$$\gamma_5^2 = \gamma_1\gamma_2\gamma_3\gamma_4\gamma_1\gamma_2\gamma_3\gamma_4 = -\gamma_1\gamma_2\gamma_3\gamma_1\gamma_2\gamma_3 = I. \qquad (5.1.24)$$

112

The Dirac equation and the algebra of gamma matrices

Moreover, γ_5 anticommutes with any of the matrices $\gamma_1, \gamma_2, \gamma_3, \gamma_4$. For example,

$$\gamma_5\gamma_1 = \gamma_1\gamma_2\gamma_3\gamma_4\gamma_1 = \gamma_1\gamma_1\gamma_2\gamma_3\gamma_4 = -\gamma_1\gamma_5, \quad \text{etc.}$$

To prove the linear independence of the matrices γ_k it is necessary to introduce the notion of *trace* (or *spur*) of a square matrix. By definition, the trace of matrix A is

$$Tr\, A = \sum_i A_{ii}. \tag{5.1.25}$$

The trace of a product of matrices has the property that it is left unchanged by a circular permutation of these matrices. Indeed,

$$Tr\, AB...YZ = A_{ij}B_{jk}...Y_{ls}Z_{si} = Z_{si}A_{ij}B_{jk}...Y_{ls} = Tr\, ZAB...Y, \tag{5.1.26}$$

where the summation convention has been used. Using property (5.1.26), it is easily seen that the trace of a matrix is invariant with respect to a matrix transformation of the type

$$A' = SAS^{-1}, \tag{5.1.27}$$

because

$$Tr\, A' = Tr\, SAS^{-1} = Tr\, S^{-1}SA = Tr\, A. \tag{5.1.28}$$

If S is a unitary matrix, then the transformation (5.1.27) is called *unitary transformation*.

Let us now define the matrices γ_i', obtained by using the matrix transformation $\gamma_i' = S\gamma_i S^{-1}$, and show that the γ_i' satisfy the same anti-commutation rule as the matrices γ_i, (5.1.12). Indeed,

$$\gamma_i'\gamma_k' + \gamma_k'\gamma_i' = (S\gamma_i S^{-1})(S\gamma_k S^{-1}) + (S\gamma_k S^{-1})(S\gamma_i S^{-1})$$

$$= S(\gamma_i\gamma_k + \gamma_k\gamma_i)S^{-1} = 2\,\delta_{ik}\,I \quad \text{q.e.d.} \tag{5.1.29}$$

Next, let us prove that the trace of a product of an odd number of γ matrices is zero. Using (5.1.12) and (5.1.26), we can write

$$Tr\, \gamma_{k_1}\gamma_{k_2}...\gamma_{k_n} = Tr\, \gamma_5\gamma_5\gamma_{k_1}\gamma_{k_2}...\gamma_{k_n}$$

$$= Tr\, \gamma_5\gamma_{k_1}\gamma_{k_2}...\gamma_{k_n}\gamma_5 = (-1)^n\, Tr\, \gamma_{k_1}\gamma_{k_2}...\gamma_{k_n},$$

and, if n is odd,

$$Tr\, \gamma_{k_1}\gamma_{k_2}...\gamma_{k_{2n+1}} = 0, \qquad (5.1.30)$$

which means that

$$Tr\, \gamma_k = 0 \; ; \quad \text{and} \quad Tr\, \gamma_{klm} = 0. \qquad (5.1.31)$$

Let us now consider a product of an even number of γ matrices, and derive a general rule for calculating its trace. Using (5.1.12), we successively have

$$Tr\, \gamma_{k_1}\gamma_{k_2}...\gamma_{k_{2n}} = Tr\, (2\delta_{k_1 k_2} - \gamma_{k_2}\gamma_{k_1})\gamma_{k_3}...\gamma_{k_{2n}}$$

$$= 2\delta_{k_1 k_2} Tr\, \gamma_{k_3}...\gamma_{k_{2n}} - Tr\, \gamma_{k_2}\gamma_{k_1}\gamma_{k_3}...\gamma_{k_{2n}}$$

$$= 2\delta_{k_1 k_2} Tr\, \gamma_{k_3}...\gamma_{k_{2n}} - Tr\, \gamma_{k_2}(2\delta_{k_1 k_3} - \gamma_{k_3}\gamma_{k_1})\gamma_{k_4}...\gamma_{k_{2n}}$$

$$= 2\delta_{k_1 k_2} Tr\, \gamma_{k_3}...\gamma_{k_{2n}} - 2\delta_{k_1 k_3} Tr\, \gamma_{k_2}\gamma_{k_4}\gamma_{k_5}...\gamma_{k_{2n}}$$

$$+ Tr\, \gamma_{k_2}\gamma_{k_3}(2\delta_{k_1 k_4} - \gamma_{k_4}\gamma_{k_1})\gamma_{k_5}...\gamma_{k_{2n}} = ...$$

$$= 2\delta_{k_1 k_2} Tr\, \gamma_{k_3}...\gamma_{k_{2n}} - 2\delta_{k_1 k_3} Tr\, \gamma_{k_2}\gamma_{k_4}...\gamma_{k_{2n}}$$

$$+ 2\delta_{k_1 k_4} Tr\, \gamma_{k_2}\gamma_{k_3}\gamma_{k_5}...\gamma_{k_{2n}} - ... + Tr\, \gamma_{k_2}...\gamma_{k_{2n-1}}\gamma_{k_1}\gamma_{k_{2n}}.$$

In the last term we substitute

$$\gamma_{k_1}\gamma_{k_{2n}} = 2\,\delta_{k_1 k_{2n}} - \gamma_{k_{2n}}\gamma_{k_1},$$

and obtain

$$Tr\, \gamma_{k_1}\gamma_{k_2}...\gamma_{k_{2n}} = 2\sum_r (-1)^r \delta_{k_1 k_r} Tr\, \gamma_{k_2}...\gamma_{k_{r-1}}\gamma_{k_{r+1}}...\gamma_{k_{2n}}$$

$$-Tr\, \gamma_{k_1}\gamma_{k_2}...\gamma_{k_{2n}},$$

and, finally

$$Tr\, \gamma_{k_1}\gamma_{k_2}...\gamma_{k_{2n}} = \sum_r (-1)^r \delta_{k_1 k_r} Tr\, \gamma_{k_2}...\gamma_{k_{r-1}}\gamma_{k_{r+1}}...\gamma_{k_{2n}}. \qquad (5.1.32)$$

The advantage of this rule is that the number of matrices under the sign of the trace on the r.h.s. is decreased by two. Here are two examples, for two and four γ matrices, respectively:

$$Tr\, \gamma_k \gamma_j = \delta_{kj} Tr\, I = 4\delta_{kj}; \qquad (5.1.33)$$

$$Tr\, \gamma_k \gamma_j \gamma_m \gamma_p = \delta_{kj} Tr\, \gamma_m \gamma_p - \delta_{km} Tr\, \gamma_j \gamma_p + \delta_{kp} Tr\, g_j \gamma_m$$

$$= 4(\delta_{kj}\delta_{mp} - \delta_{km}\delta_{jp} + \delta_{kp}\delta_{jm}). \qquad (5.1.34)$$

According to (5.1.33) and (5.1.34), we therefore have

$$Tr\,\gamma_{kl} = 0 , \quad \text{and} \quad Tr\,\gamma_5 = Tr\,\gamma_1\gamma_2\gamma_3\gamma_4 = 0. \qquad (5.1.35)$$

We conclude that, except for the unit matrix I, all matrices in (5.1.21) have null trace.

Finally, we shall show that the 16 γ matrices are *linearly independent*. Let us take the linear combination

$$M = C_0 I + C_k\gamma_k + C_{kl}\gamma_{kl} + C_{klm}\gamma_{klm} + C_5\gamma_5 = 0. \qquad (5.1.36)$$

To prove linear independence, we have to show that all coefficients C_0, C_k, C_{kl}, C_{klm}, C_5 are zero. Taking the trace of (5.1.36), and using (5.1.31) and (5.1.35), we have

$$Tr\,M = C_0 Tr\,I = 4C_0 = 0, \qquad (5.1.37)$$

and thus $C_0 = 0$. Next, we shall multiply (5.1.36) on the left by γ_h, and take the trace of the result. By means of (5.1.30), we can write

$$Tr\,(C_k\gamma_h\gamma_k + C_{klm}\gamma_h\gamma_{klm}) = 0. \qquad (5.1.38)$$

But, according to (5.1.21) and (5.1.34), for any h and $k < l < m$, one has

$$Tr\,\gamma_h\gamma_{klm} = Tr\,\gamma_h\gamma_k\gamma_l\gamma_m = 0,$$

and (5.1.33) yields

$$C_k\,Tr\,\gamma_h\gamma_k = 4\,C_k\delta_{hk} = 4\,C_h = 0. \qquad (5.1.39)$$

Following the same procedure, we obtain

$$C_{kl} = C_{klm} = C_5 = 0, \qquad (5.1.40)$$

which completes the proof.

5.2. Lagrangian formalism

Let us now show that if we choose the Lagrangian density as

$$\mathcal{L}_1 = -\overline{\psi}(\hat{\partial}\psi) - m\,\overline{\psi}\psi = -\overline{\psi}\,[(\overset{\rightarrow}{\hat{\partial}} + m)\psi] ; \quad \hat{\partial} = \gamma_k\partial_k, \qquad (5.2.1)$$

then the Euler-Lagrange equations (1.1.15) lead to both the Dirac equation and its adjoint. Indeed,

$$\frac{\partial \mathcal{L}_1}{\partial \overline{\psi}} = -(\overset{\rightarrow}{\partial} + m)\psi \; ; \quad \frac{\partial \mathcal{L}_1}{\partial \overline{\psi}_{,k}} = 0 \; ; \quad \frac{\partial \mathcal{L}_1}{\partial \psi} = -m\overline{\psi} \; ; \quad \frac{\partial \mathcal{L}_1}{\partial \psi_{,k}} = -\overline{\psi}\gamma_k,$$

so that

$$\frac{\partial \mathcal{L}_1}{\partial \overline{\psi}} - \frac{\partial}{\partial x_k}\left(\frac{\partial \mathcal{L}_1}{\partial \overline{\psi}_{,k}}\right) = -(\overset{\rightarrow}{\partial} + m)\psi = 0,$$

as well as

$$\frac{\partial \mathcal{L}_1}{\partial \psi} - \frac{\partial}{\partial x_k}\left(\frac{\partial \mathcal{L}_1}{\partial \psi_{,k}}\right) = \overline{\psi}(\overset{\leftarrow}{\partial} - m) = 0,$$

i.e. precisely equations (5.1.11) and (5.1.20).

It can be easily shown that the Lagrangian density

$$\mathcal{L}_2 = (\overline{\psi}\overset{\rightarrow}{\partial})\psi - m\overline{\psi}\psi = [\overline{\psi}(\overset{\leftarrow}{\partial} - m)]\psi \tag{5.2.2}$$

leads to the same result.

As can be seen, the difference $\mathcal{L}_2 - \mathcal{L}_1$ is a four-divergence

$$\mathcal{L}_2 - \mathcal{L}_1 = (\overline{\psi}\overset{\leftarrow}{\partial})\psi + \overline{\psi}(\overset{\rightarrow}{\partial}\psi) = \overline{\psi}\overset{\leftrightarrow}{\partial}\psi = \partial_k(\overline{\psi}\gamma_k\psi), \tag{5.2.3}$$

meaning that the two Lagrangian densities are equivalent. This fact makes it possible to use a linear combination of (5.2.1) and (5.2.2), such as

$$\mathcal{L} = \frac{1}{2}(\mathcal{L}_1 + \mathcal{L}_2) = \frac{1}{2}[(\overline{\psi}\overset{\leftarrow}{\partial})\psi - \overline{\psi}(\overset{\rightarrow}{\partial}\psi)] - m\overline{\psi}\psi. \tag{5.2.4}$$

If ψ and $\overline{\psi}$ satisfy the Dirac equation (5.1.11) and its adjoint (5.1.20), respectively, then the two Lagrangian densities, together with $\mathcal{L}$, are zero:

$$\mathcal{L}_1 = \mathcal{L}_2 = \mathcal{L} = 0. \tag{5.2.5}$$

Using (1.1.16) and (5.2.4), we are now able to write the canonical energy-momentum tensor $T_{ik}^{(c)}$. Since

$$U = \psi \; ; \quad U^+ = \overline{\psi} \; ; \quad \pi_k = \frac{\partial \mathcal{L}}{\partial \overline{\psi}_{,k}} = \frac{1}{2}\gamma_k\psi \; ; \quad \pi_k^+ = \frac{\partial \mathcal{L}}{\partial \psi_{,k}} = -\frac{1}{2}\overline{\psi}\gamma_k,$$

$$\tag{5.2.6}$$

the definition (1.5.4) gives

$$T_{ik}^{(c)} = \pi_k^+ U_{,i} + U_{,i}^+ \pi_k - \mathcal{L}\,\delta_{ik} = -\frac{1}{2}\overline{\psi}\gamma_k\psi_{,i} + \frac{1}{2}\overline{\psi}_{,i}\gamma_k\psi, \tag{5.2.7}$$

Lagrangian formalism

where we used (5.2.5) to eliminate the term $\mathcal{L}\delta_{ik}$

According to (1.8.8), the current density is

$$s_l = i(\pi_l^+ U - U^+ \pi_l) = -i\bar{\psi}\gamma_l\psi. \tag{5.2.8}$$

Multiplying this relation by the unit charge e, one obtains the electric current density.

The spin momentum density is given by (1.6.14):

$$S_{ikl} = i\left[\pi_{l(r)}^+ \tilde{S}_{(r)(p)ik}U_p + U_p^+ \tilde{S}_{(r)(p)ik}^+\pi_{l(r)}\right].$$

In the case of the spinorial field, this becomes

$$S_{(r)(p)ik} = S_{rpik} = \frac{1}{4}(\gamma_i\gamma_k)_{rp} \; ;$$

$$S_{(r)(p)ik}^+ = S_{rpik}^+ = \frac{1}{4}(\gamma_k\gamma_i)_{pr},$$

and using (1.6.15), we obtain

$$S_{ikl} = \frac{i}{4}\left[\pi_l^+(\gamma_i\gamma_k - \gamma_k\gamma_i)\,U + U^+(\gamma_k\gamma_i - \gamma_i\gamma_k)\pi_l\right]. \tag{5.2.9}$$

Denoting by

$$\gamma_i\gamma_k - \gamma_k\gamma_i = [\gamma_i, \gamma_k] = 2\,i\sigma_{ik} \tag{5.2.10}$$

and taking into account (5.2.6), we finally find

$$S_{ikl} = \frac{1}{4}\bar{\psi}(\gamma_l\sigma_{ik} + \sigma_{ik}\gamma_l)\psi. \tag{5.2.11}$$

The spatial components P_α of the momentum four-vector are found by means of (5.2.7)

$$P_\alpha = i\int T_{\alpha 4}^{(c)}\,d\vec{x} = \frac{i}{2}\int(\psi_{,\alpha}^+\psi - \psi^+\psi_{,\alpha})\,d\vec{x}, \tag{5.2.12}$$

while the Hamiltonian is given by

$$H = \int T_{44}^{(c)}\,d\vec{x} = \frac{1}{2}\int(\psi_{,4}^+\psi - \psi^+\psi_{,4})\,d\vec{x}. \tag{5.2.13}$$

Here we have also used (5.1.19), and the fact that $\gamma_4\gamma_4 = I$.

117

5.3. The free particle in the Dirac theory

We shall look for a solution of the Dirac equation for the free particle in the form of a plane, monochromatic wave

$$\psi(x) = B(\vec{p})\, e^{ipx},$$

where the one-column matrix B does not depend on x. Equation (5.1.9) then writes

$$(\hat{p} - im)\, B(\vec{p}) = 0, \tag{5.3.1}$$

where the quantities

$$\hat{p} = p_k \gamma_k \; ; \; p_4 = ip_0 = iE \; ; \; p_0 = \pm E_p = \pm\sqrt{\vec{p}^2 + m^2} \tag{5.3.2}$$

remain constant.

Let us write the explicit form of the matrices γ_k defined by (5.1.6). Since

$$\alpha_\theta = \begin{pmatrix} 0 & \sigma_\theta \\ \sigma_\theta & 0 \end{pmatrix} \; ; \; \beta = \gamma_4 = \begin{pmatrix} \epsilon & 0 \\ 0 & -\epsilon \end{pmatrix}, \tag{5.3.3}$$

where σ_θ are the three Pauli matrices

$$\sigma_1 = \begin{pmatrix} 0 & 1 \\ 1 & 0 \end{pmatrix} \; ; \; \sigma_2 = \begin{pmatrix} 0 & -i \\ i & 0 \end{pmatrix} \; ; \; \sigma_3 = \begin{pmatrix} 1 & 0 \\ 0 & -1 \end{pmatrix}, \tag{5.3.4}$$

and ϵ is the 2×2 unit matrix, we have:

$$\beta\alpha_\theta = \begin{pmatrix} \epsilon & 0 \\ 0 & -\epsilon \end{pmatrix}\begin{pmatrix} 0 & \sigma_\theta \\ \sigma_\theta & 0 \end{pmatrix} = \begin{pmatrix} 0 & \sigma_\theta \\ -\sigma_\theta & 0 \end{pmatrix};$$

$$\gamma_\theta = -i\beta\alpha_\theta = \begin{pmatrix} 0 & -i\sigma_\theta \\ i\sigma_\theta & 0 \end{pmatrix}.$$

The explicit form of γ_k is then:

$$\gamma_1 = \begin{pmatrix} 0 & 0 & 0 & -i \\ 0 & 0 & -i & 0 \\ 0 & i & 0 & 0 \\ i & 0 & 0 & 0 \end{pmatrix} \; ; \; \gamma_2 = \begin{pmatrix} 0 & 0 & 0 & -1 \\ 0 & 0 & 1 & 0 \\ 0 & 1 & 0 & 0 \\ -1 & 0 & 0 & 0 \end{pmatrix},$$

$$\tag{5.3.5}$$

$$\gamma_3 = \begin{pmatrix} 0 & 0 & -i & 0 \\ 0 & 0 & 0 & i \\ i & 0 & 0 & 0 \\ 0 & -i & 0 & 0 \end{pmatrix} \; ; \; \gamma_4 = \begin{pmatrix} 1 & 0 & 0 & 0 \\ 0 & 1 & 0 & 0 \\ 0 & 0 & -1 & 0 \\ 0 & 0 & 0 & -1 \end{pmatrix}.$$

The free particle in the Dirac theory

Consider now the column matrix (or column vector)

$$A = \begin{pmatrix} A_1 \\ A_2 \\ A_3 \\ A_4 \end{pmatrix} \tag{5.3.6}$$

and show that, for any A, the matrix

$$B(\vec{p}) = (\hat{p} + im)A \tag{5.3.7}$$

is a solution of the Dirac equation (5.3.1) for the free spinorial field. Indeed, using the energy-momentum relation (5.1.1), we have

$$(\hat{p} - im)B(\vec{p}) = (\hat{p} - im)(\hat{p} + im)A = (p^2 + m^2)A = 0.$$

Using (5.3.5), and (5.3.6), we want now to expound the elements of matrix (5.3.7). To this end, note that in the r.h.s. of (5.3.7) we have the products $\gamma_k A$ ($k = 1,2,3,4$), which are

$$\gamma_1 A = \begin{pmatrix} -iA_4 \\ -iA_3 \\ iA_2 \\ iA_1 \end{pmatrix} \; ; \quad \gamma_2 A = \begin{pmatrix} -A_4 \\ A_3 \\ A_2 \\ -A_1 \end{pmatrix} \; ;$$

$$\gamma_3 A = \begin{pmatrix} -iA_3 \\ iA_4 \\ iA_1 \\ -iA_2 \end{pmatrix} \; ; \quad \gamma_4 A = \begin{pmatrix} A_1 \\ A_2 \\ -A_3 \\ -A_4 \end{pmatrix} . \tag{5.3.8}$$

Then we can write:

$$B(\vec{p}) = \begin{pmatrix} (-ip_x - p_y)A_4 - ip_z A_3 + i(p_0 + m)A_1 \\ (-ip_x + p_y)A_3 + ip_z A_4 + i(p_0 + m)A_2 \\ (ip_x + p_y)A_2 + ip_z A_1 + i(-p_0 + m)A_3 \\ (ip_x - p_y)A_1 - ip_z A_2 + i(-p_0 + m)A_4 \end{pmatrix} .$$

If we denote

$$p_\pm = p_x \pm i p_y, \tag{5.3.9}$$

and put the elements of A in increasing order everywhere, we obtain

$$B(\vec{p}) = \begin{pmatrix} i(p_0 + m)A_1 - ip_z A_3 - ip_- A_4 \\ i(p_0 + m)A_2 - ip_+ A_3 + ip_z A_4 \\ ip_z A_1 + ip_- A_2 + i(-p_0 + m)A_3 \\ ip_+ A_1 - ip_z A_2 + i(-p_0 + m)A_4 \end{pmatrix} . \tag{5.3.10}$$

The quantization of the spinorial field

Since the elements of the column vector A are arbitrary, if $B(\vec{p})$ is a solution of the Dirac equation (5.3.1), then the column vectors $B^{(r)}(\vec{p})$, whose elements are the coefficients of A_k, are also solutions. The equation (5.3.1) being homogeneous, one can factor out i, and we are left with the following linearly independent solutions:

$$B^{(1)}(\vec{p}) = \begin{pmatrix} p_0 + m \\ 0 \\ p_z \\ p_+ \end{pmatrix} \; ; \quad B^{(2)}(\vec{p}) = \begin{pmatrix} 0 \\ p_0 + m \\ p_- \\ -p_z \end{pmatrix} \; ;$$

$$\text{(5.3.11)}$$

$$B^{(3)}(\vec{p}) = \begin{pmatrix} -p_z \\ -p_+ \\ -p_0 + m \\ 0 \end{pmatrix} \; ; \quad B^{(4)}(\vec{p}) = \begin{pmatrix} -p_- \\ p_z \\ 0 \\ -p_0 + m \end{pmatrix} .$$

It is not difficult to verify that the column vectors $B^{(r)}(\vec{p})$ ($r = 1, 2, 3, 4$) satisfy the Dirac equation (5.3.1). For example,

$$(\hat{p} - im)B^{(1)}(\vec{p}) = p_x \gamma_1 B^{(1)}(\vec{p}) + p_y \gamma_2 B^{(1)}(\vec{p})$$

$$+ p_z \gamma_3 B^{(1)}(\vec{p}) + i p_0 \gamma_4 B^{(1)}(\vec{p}) - im B^{(1)}(\vec{p})$$

$$= p_x \begin{pmatrix} -ip_+ \\ -ip_z \\ 0 \\ i(p_0 + m) \end{pmatrix} + p_y \begin{pmatrix} -p_+ \\ p_z \\ 0 \\ -(p_0 + m) \end{pmatrix} + p_z \begin{pmatrix} -ip_z \\ ip_+ \\ i(p_0 + m) \\ 0 \end{pmatrix}$$

$$+ i p_0 \begin{pmatrix} p_0 + m \\ 0 \\ -p_z \\ -p_+ \end{pmatrix} - im \begin{pmatrix} p_0 + m \\ 0 \\ p_z \\ p_+ \end{pmatrix}$$

$$= \begin{pmatrix} -ip_x p_+ - p_y p_+ - ip_z^2 + i(p_0^2 - m^2) \\ -ip_x p_z + p_y p_z + ip_z p_+ \\ ip_z(p_0 + m) - ip_0 p_z - imp_z \\ ip_x(p_0 + m) - p_y(p_0 + m) - ip_0 p_+ - imp_+ \end{pmatrix} = \begin{pmatrix} 0 \\ 0 \\ 0 \\ 0 \end{pmatrix} = 0.$$

It's easily seen that the component p_0 of the momentum four-vector can be replaced in $B^{(r)}(\vec{p})$ by either $+E_p$, or $-E_p$. The appropriate choice is obtained by requiring that the matrices $B^{(r)}(\vec{p})$ be orthogonal. Since

$$B^{(1)+}(\vec{p}) = (p_0 + m \; 0 \; p_z \; p_-),$$

we have, for example,

$$B^{(1)+}(\vec{p})B^{(4)}(\vec{p}) = -p_-(p_0 + m) + p_-(-p_0 + m). \qquad (5.3.12)$$

In order for the matrices $B^{(1)+}(\vec{p})$ and $B^{(4)}(\vec{p})$ to be orthogonal, it is necessary to chose $p_0 = +E_p$ in $B^{(1)+}(\vec{p})$, and $p_0 = -E_p$ in $B^{(4)}(\vec{p})$. The analysis of all possible products $B^{(i)+}(\vec{p})B^{(k)}(\vec{p})$ $(i \neq k)$, shows that the condition of orthogonality is fulfilled if we take $p_0 = +E_p$ in $B^{(1)}(\vec{p})$ and $B^{(2)}(\vec{p})$, and $p_0 = -E_p$ in $B^{(3)}(\vec{p})$ and $B^{(4)}(\vec{p})$. We have:

$$B^{(1)+}(\vec{p})B^{(1)}(\vec{p}) = (p_0 + m \quad 0 \quad p_z \quad p_-) \begin{pmatrix} p_0 + m \\ 0 \\ p_z \\ p_+ \end{pmatrix}$$

$$= (E_p + m)^2 + p_z^2 + p_- p_+.$$

But, according to (5.3.9),

$$p_z^2 + p_- p_+ = p_z^2 + (p_x - ip_y)(p_x + ip_y) = \vec{p}^{\,2},$$

so that

$$B^{(1)+}(\vec{p})B^{(1)}(\vec{p}) = E_p^2 + m^2 + 2mE_p + \vec{p}^{\,2} = 2E_p(E_p + m). \qquad (5.3.13)$$

Computing all products of the type $B^{(r)+}(\vec{p})B^{(r)}(\vec{p})$ (no summation over r) we find that the result is always the same, that is

$$B^{(r)+}(\vec{p})B^{(r)}(\vec{p}) = 2E_p(E_p + m). \qquad (5.3.14)$$

We can condense these results in a single relation, namely

$$B^{(i)+}(\vec{p})B^{(k)}(\vec{p}) = 2E_p(E_p + m)\,\delta_{ik}, \qquad (5.3.15)$$

which expresses the orthogonality of matrices $B^{(r)}(\vec{p})$. If we replace the matrices $B^{(r)}(\vec{p})$ by

$$u^{(r)}(\vec{p}) = \frac{B^{(r)}(\vec{p})}{\sqrt{2E_p(E_p + m)}}, \qquad (5.3.16)$$

then the new matrices $u^{(r)}(\vec{p})$ are *orthonormal*:

$$u^{(i)+}(\vec{p})u^{(k)}(\vec{p}) = \delta_{ik}. \qquad (5.3.17)$$

The quantization of the spinorial field

We can also write these matrices as

$$u^{(1)}(\vec{p}) = \frac{1}{\sqrt{2E_p(E_p+m)}} \begin{pmatrix} p_0 + m \\ 0 \\ p_z \\ p_+ \end{pmatrix}$$

$$= \frac{E_p + m}{\sqrt{2E_p(E_p+m)}} \begin{pmatrix} 1 \\ 0 \\ \lambda p_z \\ \lambda p_+ \end{pmatrix} = \kappa \begin{pmatrix} 1 \\ 0 \\ \lambda p_z \\ \lambda p_+ \end{pmatrix},$$

with

$$\lambda = \frac{1}{E_p+m} \ ; \qquad \kappa = \sqrt{\frac{E_p+m}{2E_p}}. \tag{5.3.18}$$

The new orthonormal matrices $u^{(r)}(\vec{p})$:

$$u^{(1)}(\vec{p}) = \kappa \begin{pmatrix} 1 \\ 0 \\ \lambda p_z \\ \lambda p_+ \end{pmatrix} \ ; \qquad u^{(2)}(\vec{p}) = \kappa \begin{pmatrix} 0 \\ 1 \\ \lambda p_- \\ -\lambda p_z \end{pmatrix} \ ;$$

$$\tag{5.3.19}$$

$$u^{(3)}(\vec{p}) = \kappa \begin{pmatrix} -\lambda p_z \\ -\lambda p_+ \\ 1 \\ 0 \end{pmatrix} \ ; \qquad u^{(4)}(\vec{p}) = \kappa \begin{pmatrix} -\lambda p_- \\ \lambda p_z \\ 0 \\ 1 \end{pmatrix},$$

are also solutions of the Dirac equation (5.3.1).

Let us now define the matrix U and its adjoint U^+, as

$$U = \begin{pmatrix} u^{(1)} & u^{(2)} & u^{(3)} & u^{(4)} \end{pmatrix} \ ; \qquad U^+ = \begin{pmatrix} u^{(1)+} \\ u^{(2)+} \\ u^{(3)+} \\ u^{(4)+} \end{pmatrix}. \tag{5.3.20}$$

In view of (5.3.17), we then have

$$UU^+ = U^+U = I, \tag{5.3.21}$$

because U is a unitary, non-singular matrix. Therefore, if there exists a right inverse element, then there also exists a left inverse element. We can also write this as

$$\sum_r u_{ir}u^+_{rk} = \delta_{ik}. \tag{5.3.22}$$

The free particle in the Dirac theory

Following the same procedure as in the previous chapters, we write the Fourier transform of the state function $\psi(x)$ as

$$\psi(x) = \frac{1}{(2\pi)^{3/2}} \int \psi(\vec{p})\, e^{ipx}\, d\vec{p} = \frac{1}{(2\pi)^{3/2}} \int \sum_{r=1}^{4} a_r(\vec{p}) u^{(r)}(\vec{p})\, e^{ipx}\, d\vec{p},$$

(5.3.23)

corresponding to the plane-wave expansion of the state function. The integral (5.3.23) can be split into two parts, one for each of the two possible values $p_0 = +E_p$, and $p_0 = -E_p$, as follows:

$$\psi(x) = \frac{1}{(2\pi)^{3/2}} \int \sum_{r=1}^{2} a_r(\vec{p}) u^{(r)}(\vec{p})\, e^{i(\vec{p}\cdot\vec{x} - E_p t)}\, d\vec{p}$$

$$+ \frac{1}{(2\pi)^{3/2}} \int \sum_{r=3}^{4} a_r(\vec{p}) u^{(r)}(\vec{p})\, e^{i(\vec{p}\cdot\vec{x} + E_p t)}\, d\vec{p}$$

$$= \frac{1}{(2\pi)^{3/2}} \int \sum_{r=1}^{2} a_r(\vec{p}) u^{(r)}(\vec{p})\, e^{i(\vec{p}\cdot\vec{x} - E_p t)}\, d\vec{p} \qquad (5.3.24)$$

$$+ \frac{1}{(2\pi)^{3/2}} \int \sum_{r=3}^{4} a_r(-\vec{p}) u^{(r)}(-\vec{p})\, e^{-i(\vec{p}\cdot\vec{x} - E_p t)}\, d\vec{p}$$

$$= \frac{1}{(2\pi)^{3/2}} \int \psi_+(\vec{p}) e^{i(\vec{p}\cdot\vec{x} - E_p t)}\, d\vec{p} + \frac{1}{(2\pi)^{3/2}} \int \psi_-(\vec{p})\, e^{-i(\vec{p}\cdot\vec{x} - E_p t)}\, d\vec{p},$$

or

$$\psi(x) = \psi_+(x) + \psi_-(x) = \frac{1}{(2\pi)^{3/2}} \int [\psi_+(\vec{p}) + \psi_-(-\vec{p})] e^{ipx}\, d\vec{p} \quad (5.3.25).$$

Therefore

$$\psi(\vec{p}) = \psi_+(\vec{p}) + \psi_-(\vec{p}), \qquad (5.3.26)$$

with

$$\psi_+(\vec{p}) = \sum_{r=1}^{2} a_r(\vec{p}) u^{(r)}(\vec{p}) \; ; \qquad \psi_-(\vec{p}) = \sum_{r=3}^{4} a_r(-\vec{p}) u^{(r)}(-\vec{p}). \quad (5.3.27)$$

We can also write

$$\psi_\pm(x) = \frac{1}{(2\pi)^{3/2}} \int \psi_\pm(\vec{p})\, e^{\pm ipx}\, d\vec{p}, \qquad (5.3.28)$$

The quantization of the spinorial field

where $\psi_+(\vec{p})$ refers to the states with positive energy ($p_0 = +E_p$), and $\psi_-(\vec{p})$ to those with negative energy ($p_0 = -E_p$).

To facilitate the calculation, we introduce the following notations:

$$a_1(\vec{p}) = a_{1+}(\vec{p}) \; ; \; a_2(\vec{p}) = a_{2+}(\vec{p}) \; ;$$

$$a_3(-\vec{p}) = a_{1-}(\vec{p}) \; ; \; a_4(-\vec{p}) = a_{2-}(\vec{p}),$$

and

$$u^{(1)}(\vec{p}) = v_+^{(1)}(\vec{p}) \; ; \; u^{(2)}(\vec{p}) = v_+^{(2)}(\vec{p}) \; ;$$

$$u^{(3)}(-\vec{p}) = v_-^{(1)}(\vec{p}) \; ; \; u^{(4)}(-\vec{p}) = v_-^{(2)}(\vec{p}),$$

or, written in a condensed form,

$$\begin{cases} a_s(\vec{p}) = a_{s+}(\vec{p}) \; ; \quad a_{s+2}(-\vec{p}) = a_{s-}(\vec{p}) \; ; \\ u^{(s)}(\vec{p}) = v_+^{(s)}(\vec{p}) \; ; \quad u^{(s+2)}(-\vec{p}) = v_-^{(s)}(\vec{p}) \; ; \quad (s = 1, 2). \end{cases} \qquad (5.3.29)$$

By means of (5.3.29), we can write (5.3.27) as

$$\psi_\pm(\vec{p}) = \sum_{r=1}^{2} a_{r\pm}(\vec{p}) v_\pm^{(r)}(\vec{p}), \qquad (5.3.30)$$

while the matrices (5.3.19) take the form

$$v_+^{(1)} = \kappa \begin{pmatrix} 1 \\ 0 \\ \lambda p_z \\ \lambda p_+ \end{pmatrix} \; ; \quad v_+^{(2)} = \kappa \begin{pmatrix} 0 \\ 1 \\ \lambda p_- \\ -\lambda p_z \end{pmatrix} \; ;$$

$$(5.3.31)$$

$$v_-^{(1)} = \kappa \begin{pmatrix} \lambda p_z \\ \lambda p_+ \\ 1 \\ 0 \end{pmatrix} \; ; \quad v_-^{(2)} = \kappa \begin{pmatrix} \lambda p_- \\ -\lambda p_z \\ 0 \\ 1 \end{pmatrix} .$$

In view of (5.3.28), we can write the adjoint of $\psi(x)$ both as

$$\psi^+(x) = \frac{1}{(2\pi)^{3/2}} \int \psi_+^+(\vec{p}) \, e^{ipx} \, d\vec{p} + \frac{1}{(2\pi)^{3/2}} \int \psi_-^+(\vec{p}) \, e^{-ipx} \, d\vec{p},$$

and as

$$[\psi(x)]^+ = \frac{1}{(2\pi)^{3/2}} \int [\psi_+(\vec{p})]^+ \, e^{-ipx} \, d\vec{p} + \frac{1}{(2\pi)^{3/2}} \int [\psi_-(\vec{p})]^+ \, e^{ipx} \, d\vec{p}.$$

Comparing the last two formulas, we obtain

$$[\psi_\pm(\vec{p})]^+ = \psi_\mp^\pm(\vec{p}). \tag{5.3.32}$$

Since

$$\psi_\pm^\pm(\vec{p}) = \sum_{r=1}^{2} a_{r\pm}^+(\vec{p}) v_\pm^{(r)+}(\vec{p}),$$

by means of (5.3.30) and (5.3.32) we also have

$$[v_\pm^{(r)}(\vec{p})]^+ = v_\mp^{(r)+}(\vec{p}). \tag{5.3.33}$$

Therefore, we can write:

$$\begin{cases} v_+^{(1)+}(\vec{p}) = [v_-^{(1)}(\vec{p})]^+ = [u^{(3)}(-\vec{p})]^+ = \kappa(\lambda p_z \quad \lambda p_- \quad 1 \quad 0) \; ; \\ v_+^{(2)+}(\vec{p}) = [v_-^{(2)}(\vec{p})]^+ = [u^{(4)}(-\vec{p})]^+ = \kappa(\lambda p_+ \quad -\lambda p_z \quad 0 \quad 1) \; ; \\ v_-^{(1)+}(\vec{p}) = [v_+^{(1)}(\vec{p})]^+ = [u^{(1)}(\vec{p})]^+ = \kappa(1 \quad 0 \quad \lambda p_z \quad \lambda p_-) \; ; \\ v_-^{(2)+}(\vec{p}) = [v_+^{(2)}(\vec{p})]^+ = [u^{(2)}(\vec{p})]^+ = \kappa(0 \quad 1 \quad \lambda p_+ \quad -\lambda p_z), \end{cases} \tag{5.3.34}$$

that is

$$v_+^{(r)+}(\vec{p}) = [u^{(r+2)}(-\vec{p})]^+ \; ; \quad v_-^{(r)+}(\vec{p}) = [u^{(r)}(\vec{p})]^+. \tag{5.3.35}$$

We also have

$$v_+^{(s)+}(\vec{p}) \, v_-^{(t)}(\vec{p}) = [u^{(s+2)}(-\vec{p})]^+ \, u^{(t+2)}(-\vec{p}) = \delta_{st} \; ;$$

$$v_-^{(s)+}(\vec{p}) \, v_+^{(t)}(\vec{p}) = [u^{(s)}(\vec{p})]^+ \, u^{(t)}(\vec{p}) = \delta_{st},$$

or

$$v_\pm^{(s)+}(\vec{p}) \, v_\mp^{(t)}(\vec{p}) = \delta_{st} \; ; \quad (s,t = 1,2). \tag{5.3.36}$$

Relations (5.3.34) also yield

$$v_+^{(s)+}(\vec{p}) \, v_+^{(t)}(-\vec{p}) = [u^{(s+2)}(-\vec{p})]^+ \, u^{(t)}(-\vec{p}) = 0 \; ;$$

$$v_-^{(s)+}(\vec{p}) \, v_-^{(t)}(-\vec{p}) = [u^{(s)}(\vec{p})]^+ \, u^{(t+2)}(\vec{p}) = 0 \; ;$$

that is

$$v_\pm^{(s)+}(\vec{p}) \, v_\pm^{(s)}(-\vec{p}) = 0. \tag{5.3.37}$$

On the other hand, taking into account (5.3.30), we have:

$$\psi_\pm(x) = \frac{1}{(2\pi)^{3/2}} \int a_{s\pm}(\vec{p}) \, v_\pm^{(s)}(\vec{p}) \, e^{\pm ipx} \, d\vec{p} \; ; \tag{5.3.38}$$

125

The quantization of the spinorial field

$$\psi_{\pm}^{\pm}(x) = \frac{1}{(2\pi)^{3/2}} \int a_{s\pm}^{+}(\vec{p}) \, v_{\pm}^{(s)+}(\vec{p}) \, e^{\pm ipx} \, d\vec{p} \; ; \qquad (5.3.39)$$

$$\overline{\psi}_{\pm}(x) = \psi_{\pm}^{\pm}(x)\gamma_4 = \frac{1}{(2\pi)^{3/2}} \int a_{s\pm}^{+}(\vec{p}) \, \overline{v}_{\pm}^{(s)}(\vec{p}) \, e^{\pm ipx} \, d\vec{p} \;, \qquad (5.3.40)$$

where

$$\overline{v}_{\pm}^{(s)}(\vec{p}) = v_{\pm}^{(s)+}(\vec{p}) \, \gamma_4. \qquad (5.3.41)$$

In view of (5.3.5), (5.3.31) and (5,3.41), we obtain

$$\begin{cases} \overline{v}_{+}^{(1)}(\vec{p}) = \kappa(\lambda p_z \quad \lambda p_- \quad -1 \quad 0) \; ; \\ \overline{v}_{+}^{(2)}(\vec{p}) = \kappa(\lambda p_+ \quad -\lambda p_z \quad 0 \quad -1) \; ; \\ \overline{v}_{-}^{(1)}(\vec{p}) = \kappa(1 \quad 0 \quad -\lambda p_z \quad -\lambda; p_-) \; ; \\ \overline{v}_{-}^{(2)}(\vec{p}) = \kappa(0 \quad 1 \quad -\lambda p_+ \quad \lambda p_z) \; . \end{cases} \qquad (5.3.42)$$

Using (5.3.42) and (5.3.34), we also have

$$\overline{v}_{+}^{(r)}(\vec{p}) = -[u^{(r+2)}(\vec{p})]^{+} \; ; \qquad \overline{v}_{-}^{(r)}(\vec{p}) = [u^{(r)}(-\vec{p})]^{+}. \qquad (5.3.43)$$

Using (5.3.2) and (5.3.18), let us now calculate

$$[u^{(s)}(\vec{p})]^{+} \, u^{(t)}(-\vec{p}) = \kappa^2(1 - \lambda^2 \vec{p}^2)\delta_{st}$$

$$= \frac{E_p + m}{2E_p}\Big[1 - \frac{\vec{p}^2}{(E_p + m)^2}\Big]\delta_{st}$$

$$= \frac{1}{2E_p(E_p + m)}(E_p^2 + 2mE_p + m^2 - \vec{p}^2)\delta_{st} = \frac{m}{E_p}\delta_{st}.$$

Consequently,

$$[u^{(s)}(\vec{p})]^{+} \, u^{(t)}(-\vec{p}) = [u^{(s+2)}(\vec{p})]^{+} \, u^{(t+2)}(-\vec{p}) = \frac{m}{E_p}\delta_{st}, \qquad (5.3.44)$$

and then, using (5.3.33),

$$\begin{cases} [v_{+}^{(s)}(\vec{p})]^{+} \, v_{+}^{(t)}(-\vec{p}) = v_{+}^{(s)+}(\vec{p}) \, v_{+}^{(t)}(-\vec{p}) = \frac{m}{E_p}\delta_{st} \; ; \\ [v_{-}^{(s)}(-\vec{p})]^{+} v_{-}^{(t)}(\vec{p}) = v_{-}^{(s)+}(-\vec{p}) v_{-}^{(t)}(\vec{p}) = \frac{m}{E_p}\delta_{st}. \end{cases} \qquad (5.3.45)$$

In view of (5.3.29), and (5.3.43), we also have:

$$\begin{cases} \overline{v}_{+}^{(s)}(\vec{p})v_{-}^{(t)}(\vec{p}) = -[u^{(s+2)}(\vec{p})]^{+} \, u^{(t+2)}(-\vec{p}) = -\frac{m}{E_p}\delta_{st} \; ; \\ \overline{v}_{-}^{(s)}(\vec{p})v_{+}^{(t)}(\vec{p}) = [u^{(s)}(-\vec{p})]^{+} \, u^{(t)}(\vec{p}) = \frac{m}{E_p}\delta_{st}, \end{cases}$$

126

The free particle in the Dirac theory

or, in a condensed form,

$$\bar{v}_{\pm}^{(s)}(\vec{p})\, v_{\mp}^{(t)}(\vec{p}) = \mp \frac{m}{E_p}\delta_{st}. \tag{5.3.46}$$

Recalling that the matrices $u^{(1)}$, $u^{(2)}$ and $v_{+}^{(1)}$, $v_{+}^{(2)}$ satisfy the Dirac equation for $p_0 = +E_p$, while matrices $u^{(3)}$, $u^{(3)}$ and $v_{-}^{(1)}$, $v_{-}^{(2)}$ are solutions of the Dirac equation for $p_0 = -E_p$, we can write

$$(\vec{p}\cdot\vec{\gamma} + iE_p\gamma_4 - im)\,u^{(r)} = \begin{cases} 0 & ,\ r = 1,2\,; \\ 2iE_p\gamma_4 u^{(r)} & ,\ r = 3,4\,, \end{cases}$$

$$(\vec{p}\cdot\vec{\gamma} - iE_p\gamma_4 - im)\,u^{(r)} = \begin{cases} -2iE_p\gamma_4 u^{(r)} & ,\ r = 1,2\,; \\ 0 & ,\ r = 3,4\,, \end{cases}$$

or

$$\frac{\gamma_4}{2iE_p}(\vec{p}\cdot\vec{\gamma} + iE_p\gamma_4 - im)\,u^{(r)} = \begin{cases} 0 & ,\ r = 1,2\,; \\ u^{(r)} & ,\ r = 3,4\,, \end{cases} \tag{5.3.47}$$

$$-\frac{\gamma_4}{2iE_p}(\vec{p}\cdot\vec{\gamma} - iE_p\gamma_4 - im)\,u^{(r)} = \begin{cases} u^{(r)} & ,\ r = 1,2\,; \\ 0 & ,\ r = 3,4\,, \end{cases} \tag{5.3.48}$$

after adding and subtracting the term $iE_p\gamma_4$. If we denote

$$\Lambda_- = -\frac{i\gamma_4}{2E_p}(\vec{p}\cdot\vec{\gamma} + iE_p\gamma_4 - im)\,; \tag{5.3.49}$$

$$\Lambda_+ = \frac{i\gamma_4}{2E_p}(\vec{p}\cdot\vec{\gamma} - iE_p\gamma_4 - im)\,; \tag{5.3.50}$$

then (5.3.47) and (5.3.48) yield

$$\Lambda_- u^{(r)} = \begin{cases} 0 & ,\ r = 1,2\,; \\ u^{(r)} & ,\ r = 3,4\,, \end{cases} \tag{5.3.51}$$

$$\Lambda_+ u^{(r)} = \begin{cases} u^{(r)} & ,\ r = 1,2\,; \\ 0 & ,\ r = 3,4\,, \end{cases} \tag{5.3.52}$$

The operators Λ_- and Λ_+ defined by (5.3.51) and (5.3.52) have the following properties:

a) $\Lambda_+ + \Lambda_- = I$ (unity operator).

b) Both operators are Hermitian: $\Lambda_{\pm}^{\dagger} = \Lambda_{\pm}$. For example,

$$\Lambda_-^{\dagger} = \left[\frac{\gamma_4}{2iE_p}(\vec{p}\cdot\vec{\gamma} + iE_p\gamma_4 - im)\right]^{\dagger} = (\vec{p}\cdot\vec{\gamma} + iE_p\gamma_4 - im)^{\dagger}\left(\frac{\gamma_4}{2iE_p}\right)^{\dagger}$$

$$= -\frac{1}{2iE_p}(\vec{p}\cdot\vec{\gamma} - iE_p\gamma_4 + im)\gamma_4 = -\frac{\gamma_4}{2iE_p}(-\vec{p}\cdot\vec{\gamma} - iE_p\gamma_4 + im) = \Lambda_-.$$

c) Operators Λ_+ and Λ_- are orthogonal, that is $\Lambda_+\Lambda_- = 0$. Indeed,

$$\Lambda_+\Lambda_- = \frac{-\gamma_4}{2iE_p}(\vec{p}\cdot\vec{\gamma} - iE_p\gamma_4 - im)\frac{\gamma_4}{2iE_p}(\vec{p}\cdot\vec{\gamma} + iE_p\gamma_4 - im)$$

$$= -\frac{1}{4E_p^2}(\vec{p}\cdot\vec{\gamma} + iE_p\gamma_4 + im)(\vec{p}\cdot\vec{\gamma} + iE_p\gamma_4 - im)$$

$$= -\frac{1}{4E_p^2}\Big[(\vec{p}\cdot\vec{\gamma})^2 + iE_p\vec{p}\cdot\vec{\gamma}\gamma_4 - im\vec{p}\cdot\vec{\gamma} + iE_p\gamma_4\vec{p}\cdot\vec{\gamma}$$

$$-E_p^2 + mE_p\gamma_4 + im\vec{p}\cdot\vec{\gamma} - mE_p\gamma_4 + m^2\Big].$$

Since

$$(\vec{p}\cdot\vec{\gamma})^2 = (p_x\gamma_1 + p_y\gamma_2 + p_z\gamma_3)(p_x\gamma_1 + p_y\gamma_2 + p_z\gamma_3) = p_x^2 + p_y^2 + p_z^2 = \vec{p}^2,$$

and using (5.3.2), after some cancellations we are left with

$$\Lambda_+\Lambda_- = -\frac{1}{4E_p^2}(\vec{p}^2 + m^2 - E_p^2) = 0.$$

d) The two operators are idempotent, i.e. $\Lambda_\pm^2 = \Lambda_\pm$. For example:

$$\Lambda_+^2 = \Big[\frac{\gamma_4}{2iE_p}(\vec{p}\cdot\vec{\gamma} - iE_p\gamma_4 - im)\Big]\Big[\frac{\gamma_4}{2iE_p}(\vec{p}\cdot\vec{\gamma} - iE_p\gamma_4 - im)\Big]$$

$$= \frac{1}{4E_p^2}(\vec{p}\cdot\vec{\gamma} + iE_p\gamma_4 + im)(\vec{p}\cdot\vec{\gamma} - iE_p\gamma_4 - im)$$

$$= \frac{1}{4E_p^2}\Big[(\vec{p}\cdot\vec{\gamma})^2 - iE_p(\vec{p}\cdot\vec{\gamma})\gamma_4 - im(\vec{p}\cdot\vec{\gamma})$$

$$+iE_p\gamma_4(\vec{p}\cdot\vec{\gamma}) + E_p^2 + mE_p\gamma_4 + im(\vec{p}\cdot\vec{\gamma}) + mE_p\gamma_4 + m^2\Big]$$

$$= \frac{1}{2E_p^2}(iE_p\gamma_4\vec{p}\cdot\vec{\gamma} + mE_p\gamma_4 + E_p^2) = -\frac{\gamma_4}{2iE_p}(\vec{p}\cdot\vec{\gamma} - iE_p\gamma_4 - im) = \Lambda_+.$$

Properties (b) and (d) show that $\Lambda_\pm$ are projection operators. According to the last property, the four spinors $u^{(r)}$ form the basis of a four-dimensional space. In this case, the projection operators Λ_+ and Λ_- separate the space into two bi-dimensional orthogonal subspaces

The free particle in the Dirac theory

(see properties (a) and (c)), corresponding to positive and negative energies, respectively.

Multiplying (5.3.48) on the right by $-2iE_p u^{(r)+}\gamma_4$ and summing over r, we obtain

$$\gamma_4(\vec{p}\cdot\vec{\gamma} - iE_p\gamma_4 - im)\sum_{r=1}^{4} u^{(r)}u^{(r)+}\gamma_4 = -2iE_p\sum_{r=1}^{2} u^{(r)}u^{(r)+}\gamma_4.$$

On the l.h.s. we take into account that $\sum_{r=1}^{4} u^{(r)}u^{(r)+} = I$, while on the r.h.s. we observe that, for $r = 1, 2$, (5.3.29),(5.3.35), and (5.3.41) yield

$$u^{(r)} = v_+^{(r)}\ ;\quad u^{(r)+} = v_-^{(r)+};\quad u^{(r)+}\gamma_4 = \bar{v}_-^{(r)}$$

Then we can write

$$\gamma_4(\vec{p}\cdot\vec{\gamma} - iE_p\gamma_4 - im)\gamma_4 = -2iE_p\sum_{r=1}^{2} v_+^{(r)}(\vec{p})\bar{v}_-^{(r)}(\vec{p}).$$

Using the anti-commutation relation (5.1.12), we obtain

$$\vec{p}\cdot\vec{\gamma} + iE_p\gamma_4 + im = \hat{p} + im = 2iE_p\sum_{r=1}^{2} v_+^{(r)}(\vec{p})\bar{v}_-^{(r)}(\vec{p}),$$

which leads to

$$\sum_{r=1}^{2} v_+^{(r)}(\vec{p})\bar{v}_-^{(r)}(\vec{p}) = \frac{-i\hat{p} + m}{2E_p}. \tag{5.3.53}$$

Similarly, multiplying (5.3.47), on the right, by $2iE_p u^{(r)+}\gamma_4$, then summing over s, one gets:

$$\gamma_4(\vec{p}\cdot\vec{\gamma} + iE_p\gamma_4 - im)\sum_{r=1}^{4} u^{(r)}u^{(r)+}\gamma_4 = 2iE_p\sum_{r=3}^{4} u^{(r)}u^{(r)+}\gamma_4.$$

In view of (5.3.29), (5.3.35), (5.3.41), and using the relation $\sum u^{(r)}u^{(r)+} = I$, we have

$$\sum_{r=3}^{4} u^{(r)}(\vec{p})u^{(r)+}(\vec{p}) = \sum_{r=1}^{2} u^{(r+2)}(\vec{p})u^{(r+2)+}(\vec{p}) =$$

$$\sum_{r=1}^{2} v_-^{(r)}(-\vec{p})v_+^{(r)+}(-\vec{p}),$$

The quantization of the spinorial field

and, using anti-commutation relation (5.1.12),

$$-\vec{p}\cdot\vec{\gamma}+iE_p\gamma_4-im = 2iE_p\sum_{r=1}^{2}v_-^{(r)}(-\vec{p})\overline{v}_+^{(r)+}(-\vec{p}).$$

Replacing $\vec{p}$ by $-\vec{p}$, we obtain

$$\vec{p}\cdot\vec{\gamma}+iE_p\gamma_4-im = \hat{p}-im = 2iE_p\sum_{r=1}^{2}v_-^{(r)}(\vec{p})\overline{v}_+^{(r)}(\vec{p}),$$

and finally

$$\sum_{r=1}^{2}v_-^{(r)}(\vec{p})\overline{v}_+^{(r)}(\vec{p}) = \frac{-i\hat{p}-m}{2E_p}. \tag{5.3.54}$$

It is obvious that relation (5.3.53) corresponds to one type of particles (electrons), and relation (5.3.54) to the other type (positrons). This connection will be explored later.

5.4. Energy, momentum, charge, and spin of the free spinorial field in momentum representation

Energy. In view of (5.2.13), we have

$$H = H_{1+} + H_{1-} + H_{2+} + H_{2-} , \tag{5.4.1}$$

where we denoted

$$H_{1\pm} = \frac{1}{2}\int (\psi_{\pm,4}^+\psi_\pm - \psi_\pm^+\psi_{\pm,4})\,d\vec{x} ; \tag{5.4.2}$$

$$H_{2\pm} = \frac{1}{2}\int (\psi_{\pm,4}^+\psi_\mp - \psi_\pm^+\psi_{\mp,4})\,d\vec{x} . \tag{5.4.3}$$

Using (5.3.38) and (5.3.39), we can write

$$\psi_{\pm,4}(x) = \frac{1}{(2\pi)^{3/2}}\int \mp E_p a_{s\pm}(\vec{p})v_\pm^{(s)}(\vec{p})\,e^{\pm ipx}\,d\vec{p} ; \tag{5.4.4}$$

$$\psi_{\pm,4}^+(x) = \frac{1}{(2\pi)^{3/2}}\int \mp E_p a_{s\pm}^+(\vec{p})v_\pm^{(s)+}(\vec{p})\,e^{\pm ipx}\,d\vec{p} . \tag{5.4.5}$$

Using (5.3.38), (5.3.39), (5,4,4), and (5.4.5), we shall first calculate

$$H_{1+} = \frac{1}{2}\int (\psi_{+,4}^+\psi_+ - \psi_+^+\psi_{+,4})\,d\vec{x}$$

$$= \frac{1}{2}\frac{1}{(2\pi)^3}\int \left[-E_p a_s^+(\vec{p})v_+^{(s)+}(\vec{p})a_{r+}(\vec{p}')v_+^{(r)}(\vec{p}')\right.$$

$$\left.+E_p' a_{s+}^+(\vec{p})v_+^{(s)+}(\vec{p})a_{r+}(\vec{p}')v_+^{(r)}(\vec{p}')\right] e^{i(p+p')x}\, d\vec{x}\, d\vec{p}\, d\vec{p}'.$$

Integrating with respect to $\vec{x}$ and then $\vec{p}'$, we find

$$H_{1+} = \frac{1}{2}\int \left[-E_p a_{s+}^+(\vec{p})a_{r+}(-\vec{p})v_+^{(s)+}(\vec{p})v_+^{(r)}(-\vec{p})\right.$$

$$\left.+E_p a_{s+}^+(\vec{p})a_{r+}(-\vec{p})v_+^{(s)+}(\vec{p})v_+^{(r)}(-\vec{p})\right] e^{-2iE_p t}\, d\vec{p},$$

and therefore

$$H_{1+} = 0. \qquad (5.4.6)$$

In a similar way it can be shown that

$$H_{1-} = 0. \qquad (5.4.7)$$

Furthermore, we have

$$H_{2+} = \frac{1}{2}\int (\psi_{+,4}^+\psi_- - \psi_+^+\psi_{-,4})\, d\vec{x}$$

$$= \frac{1}{2}\frac{1}{(2\pi)^3}\int \left[-E_p a_{s+}^+(\vec{p})v_+^{(s)+}(\vec{p})a_{r-}(\vec{p}')v_-^{(r)}(\vec{p}')\right.$$

$$\left.-E_p' a_{s+}^+(\vec{p})v_+^{(s)+}(\vec{p})a_{r-}(\vec{p}')v_-^{(r)}(\vec{p}')\right] e^{i(p-p')x}\, d\vec{x}\, d\vec{p}\, d\vec{p}'.$$

Integrating with respect to $\vec{x}$ and $\vec{p}'$, and using (5.3.36), we have

$$H_{2+} = \frac{1}{2}\int \left[-E_p a_{s+}^+(\vec{p})a_{r-}(\vec{p}) - E_p a_{s+}^+(\vec{p})a_{r-}(\vec{p})\right]v_+^{(s)+}(\vec{p})v_-^{(r)}(\vec{p})\, d\vec{p}$$

$$= -\int E_p\, a_{r+}^+(\vec{p})a_{r-}(\vec{p})\, d\vec{p}. \qquad (5.4.8)$$

Using the same procedure, we get

$$H_{2-} = \int E_p\, a_{r-}^+(\vec{p})a_{r+}(\vec{p})\, d\vec{p}. \qquad (5.4.9)$$

Substituting (5.4.6)-(5.4.9) in (5.4.1), we finally obtain the Hamiltonian of the free spinorial field in momentum representation

$$H = \int E_p\left[a_{r-}^+(\vec{p})a_{r+}(\vec{p}) - a_{r+}^+(\vec{p})a_{r-}(\vec{p})\right] d\vec{p}\,; \quad (r=1,2) \quad (5.4.10)$$

The quantization of the spinorial field

Momentum. The spatial components P_α of the momentum four-vector are given by (5.2.12). We can write

$$P_\alpha = P_{\alpha 1+} + P_{\alpha 1-} + P_{\alpha 2+} + P_{\alpha 2-},$$

where

$$P_{\alpha 1\pm} = \frac{i}{2} \int (\psi_{\pm,\alpha}^{+} \psi_\pm - \psi_\pm^{+} \psi_{\pm,\alpha}) \, d\vec{x} \; ;$$

$$P_{\alpha 2\pm} = \frac{i}{2} \int (\psi_{\pm,\alpha}^{+} \psi_\mp - \psi_\pm^{+} \psi_{\mp,\alpha}) \, d\vec{x} \; .$$

The procedure outlined above leads to

$$P_\alpha = \int p_\alpha [a_{r-}^{+}(\vec{p}) a_{r+}(\vec{p}) - a_{r+}^{+}(\vec{p}) a_{r-}(\vec{p})] \, d\vec{p} \, , \quad (r = 1, 2). \quad (5.4.11)$$

Charge. By means of (5.2.8), we can write

$$Q = -ie \int s_4 \, d\vec{x} = -e \int \overline{\psi} \gamma_4 \psi \, d\vec{x}$$

$$= -e \int \psi^{+} \psi \, d\vec{x} = Q_{1+} + Q_{1-} + Q_{2+} + Q_{2-}, \quad (5.4.12)$$

with

$$Q_{1\pm} = -e \int \psi_\pm^{+} \psi_\pm \, d\vec{x} \; ; \quad (5.4.13)$$

$$Q_{2\pm} = -e \int \psi_\pm^{+} \psi_\mp \, d\vec{x} \; ; \quad (5.4.14)$$

Let us start with Q_{1+}. Following the same steps, we obtain

$$Q_{1+} = -\frac{e}{(2\pi)^3} \int a_{s+}^{+}(\vec{p}) v_+^{(s)+}(\vec{p}) a_{r+}(\vec{p}') v_+^{(r)}(\vec{p}') \, e^{i(p+p')x} \, d\vec{x} \, d\vec{p} \, d\vec{p}',$$

or, after integration with respect to $\vec{x}$ and $\vec{p}'$

$$Q_{1+} = -e \int a_{s+}^{+}(\vec{p}) a_{r+}(-\vec{p}) v_+^{(s)+}(\vec{p}) v_+^{(r)}(-\vec{p}) \, d\vec{p}$$

Therefore, in view of (5.3.37), we obtain

$$Q_{1+} = 0. \quad (5.4.15)$$

Similarly, we get

$$Q_{1-} = 0. \quad (5.4.16)$$

We also have

$$Q_{2+} = -\frac{e}{(2\pi)^3} \int a_{s+}^+(\vec{p})v_+^{(s)+}(\vec{p})a_{r-}(\vec{p}')v_-^{(r)}(\vec{p}') e^{i(p-p')x} \, d\vec{x} \, d\vec{p} \, d\vec{p}',$$

or, after integration with respect to $\vec{x}$ and $\vec{p}'$

$$Q_{2+} = -e \int a_{r+}^+(\vec{p})a_{r-}(\vec{p}) \, d\vec{p}. \tag{5.4.17}$$

Similarly, we get

$$Q_{2-} = -e \int a_{r-}^+(\vec{p})a_{r+}(\vec{p}) \, d\vec{p}. \tag{5.4.18}$$

Substituting all these results in (5.4.12), we arrive at

$$Q = -e \int [a_{r+}^+(\vec{p})a_{r-}(\vec{p}) + a_{r-}^+(\vec{p})a_{r+}(\vec{p})] \, d\vec{p}, \quad (r=1,2). \tag{5.4.19}$$

Spin. We use the property that the tensor

$$I_{\alpha\beta} = \int S_{\alpha\beta4} \, d\vec{x},$$

as any antisymmetric tensor, can be associated with an axial vector $\vec{S}$ whose components are

$$S_1 = \int S_{234} \, d\vec{x} \; ; \quad S_2 = \int S_{314} \, d\vec{x} \; ; \quad S_3 = S_z = \int S_{124} \, d\vec{x}.$$

Suppose that the propagation takes place along the z-axis, i.e.

$$p_x = p_y = 0 \; ; \qquad p_+ = p_- = 0. \tag{5.4.20}$$

Using (5.2.11), we then have for S_z

$$S_z = \frac{1}{4} \int \overline{\psi}(\gamma_4\sigma_{12} + \sigma_{12}\gamma_4)\psi \, d\vec{x} = \frac{1}{2} \int \psi^+\sigma_{12}\psi \, d\vec{x}, \tag{5.4.21}$$

where, according to (5.2.10), σ_{12} is

$$\sigma_{12} = -i\gamma_1\gamma_2 = \frac{1}{2i}(\gamma_1\gamma_2 - \gamma_2\gamma_1). \tag{5.4.22}$$

The quantization of the spinorial field

Since γ_1 and γ_2 anticommute, we have

$$\sigma_{12} = \tau_3 = \begin{pmatrix} 1 & 0 & 0 & 0 \\ 0 & -1 & 0 & 0 \\ 0 & 0 & 1 & 0 \\ 0 & 0 & 0 & -1 \end{pmatrix}, \tag{5.4.23}$$

where τ_3 is a new notation. To express S_z in momentum representation, let us take advantage of the fact that, under condition (5.4.20), the matrices $v_{\pm}^{(1)}$ and $v_{\pm}^{(2)}$ given by (5.3.31) become

$$v_{+}^{(1)} = \kappa \begin{pmatrix} 1 \\ 0 \\ \lambda p_z \\ 0 \end{pmatrix} ; \qquad v_{+}^{(2)} = \kappa \begin{pmatrix} 0 \\ 1 \\ 0 \\ -\lambda p_z \end{pmatrix} ;$$

$$\tag{5.4.24}$$

$$v_{-}^{(1)} = \kappa \begin{pmatrix} \lambda p_z \\ 0 \\ 1 \\ 0 \end{pmatrix} ; \qquad v_{-}^{(2)} = \kappa \begin{pmatrix} 0 \\ -\lambda p_z \\ 0 \\ 1 \end{pmatrix} ,$$

while the adjoint matrices are

$$\begin{cases} v_{+}^{(1)+} = \kappa(\lambda p_z \quad 0 \quad 1 \quad 0) ; & v_{+}^{(2)+} = \kappa(0 \quad -\lambda p_z \quad 0 \quad 1) ; \\ v_{-}^{(1)+} = \kappa(1 \quad 0 \quad \lambda p_z \quad o) ; & v_{-}^{(2)+} = \kappa(0 \quad 1 \quad 0 \quad -\lambda p_z). \end{cases}$$

$$\tag{5.4.25}$$

It can also easily be seen that

$$\tau_3 v_{\pm}^{(1)} = v_{\pm}^{(1)} ; \qquad \tau_3 v_{\pm}^{(2)} = -v_{\pm}^{(2)}. \tag{5.4.26}$$

We now have all the elements necessary to calculate the spin angular momentum. As usual, we write

$$S_z = S_{1+} + S_{1-} + S_{2+} + S_{2-}, \tag{5.4.27}$$

where

$$S_{1\pm} = \frac{1}{2} \int \psi_{\pm}^{+} \tau_3 \psi_{\pm} \, d\vec{x} ; \tag{3.4.28}$$

$$S_{2\pm} = \frac{1}{2} \int \psi_{\pm}^{+} \tau_3 \psi_{\mp} \, d\vec{x} . \tag{3.4.29}$$

134

Energy, momentum, charge and spin

Using (5.3.28) and (5.3.29), we have:

$$S_{1+} = \frac{1}{2}\frac{1}{(2\pi)^3}\int a_{s+}^+(\vec{p})a_{r+}(\vec{p}')v_+^{(s)+}(\vec{p})\tau_3 v_+^{(r)}(\vec{p}')\,e^{i(p+p')x}\,d\vec{x}\,d\vec{p}\,d\vec{p}',$$

and, after integration with respect to $\vec{x}$ and $\vec{p}'$,

$$S_{1+} = \frac{1}{2}\int a_{s+}^+(\vec{p})a_{r+}(-\vec{p})v_+^{(s)+}(\vec{p})\tau_3 v_+^{(r)}(-\vec{p})\,e^{-2iE_p t}\,d\vec{p}.$$

But, according to (5.4.26), the operator τ_3 acting on $v_+^{(r)}$ can only produce a sign change. Therefore, under the integral sign we have the product $v_+^{(s)+}(\vec{p})v_+^{(r)}(-\vec{p})$ which, by virtue of (5.3.37), is null, so that

$$S_{1+} = 0. \tag{5.4.30}$$

Similarly, we obtain

$$S_{1-} = 0. \tag{5.4.31}$$

Further, S_{2+} writes

$$S_{2+} = \frac{1}{2}\frac{1}{(2\pi)^3}\int a_{s+}^+(\vec{p})a_{r-}(\vec{p}')v_+^{(s)}(\vec{p})\tau_3 v_-^{(r)}(\vec{p}')\,e^{i(p-p')x}\,d\vec{x}\,d\vec{p}\,d\vec{p}'$$

$$= \frac{1}{2}\int a_{s+}^+(\vec{p})a_{r-}(\vec{p})v_+^{(s)+}(\vec{p})\tau_3 v_-^{(r)}(\vec{p})\,d\vec{p},$$

where we have carried out the integrations with respect to $\vec{x}$ and $\vec{p}'$. Using (5.3.36) and (5.4.26), we can show that

$$v_+^{(1)+}\tau_3 v_-^{(1)} = v_+^{(1)+}v_-^{(1)} = I \;;\quad v_+^{(2)+}\tau_3 v_-^{(2)} = -v_+^{(2)+}v_-^{(2)} = -I, \tag{5.4.32}$$

and S_{2+} writes

$$S_{2+} = \frac{1}{2}\int [a_{1+}^+(\vec{p})a_{1-}(\vec{p}) - a_{2+}^+(\vec{p})a_{2-}(\vec{p})]\,d\vec{p}. \tag{5.4.33}$$

We also have

$$S_{2-} = \frac{1}{2}\frac{1}{(2\pi)^3}\int a_{s-}^+(\vec{p})a_{r+}(\vec{p}')v_-^{(s)+}(\vec{p})\tau_3 v_+^{(r)}(\vec{p}')\,e^{-i(p-p')x}\,d\vec{x}\,d\vec{p}\,d\vec{p}'$$

$$= \frac{1}{2}\int [a_{1-}^+(\vec{p})a_{1+}(\vec{p})v_-^{(1)+}(\vec{p})v_+^{(1)}(\vec{p}) \tag{5.4.34}$$

$$- a_{2-}^+(\vec{p})a_{2+}(\vec{p})v_-^{(2)+}(\vec{p})v_+^{(2)}(\vec{p})] \, d\vec{p}$$

$$= \frac{1}{2}\int [a_{1-}^+(\vec{p})a_{1+}(\vec{p}) - a_{2-}^+(\vec{p})a_{2+}(\vec{p})] \, d\vec{p}.$$

Substituting (5.4.30), (5.4.31), (5.4.33), and (5.4.34) in (5.4.27), we finally obtain the z-component of the spin momentum

$$S_z = \frac{1}{2}\int [a_{1+}^+(\vec{p})a_{1-}(\vec{p}) - a_{2+}^+(\vec{p})a_{2-}(\vec{p})$$

$$+a_{1-}^+(\vec{p})a_{1+}(\vec{p}) - a_{2-}^+(\vec{p})a_{2+}(\vec{p})] \, d\vec{p}. \tag{5.4.35}$$

5.5. Anti-commutators of the free spinorial field

a) As we have seen in Chapter II, field quantization can be accomplished using commutation or anti-commutation relations. This time we have to resort to anti-commutation relations of the type

$$\{a_{s+}(\vec{p}), a_{r-}^+(\vec{p}')\} = \delta_{sr}\delta(\vec{p}-\vec{p}') \; ; \tag{5.5.1}$$

$$\{a_{s+}^+(\vec{p}), a_{r-}(\vec{p}')\} = \delta_{sr}\delta(\vec{p}-\vec{p}') \; ; \tag{5.5.2}$$

$$\{a_{s\pm}(\vec{p}), a_{r\pm}(\vec{p}')\} = \{a_{s\pm}(\vec{p}), a_{r\pm}^+(\vec{p}')\} = \{a_{s\pm}^+(\vec{p}), a_{r\pm}^+(\vec{p}')\}$$

$$= \{a_{s+}(\vec{p}), a_{r-}(\vec{p}')\} = \{a_{s+}^+(\vec{p}), a_{r-}^+(\vec{p}')\} = 0 \; ; \; (s,r=1,2), \tag{5.5.3}$$

because only this way we can insure the positive definite character of (5.4.10). Indeed, taking advantage of (5.5.2), we have

$$H = \int E_p[a_{r-}(\vec{p})a_{r+}^+(\vec{p}) + a_{r-}^+(\vec{p})a_{r+}(\vec{p})] \, d\vec{p} + E_0 \; ; \quad (r=1,2), \tag{5.5.4}$$

where

$$E_0 = -\delta(0)\int E_p \, d\vec{p} \tag{5.5.5}$$

is the vacuum energy level, which can be chosen as the origin of the measured energy scale.

A similar reasoning allows one to write the charge, given by (5.4.19), as

$$Q = e\int [a_{r-}(\vec{p})a_{r+}^+(\vec{p}) - a_{r-}^+(\vec{p})a_{r+}(\vec{p})] \, d\vec{p} + Q_0; \; (r=1,2), \tag{5.5.6}$$

where

$$Q_0 = -e\delta(0) \int d\vec{p}$$

is the electric charge of the vacuum. By virtue of (5.4.11), we also have

$$P_\alpha = \int p_\alpha [a_{r-}(\vec{p})a_{r+}^+(\vec{p}) + a_{r-}^+(\vec{p})a_{r+}(\vec{p})] \, d\vec{p} + P_{\alpha 0}; \quad (r = 1, 2), \quad (5.5.7)$$

with

$$P_{\alpha 0} = -\delta(0) \int p_\alpha \, d\vec{p}.$$

The analysis of H, Q, and P_α given by (5.5.4), (5.5.6) and (5.5.7), respectively, shows that the quantities

$$N_{+r}(\vec{p}) = a_{r-}(\vec{p})a_{r+}^+(\vec{p}) \; ; \tag{5.5.8}$$
$$N_{-r}(\vec{p}) = a_{r-}^+(\vec{p})a_{r+}(\vec{p}) \; , \quad (r = 1, 2 \; ; \quad \text{no summation} \,) \tag{5.5.9}$$

are operators that generate particle numbers. Therefore, we are dealing with two types of particles, denoted by $N_{+r}(\vec{p})$ and $N_{-r}(\vec{p})$. It follows that $a_{r+}^+(\vec{p})$ is the annihilation operator of a particle with momentum $\vec{p}$ and charge $+e$, while $a_{r-}(\vec{p})$ is the creation operator for the same particle. It also follows that $a_{r+}(\vec{p})$ is the annihilation operator of a particle with momentum $\vec{p}$ and charge $-e$, whereas $a_{r-}^+(\vec{p})$ is the creation operator of such particles. Note also that the states with the same charge have the same sign for the energy as well.

We can therefore assert the fact that the operators a_{r+}, a_{r-}^+ are associated with the electron, and the operators a_{r+}^+, a_{r-} correspond to the positron. In the configuration representation, ψ_+, and $\overline{\psi}_-$ are associated with the electrons, while $\overline{\psi}_+$, and ψ_- stand for positrons.

Taking into account (5.5.2) and (5.5.9) (and omitting, as we always have, the vacuum terms), the spin (5.4.35) writes

$$S_z = \frac{1}{2} \int [-a_{1-}(\vec{p})a_{1+}^+(\vec{p}) + a_{2-}(\vec{p})a_{2+}^+(\vec{p})$$

$$+a_{1-}^+(\vec{p})a_{1+}(\vec{p}) - a_{2-}^+(\vec{p})a_{2+}(\vec{p})] \, d\vec{p} \tag{5.5.10}$$

$$= \frac{1}{2} \int [-N_{+1}(\vec{p}) + N_{+2}(\vec{p}) + N_{-1}(\vec{p}) - N_{-2}(\vec{p})] \, d\vec{p},$$

meaning that for each of the two types of particles, $N_{+r}(\vec{p})$ and $N_{-r}(\vec{p})$, the projection of the spin along the z-axis is either $+\frac{1}{2}$, or $-\frac{1}{2}$.

The quantization of the spinorial field

b) Let us take the ik component of the anti-commutator $\{\psi(x), \overline{\psi}(y)\}$:

$$\{\psi_i(x), \overline{\psi}_k(y)\} = \{\psi_{i+}(x), \overline{\psi}_{k+}(y)\} + \{\psi_{i-}(x), \overline{\psi}_{k-}(y)\}$$

$$+\{\psi_{i-}(x), \overline{\psi}_{k+}(y)\} + \{\psi_{i+}(x), \overline{\psi}_{k-}(y)\}. \tag{5.5.11}$$

Taking into account (5.3.38) and (5.3.40), we can write

$$\{\psi_{i+}(x), \overline{\psi}_{k+}(y)\}$$

$$= \frac{1}{(2\pi)^3} \int \{a_{s+}(\vec{p}), a_{r+}^+(\vec{p}')\} v_{i+}^{(s)}(\vec{p}) \, \overline{v}_{k+}^{(r)}(\vec{p}') \, e^{i(px+p'y)} \, d\vec{p} \, d\vec{p}' = 0,$$

$$\tag{5.5.12}$$

by virtue of (5.5.3). Similarly, it can be shown that

$$\{\psi_{i-}(x), \overline{\psi}_{k-}(y)\} = 0. \tag{5.5.13}$$

Using (5.5.2), we also have

$$\{\psi_{i-}(x), \overline{\psi}_{k+}(y)\}$$

$$= \frac{1}{(2\pi)^3} \int \{a_{s-}(\vec{p}), a_{r+}^+(\vec{p}')\} v_{i-}^{(s)}(\vec{p}) \, \overline{v}_{k+}^{(r)}(\vec{p}') \, e^{-i(px-p'y)} \, d\vec{p} \, d\vec{p}'$$

$$\tag{5.5.14}$$

$$= \frac{1}{(2\pi)^3} \int v_{i-}^{(s)}(\vec{p}) v_{k+}^{(s)}(\vec{p}) \, e^{-ip(x-y)} \, d\vec{p},$$

where we have performed the summation over r, and the integration with respect to $\vec{p}'$.

However, according to (5.3.53) and (5.3.54),

$$v_{i\pm}^{(s)}(\vec{p}) \, \overline{v}_{k\mp}^{(s)}(\vec{p}) = \frac{(-i\hat{p} \pm m)_{ik}}{2E_p} ; \quad (s = 1, 2), \tag{5.5.15}$$

so that

$$\{\psi_{i-}(x), \overline{\psi}_{k+}(y)\} = \frac{1}{(2\pi)^3} \int \frac{(\hat{\partial} - m)_{ik}}{2E_p} e^{-ip(x-y)} \, d\vec{p},$$

where ∂ indicates the partial derivative with respect to x. Denoting (see Appendix, Section 8)

$$S_{ik\pm}(x - y) = \pm \frac{1}{(2\pi)^3 i} \int \frac{(\hat{\partial} - m)_{ik}}{2E_p} e^{\pm ip(x-y)} \, d\vec{p}, \tag{5.5.16}$$

138

Anti-commutators of the free spinorial field

we have
$$\{\psi_{i\pm}(x), \overline{\psi}_{k\mp}(y)\} = -i\, S_{ik\pm}(x-y). \tag{5.5.17}$$

Expression (5.5.16) can also be written as
$$S_{ik\pm}(x-y) = (\hat{\partial}-m)_{ik}\left[\pm\frac{1}{(2\pi)^3 i}\int\frac{1}{2E_p}e^{\pm ip(x-y)}\,d\vec{p}\right]$$
$$= (\hat{\partial}-m)_{ik}D_{m\pm}(x-y). \tag{5.5.18}$$

One notices that $S_{ik-}(x-y)$ is a particular solution of the Dirac equation (5.1.11). Indeed,
$$(\hat{\partial}+m)S_-(x-y) = (\hat{\partial}+m)(\hat{\partial}-m)D_{m-}(x-y)$$
$$= -(p^2+m^2)D_{m-}(x-y) = 0.$$

On the other hand,
$$\{\psi_{i+}(x), \overline{\psi}_{k-}(y)\} = \frac{1}{(2\pi)^3}\int\{a_{s+}(\vec{p}), a_{r-}^+(\vec{p}')\}$$
$$\times v_{i+}^{(s)}(\vec{p})\,\overline{v}_{k-}^{(r)}(\vec{p}')\,e^{i(px-p'y)}\,d\vec{p}\,d\vec{p}'$$
$$= \frac{1}{(2\pi)^3}\int v_{i+}^{(s)}(\vec{p})\,\overline{v}_{k-}^{(s)}(\vec{p})\,e^{ip(x-y)}\,d\vec{p}.$$

Using again (5.3.53), we obtain
$$\{\psi_{i+}(x), \overline{\psi}_{k-}(y)\} = \frac{1}{(2\pi)^3}\int\frac{(-\hat{\partial}+m)_{ik}}{2E_p}e^{ip(x-y)}\,d\vec{p}$$
$$= -i\,(\hat{\partial}-m)_{ik}\left[\frac{1}{(2\pi)^3 i}\int\frac{1}{2E_p}e^{ip(x-y)}\,d\vec{p}\right] \tag{5.5.19}$$
$$= -i\,(\hat{\partial}-m)_{ik}D_{m+}(x-y) = -i\,S_{ik+}(x-y).$$

As can be seen, $S_{ik+}(x-y)$ is also a particular solution of the Dirac equation (5.1.11)

The above relations (5.5.18) and (5.5.19) can be written in the following condensed form
$$S_{ik\pm}(x-y) = (\hat{\partial}-m)\,D_{m\pm}(x-y). \tag{5.5.20}$$

139

The quantization of the spinorial field

By means of (5.5.12), (5.5.13), (5.5.17), and (5.5.18), the anti-commutator (5.5.11) becomes

$$\{\psi_i(x), \overline{\psi}_k(y)\} = -i\, S_{ik}(x-y), \qquad (5.5.21)$$

where we denoted

$$S_{ik}(x-y) = S_{ik+}(x-y) + S_{ik-}(x-y)$$

$$= (\hat{\partial} - m)_{ik}[D_{m+}(x-y) + D_{m-}(x-y)] \qquad (5.5.22)$$

$$= (\hat{\partial} - m)_{ik} D_m(x-y).$$

Similarly, it can be shown that

$$\{\psi(x), \psi(y)\} = \{\overline{\psi}(x), \overline{\psi}(y)\} = 0. \qquad (5.5.23)$$

5.6. Products of spinorial operators

Consider the product

$$\psi(x)\overline{\psi}_k(y) = \psi_{i+}(x)\overline{\psi}_{k+}(y) + \psi_{i-}(x)\overline{\psi}_{k-}(y)$$

$$+\psi_{i+}(x)\overline{\psi}_{k-}(y) + \psi_{i-}(x)\overline{\psi}_{k+}(y). \qquad (5.6.1)$$

Since the operators $\psi(x)$ and $\overline{\psi}_k(y)$ satisfy an anti-commutation relation, we shall define the *normal product* of these operators as

$$N[\psi(x)\overline{\psi}_k(y)] = \psi_{i+}(x)\overline{\psi}_{k+}(y) + \psi_{i-}(x)\overline{\psi}_{k-}(y)$$

$$- \overline{\psi}_{k-}(y)\psi_{i+}(x) + \psi_{i-}(x)\overline{\psi}_{k+}(y). \qquad (5.6.2)$$

The difference between the last two relations, with the help of (5.5.19), yields

$$\psi(x)\overline{\psi}_k(y) = N[\psi(x)\overline{\psi}_k(y)] - iS_{ik+}(x-y). \qquad (5.6.3)$$

The *time-ordered product* of the operators $\psi(x)$ and $\overline{\psi}_k(y)$ is defined by

$$T[\psi_i(x)\overline{\psi}_k(y)] = \begin{cases} \psi_i(x)\overline{\psi}_k(y) &, \quad x_0 > y_0 ; \\ -\overline{\psi}_k(y)\psi_i(x) &, \quad x_0 < y_0 , \end{cases} \qquad (5.6.4)$$

or,

$$T[\psi(x)\overline{\psi}_k(y)] = \theta(x_0 - y_0)\psi_i(x)\overline{\psi}_k(y) - \theta(y_0 - x_0)\overline{\psi}_k(y)\psi_i(x). \qquad (5.6.5)$$

Let us now take the difference between (5.6.5) and (5.6.3). Recalling (5.5.21) and (5.5.22), we have:

$$T[\psi_i(x)\overline{\psi}_k(y)] - N[\psi_i(x)\overline{\psi}_k(y)]$$

$$= [\theta(x_0 - y_0) - 1]\psi_i(x)\overline{\psi}_k(y) - \theta(y_0 - x_0)\overline{\psi}_k(y)\psi_i(x) - iS_{ik+}(x-y)$$

$$\tag{5.6.6}$$

$$= i\theta(y_0 - x_0)[S_{ik+}(x-y) + S_{ik-}(x-y)] - iS_{ik+}(x-y) = -S_{ik}^{(c)}(x-y),$$

where by $S_{ik}^{(c)}(x-y)$ we denoted the causal function

$$S_{ik-}^{(c)}(x-y) = i[\theta(x_0 - y_0)S_{ik+}(x-y) - \theta(y_0 - x_0)S_{ik-}(x-y)]. \tag{5.6.7}$$

If we set $y = 0$, $x_0 = t$ in (5.6.7) and use (5.5.20), (3.5.12), we find

$$S_{ik}^{(c)}(x) = i(\hat{\partial} - m)_{ik}[\theta(t)D_{m+}(x) - \theta(-t)D_{m-}(x)]$$

$$= (\hat{\partial} - m)_{ik}\frac{1}{(2\pi)^3}\int \frac{1}{2E_p} e^{i\vec{p}\cdot\vec{x}-E_p|t|)} d\vec{p} = (\hat{\partial} - m)_{ik}D_m^{(c)}(x). \tag{5.6.8}$$

The vacuum expectation value of the normal product of operators $\psi_i(x)$ and $\overline{\psi}_k(y)$ is zero:

$$< N[\psi_i(x)\overline{\psi}_k(y)] >_0 = 0. \tag{5.6.9}$$

Therefore, by taking the vacuum expectation value of (5.6.3) we get

$$< \psi_i(x)\overline{\psi}_k(y) >_0 = -i\, S_{ik+}(x-y) = \underbrace{\psi_i(x)\overline{\psi}_k(y)}, \tag{5.6.10}$$

where the symbol $\underbrace{}$ denotes, as we know, the vacuum contraction of the two operators.

Let us now take the vacuum expectation value of (5.6.6). Using (5.6.9), we can write

$$< T[\psi_i(x)\overline{\psi}_k(y)] >_0 = -iS_{ik}^{(c)}(x-y) = \overbrace{\psi_i(x)\overline{\psi}_k(y)}, \tag{5.6.11}$$

where $\frown$ stands for the *time-ordered contraction* (or *T-contraction*) of the operators $\psi_i(x)$ and $\psi_k(y)$.

Summarizing these results, we can write relations (5.6.3) and (5.6.6) as follows:

$$\psi_i(x)\overline{\psi}_k(y) = N[\psi_i(x)\overline{\psi}_k(y)] + \underbrace{\psi_i(x)\overline{\psi}_k(y)} \; ; \tag{5.6.12}$$

The quantization of the spinorial field

$$T[\psi_i(x)\overline{\psi}_k(y)] = N[\psi_i(x)\overline{\psi}_k(y)] + \overbrace{\psi_i(x)\overline{\psi}_k(y)}. \qquad (5.6.13)$$

Observation. Wick's theorems remain valid in the case of the spinorial field, on the condition that the order of the factors in the operator products does not change. Otherwise, any permutation of two spinors implies a change of sign. When bosonic operators are accompanied by fermionic operators, we take into account the fact that bosonic and fermionic operators commute.

CHAPTER VI

GENERAL PROBLEMS OF FIELD INTERACTIONS

*The reason Dick's [Richard Feynman] physics
was so hard for ordinary people to grasp was
that he did not use equations... It was no
wonder that people who had spent their lives
solving equations were baffled by him. Their
minds were analytical; his was pictorial.*

Freeman Dyson

6.1. Generalities

In the previous chapters we studied the quantization of the "classical"
free fields: scalar, electromagnetic, and spinorial. In reality, the notion
of "free field" is a fiction, since there are neither isolated fields, nor
free particles in Nature. Therefore, to study real physical phenomena
we have to consider the quantum fields in mutual interaction.

The general methods used to describe field interactions are taken
from the classical field theory. These procedures are essentially based
on the Lagrangian density composition: to the sum of Lagrangian
densities of the free fields one adds an *interaction Lagrangian density*.
The choice of the interaction Lagrangian density is very important,
and we shall carefully discuss it in Section 6.3.

Since we want to study field interactions, it is advisable to use
the interaction representation (or *Dirac representation*; in this representation, both the operators and the state vectors evolve in time).
In this case, the time evolution of the state vector is given by the
time-dependent Schrödinger equation

$$i \frac{\partial}{\partial t} |\psi(t) >= H|\psi(t) > . \qquad (6.1.1)$$

Here both the state vector $|\psi(t) >$ and the interaction Hamiltonian H
are expressed in interaction representation.

6.2. The S-matrix

Consider a quantum system which is described by the state vector $|\psi(t_0) >$ at time t_0, and by $|\psi(t) >$ at time t. The connection between $|\psi(t_0) >$ and $|\psi(t) >$ can be written as

$$|\psi(t) >= S(t, t_0) |\psi(t_0) >, \tag{6.2.1}$$

where $S(t, t_0)$ is an *operator*.

Let us now require that the operator $S(t, t_0)$ preserve the norm of the state vector, that is

$$< \psi(t)|\psi(t) >=< \psi(t_0)|S^+(t, t_0) S(t, t_0)|\psi(t_0) >=< \psi(t_0)|\psi(t_0) >$$

which yields

$$S^+(t, t_0) S(t, t_0) = S(t, t_0) S^+(t, t_0) = I, \tag{6.2.2}$$

where I is the unit operator. This shows that $S(t, t_0)$ is a *unitary operator*. Substituting (6.2.1) in (6.1.1), we have

$$i \frac{\partial}{\partial t} S(t, t_0) = H S(t, t_0), \tag{6.2.3}$$

with the obvious condition

$$S(t_0, t_0) = I. \tag{6.2.4}$$

By definition,

$$\lim_{\substack{t_0 \to -\infty \\ t \to +\infty}} S(t, t_0) = S. \tag{6.2.5}$$

is called *S-operator*, or *S-matrix*, as well as *scattering matrix*. The term "matrix" comes from the fact that in quantum field theory one usually employs the matrix representation of this operator. The importance of the S-matrix and its role in interaction processes will come to light a little further in this chapter.

Note that relation (6.2.5) can formally be written as

$$S(+\infty, -\infty) = S. \tag{6.2.6}$$

Equation (6.2.1) then gives

$$|\psi(+\infty) >= S |\psi(-\infty) > . \tag{6.2.7}$$

The S-matrix

$$* \quad * \quad *$$

From (6.2.5) it follows that, in order to determine the operator S, one must take a double limit. We shall perform this operation in two steps: first take the limit for $t_0 \to -\infty$, and then the limit $t \to +\infty$.

a) We shall conveniently denote

$$S(t) = \lim_{t_0 \to -\infty} S(t, t_0) = S(t, -\infty). \qquad (6.2.8)$$

This operator must also satisfy equation (6.2.3), and condition (6.2.4) as well, that is

$$\frac{\partial}{\partial t} S(t) = -i\, H\, S(t) ; \quad S(-\infty) = I. \qquad (6.2.9)$$

Integrating $(6.2.9)_1$ between the limits $-\infty$ and t, we find

$$S(t) - S(-\infty) = -i \int_{-\infty}^{t} H(t_1)\, S(t_1)\, dt_1,$$

or, in view of $(6.2.9)_2$

$$S(t) = I - i \int_{-\infty}^{t} H(t_1)\, S(t_1)\, dt_1. \qquad (6.2.10)$$

Equation (6.2.10) is an inhomogeneous Voltera-type equation, with the nucleus $-i\, H(t)$, and can be solved using *iteration method*. Substituting $S(t)$ under the integral sign in (6.2.10), we have

$$S(t) = 1 + (-i) \int_{-\infty}^{t} H(t_1)\, dt_1$$

$$+ (-i)^2 \int_{-\infty}^{t} \int_{-\infty}^{t_1} H(t_1)\, H(t_2)\, S(t_2)\, dt_1\, dt_2,$$

with the obligatory condition that $t \geq t_1 \geq t_2$. Repeating this procedure, we find

$$S(t) = 1 + (-i) \int_{-\infty}^{t} H(t_1)\, dt_1$$

$$+ (-i)^2 \int_{-\infty}^{t} \int_{-\infty}^{t_1} H(t_1)\, H(t_2)\, S(t_2)\, dt_1\, dt_2$$

$$+ (-i)^3 \int_{-\infty}^{t} \int_{-\infty}^{t_1} \int_{-\infty}^{t_2} H(t_1)\, H(t_2)\, H(t_3)\, S(t_3)\, dt_1\, dt_2\, dt_3,$$

145

General problems of field interactions

where $t \geq t_1 \geq t_2 \geq t_3$. If this process is repeated an infinite number of times, we realize that $S(t)$ can be written as a series

$$S(t) = \sum_{n=0}^{\infty} S^{(n)}(t), \tag{6.2.11}$$

where

$$S^{(n)}(t) = (-i)^n \int_{-\infty}^{t} \int_{-\infty}^{t_1} \cdots \int_{-\infty}^{t_{n-1}} H(t_1)\, H(t_2)...H(t_{n-1})\, H(t_n)$$

$$\times\, dt_1\, dt_2\, ...\, dt_{n-1}\, dt_n, \tag{6.2.12}$$

with $t \geq t_1 \geq t_2 \geq ... \geq t_{n-1} \geq t_n$.

b) The passing to the limit $t \to \infty$ is complicated by the chronology of the limits in (6.2.12). This difficulty can be eliminated by writing (6.2.12) in a more symmetrical form, by using the concept of *time-ordered product* (see Section 3.5).

To this end, let us prove the validity of the formula

$$\int_a^b \int_a^x f(x)\, f(y)\, dx\, dy = \frac{1}{2!} \int_a^b \int_a^b T[f(x)\, f(y)]\, dx\, dy\ ;\ (x \geq y). \tag{6.2.13}$$

Suppose that $F(x)$ is the primitive function of $f(x)$, given by

$$F(x) = \int_a^x f(y)\, dy\ ;\quad F(a) = 0\ ;\quad F'(x) = f(x). \tag{6.2.14}$$

Then,

$$\int_a^b \int_a^x f(x)\, f(y)\, dx\, dy = \int_a^b f(x)\left\{ \int_a^x f(y)\, dy \right\} dx$$

$$= \int_a^b F'(x)\, F(x)\, dx = \frac{1}{2} F^2(x)\Big|_a^b,$$

so that

$$\int_a^b \int_a^x f(x)\, f(y)\, dx\, dy = \frac{1}{2} F^2(b). \tag{6.2.15}$$

Since $y \leq x$, we have

$$\frac{1}{2!} \int_a^b \int_a^b T[f(x)\, f(y)]\, dx\, dy == \frac{1}{2!} \int_a^b f(x)\, dx \int_a^b f(y)\, dy = \frac{1}{2} F^2(b), \tag{6.2.16}$$

and formula (6.2.13) immediately follows by comparing (6.2.15) with (6.2.14).

Generalizing (6.2.13), we are allowed to write (6.2.12) as

$$S^{(n)}(t) = \frac{(-i)^n}{n!} \int_{-\infty}^t \int_{-\infty}^t \cdots \int_{-\infty}^t T[H(t_1)\, H(t_2)\ldots H(t_n)]$$

$$\times dt_1\, dt_2\, \ldots\, dt_n. \tag{6.2.17}$$

This expression can be obtained using the complete induction method, and we leave this proof to the reader.

Now we can take the limit $t \to +\infty$ in (6.2.12). By analogy with (6.2.11), we have

$$S = \sum_{n=0}^{\infty} S^{(n)}, \tag{6.2.18}$$

where

$$S^{(n)} = \frac{(-i)^n}{n!} \int_{-\infty}^{+\infty} \cdots \int_{-\infty}^{+\infty} T[H(t_1)\, H(t_2)\ldots H(t_n)]\, dt_1\ldots dt_n. \tag{6.2.19}$$

Using the series expansion of the exponential function, we can formally write the S-matrix as

$$S = T\left[e^{-i \int_{-\infty}^{+\infty} H(t)\, dt}\right]. \tag{6.2.20}$$

$$*\quad *\quad *$$

Since the energy densities are frequently used in QFT, we shall resort to the definition

$$H(t) = \int \mathcal{H}(x)\, d\vec{x}, \tag{6.2.21}$$

where $\mathcal{H}(x) = \mathcal{H}(\vec{x}, t)$ is the interaction Hamiltonian density, and the integral extends over the whole three-dimensional space. Therefore

$$\int_{-\infty}^{+\infty} H(t)\, dt = \int \mathcal{H}(x)\, d\vec{x}\, dt = \frac{1}{i} \int \mathcal{H}\, dx,$$

and (6.2.19) becomes

$$S^{(n)} = \frac{(-1)^n}{n!} \int \cdots \int T[\mathcal{H}(x_1)\ldots\mathcal{H}(x_n)]\, dx_1\ldots dx_n. \tag{6.2.22}$$

147

Here we have n quadruple integrals, each of them extending over the four-dimensional space. Clearly, the following notations have been used:

$$dx_1 = (d\vec{x})_1 \, d(it_1) \; ; ... \; dx_n = (d\vec{x})_n \, d(it_n).$$

* * *

The S-matrix theory has been prompted by a practical problem: the scattering of particles in microcosmos. Here one searches for the final state $|\psi(+\infty) >$ of a beam of particles, resulting from a known initial beam, described by $|\psi(-\infty) >$, after diffusion off a scattering centre. Suppose that the initial system is a parallel and monochromatic particle beam. (According to the energy-momentum relation, here *monochromatic* means *monoenergetic*, while *parallel* shows that all particles have their momenta oriented in the same direction). We also assume that the scattering potential acts over a finite region of space. All these assumptions insure that in the final state we have again free particles, but with different characteristics (energy, momentum, etc.).

If we denote by $|k >$ the *pure state* vectors (i.e. all particles have the same characteristics), and since according to our assumptions the initial state is pure, then we can write

$$|\psi(-\infty) >= |i > .$$

The final state is a superposition of such states, therefore

$$\psi(+\infty) >= \int c(f) \, |f > \, df,$$

or, in view of (6.2.7),

$$S|i >= \int c(f) \, |f > \, df. \tag{6.2.23}$$

In this case

$$c(f) =< f|S|i >, \tag{6.2.24}$$

and the probability density for the system to be in the final state $|f >$, if initially it was in state $|i >$, writes

$$w(i \to f) = c^+(f) \, c(f) = | < f|S|i > |^2. \tag{6.2.25}$$

As one observes, the r.h.s. of (6.2.24) yields the matrix representation of S. This means that the elements of the S-matrix play an essential role in the study of transitions from initial to final states in scattering processes.

6.3. Choice of the interaction Lagrangian density

As we have seen in the previous section, in order to determine the transition probability, it is necessary to know the interaction Lagrangian density $\mathcal{L}$. Its form can be postulated according to the following criteria:

a) The interaction Lagrangian density, as well as the other terms in the total Lagrangian density, must be an invariant with respect to the transformation group corresponding to the symmetry of the studied system. (First of all, we ask for the relativistic invariance).

b) Each term in the interaction Lagrangian density must contain the characteristic functions of *all* interacting fields. (We shall consider only the case where the number of interacting fields is two).

c) The interaction Lagrangian density constructed following the points (a) and (b) has to lead to results in agreement with reality.

d) Since explicit calculation is usually a very difficult task (or even impossible), the mathematical expression for the interaction Lagrangian density must be as simple as possible.

Let us illustrate the above requirements by means of the following examples:

1°. If the interacting systems are the scalar and vectorial fields φ and A_k, respectively, the simplest invariant can be chosen as $A_k \varphi_{,k}$, leading to the following interaction Lagrangian density

$$\mathcal{L} = c_1 A_k \varphi_{,k},$$

where c_1 is a constant.

2°. The simplest invariants describing the interaction between the scalar and spinorial fields φ and ψ, respectively, are $\varphi \bar{\psi} \psi$ and $\bar{\psi} \gamma_k \psi \varphi_{,k}$, which suggests the combination

$$\mathcal{L} = c_2' \varphi \bar{\psi} \psi + c_2'' \bar{\psi} \gamma_k \psi \varphi_{,k}.$$

If the phenomenon is satisfactory described by one of the terms, the other term can be safely discarded.

3°. The interaction between the electromagnetic and Dirac fields, A_k and ψ, respectively, can be described by the Lagrangian density

$$\mathcal{L} = c_3' A_k \bar{\psi} \gamma_k \psi + c_3'' H_{km} \bar{\psi} \gamma_{km} \psi,$$

with $H_{km} = A_{m,k} - A_{k,m}$; $\gamma_{km} = \gamma_k \gamma_m$.

In order to compare various interactions, we have to bear in mind that all the interacting fields are normalized, so that they satisfy the quantization relation (2.5.15). Once this condition is satisfied, the constants c_1, c_2', c_2'', c_3', and c_3'' will provide information about the *interaction intensity*. These coefficients are called *coupling constants*.

We shall focus our attention on the type of interaction described at 3°. Its study, together with the quantization of the free electromagnetic and Dirac fields (detailed in the Chap. IV and V), is the object of *Quantum Electrodynamics*.

$$* \quad * \quad *$$

To put the electromagnetic interaction in context, we give some examples of other interactions.

The interaction considered in the case 2°, where nucleons are quanta of the spinorial field, while π-mesons are quanta of the scalar field, is specific to nuclear interactions, which are more intense than the electromagnetic ones. These are called *strong interactions*.

Those interactions involving spinorial fields with zero-mass quanta (e.g. neutrinos) characterize many transformations of elementary particles (like the beta-decay). These interactions belong to the class of *weak interactions*, which are less intense than the electromagnetic interactions.

The least intense interactions are those involving the gravitational field (gravitational interactions), whose importance is mostly theoretical.

To conclude, we can order the interactions according to their intensity as: gravitational, weak, electromagnetic and strong.

$$* \quad * \quad *$$

Let us now return to finding the appropriate interaction Lagrangian density for quantum electrodynamics. To this end, we shall make use of the classical theory of electromagnetism. The action term describing the interaction, in accordance with the Maxwellian theory, is

$$\frac{1}{i} \int \mathcal{L} \, d\Omega = \frac{1}{i} \int s_k \, A_k \, d\Omega. \tag{6.3.1}$$

Here we have used (1.1.3). According to (5.2.8), the current density four-vector s_k is

$$s_k = -ie \, \overline{\psi} \, \gamma_k \, \psi. \tag{6.3.2}$$

The last two relations give

$$\mathcal{L} = -i \, e \, \overline{\psi} \, \hat{A} \, \psi, \tag{6.3.3}$$

where

$$\hat{A} = \gamma_k A_k. \tag{6.3.4}$$

Choice of the interaction Lagrangian density

The interaction Lagrangian density (6.3.3) coincides with the first term in example 3°. Almost all quantum electrodynamics is based on this Lagrangian density. We shall use the second term of 3° only when needed.

As can be seen, $c'_3 = ie$, meaning that the absolute value of the coupling constant equals the elementary charge.

*   *   *

The interaction Lagrangian density is needed to determine the interaction Hamiltonian density $\mathcal{H}$, appearing in the expression for the S-matrix. To this end, we remember relations (1.5.4)

$$T_{ik} = \pi^+_{k(r)} U_{(r),i} + U^+_{(r),i} \pi_{k(r)} - \mathcal{L}\delta_{ik}$$

and (1.1.16)

$$\frac{\partial \mathcal{L}}{\partial U_{(r),k}} = \pi^+_{k(r)} \qquad \frac{\partial \mathcal{L}}{\partial U^+_{(r),k}} = \pi_{k(r)}.$$

Since $\mathcal{L}$ given by (6.3.3) does not depend on the derivatives of ψ and $\hat{A}$, we have $\pi_{k(r)} = \pi^+_{k(r)} = 0$, so that

$$\mathcal{H} = T_{44} = -\mathcal{L} = i\,e\,\overline{\psi}\,\hat{A}\,\psi.$$

As we already know (see Section 3.5), classical expressions are translated in QFT under the sign of the normal product, so that the interaction Hamiltonian density for quantum electrodynamics finally reads

$$\mathcal{H} = i\,e\,N\,(\overline{\psi}\,\hat{A}\,\psi). \tag{6.3.5}$$

By means of (6.3.5), relation (6.2.22) becomes:

$$S^{(n)} = \frac{(-ie)^n}{n!} \int \ldots \int T\{N[\overline{\psi}(x_1)\,\hat{A}(x_1)\,\psi(x_1)]$$

$$\times \ldots N[\overline{\psi}(x_n)\,\hat{A}(x_n)\,\psi(x_n)]\}\,dx_1\,\ldots\,dx_n. \tag{6.3.6}$$

Naturally, the S-matrix is given by (6.2.18), i.e. $S = \sum_{n=0}^{\infty} S^{(n)}$.

Observing that $S^{(n)}$ involves the time-ordered product of n normal products, we can apply the third Wick's theorem (see Section 3.7, and the **Observation** on p.142). To summarize, we shall use (4.4.17), (5.6.11), (A.128), and (A.130) in order to derive the T-contractions of the operators entering our example:

$$\overbrace{A_i(x)\,A_k}(y) = \frac{1}{i}\delta_{ik}\,D^{(c)}_0(x - y)$$

$$= \frac{\delta_{ik}}{(2\pi)^4 i} \lim_{\alpha \to 0} \int \frac{e^{ik(x-y)}}{k^2 - i\alpha} \, d\vec{k} \, dk_0; \quad \alpha > 0;$$

$$\overbrace{\psi_i(x) \overline{\psi}_k(y)} = -S_{ik}^{(c)}(x-y)$$

$$= -\frac{1}{(2\pi)^4 i} \lim_{\alpha \to 0} \int \frac{(i\hat{p} - m)_{ik} \, e^{ip(x-y)}}{p^2 + m^2 - i\alpha} \, d\vec{p} \, dp_0; \qquad (6.3.7)$$

$$\overbrace{\psi(x) A(y)} = \overbrace{\overline{\psi}(x) A(y)} = 0.$$

Some operators are contraction-free, and we remind the reader the expressions (4.2.2), (4.2.21), (5.3.38), and (5.3.40):

$$A_{j\pm}(x) = \frac{1}{(2\pi)^{3/2}} \int \frac{1}{\sqrt{2\omega}} e_j^l \, a_{l\pm}(\vec{k}) \, e^{\pm ikx} \, d\vec{k};$$

$$\psi_\pm(x) = \frac{1}{(2\pi)^{3/2}} \int a_{s\pm}(\vec{p}) \, v_\pm^{(s)}(\vec{p}) \, e^{\pm ipx} \, d\vec{p}; \qquad (6.3.8)$$

$$\overline{\psi}_\pm(x) = \frac{1}{(2\pi)^{3/2}} \int a_{s\pm}^+(\vec{p}) \, \overline{v}_\pm^{(s)}(\vec{p}) \, e^{\pm ipx} \, d\vec{p}.$$

6.4. The Feynman-Dyson diagrams

The general expression for the S-matrix, obtained by means of Wick's theorem, is very complicated and tedious for concrete calculations. For this reason, it is necessary to study the terms individually, or in groups, by analyzing the factors composing each of them. This investigation is facilitated by associating some graphic symbols with the aforementioned factors, and by defining some rules of assembling these symbols, which leads to a certain term of the S-matrix expansion. These symbols are called *Feynman-Dyson diagrams* or *graphs*, while the prescriptions for their usage, including the correspondence with the S-matrix terms, are known as *Feynman-Dyson rules* [1] [2]. The Nobel Prize in Physics 1965 was awarded jointly to Sin-Itiro Tomonaga, Julian Schwinger and Richard P. Feynman *"for their fundamental work in quantum electrodynamics, with deep-ploughing consequences for the physics of elementary particles"*.

---

[1] R.P.Feynman, Phys. Rev. vol.76, 1949, p.769; vol.80, 1950, p.440.
[2] F.I.Dyson, Phys. Rev. vol.75, 1949, p.1736.

The Feynman-Dyson diagrams

The above mentioned correspondences are summarized in the GRAPH TABLE. We detail below the use of this Table, according to Feynman-Dyson rules and diagrams.

a) In the first column are inserted those factors that can possibly appear in some term of the S-matrix expansion. The first five rows contain the non-contracted operators: $\psi_\pm$, $\overline{\psi}_\pm$, and $A_\pm$. As we know (see Chap.IV, and Chap.V), these are creation and annihilation operators, used to produce the annihilation of an initial state, and create the final state, as a result of the interaction.

b) On rows 6 and 7 are found the operators under time-ordered contraction. They express the creation of a particle (electron, positron, or photon) at the event y (a four-dimensional spacetime point), and the annihilation of the same particle at the event x. They are called *internal lines* or *propagators*. These processes take place during the interaction, but they cannot be observed (we are not able to follow these microscopic processes, but only detect their result).

c) Any term in $S^{(n)}$ (see (6.3.6)) contains n quadruple integrals. Due to the interaction Hamiltonian density (6.3.5), there are always three operators (two fermionic, $\overline{\psi}$, and ψ, and one bosonic A_i) that depend on the same event. The event where two fermion and one photon lines meet is called *vertex*. Each term in $S(n)$ yields n quadruple integrals, i.e. to each vertex corresponds such an integral. If we attach the factor ie to each vertex, then the n vertices of $S(n)$ imply the factor $(ie)^n$, as it appears in (6.3.6). These observations are synthesized on row 8 (columns 1 and 2), where the dot under the integral substitutes the field operators which depend on the indicated integration variable. According to Wick's theorems, any element of the $S^{(n)}$ expansion can be constructed using the symbols shown in the first column, the only necessary adjustment being the factor $1/n!$

d) To each element of $S^{(n)}$, from the first column, one associates certain graphic symbols, as shown in column 3. Here are the rules of using and/or reading these symbols:

- The time-axis is oriented vertically and directed upwards. This means that all graphs and diagrams are read in chronological order, from the bottom to the top.
- A point corresponds to an event.
- An upwards directed line represents the propagation of an electron, while a downwards directed line describes the propagation of a positron.
- A dashed line corresponds to the propagation of a photon.
- The event (point) in which a line ends, stands for the annihilation of the particle represented by that line, while the event from which a

153

line starts describes creation of the particle associated with that line.

e) We usually need the elements of S-matrix in momentum representation. The transition is performed by means of (6.3.7) and (6.3.8), and is facilitated by the following indications. The first 5 rows of the 4th column do not contain integrals with respect to $\vec{p}$ and $\vec{k}$, as should be expected from (6.3.8), because the momenta of the initial (incident) and final (emergent) particles are supposed to be known. (The momentum is arbitrary, but fixed).

Indeed, let us calculate the non-zero matrix elements of the operators corresponding to the electromagnetic and spinorial fields given by (6.3.8). These elements are non-zero only between states for which the number of particles differs by one, for example the vacuum state and the state with one particle with given momentum $\vec{k}$ $(\vec{p})$, and polarization l (s). To this end, it is necessary to know the commutators (anti-commutators) of $A_{j\pm}(x)$, $\psi_\pm(x)$, and $\overline{\psi}_\pm$ with $a_{l\pm}(\vec{k})$, $a_{s\pm}(\vec{p})$, and $a_{s\pm}^+(\vec{p})$, respectively. It then follows that the creation operator $A_{j-}(x)$ has non-zero matrix elements between the vacuum state $|0>$ and the state with one particle of momentum $\vec{k}'$ and polarization l', i.e. $<0|a_{l'+}(\vec{k}')$, according to Fock's representation (see Section 3.6). Using the commutation relations (4.4.4), and assuming that the vacuum state is normalized $(<0|0>=1)$, we have

$$< 0|a_{l'+}(\vec{k}')A_{j-}(x)|0 >$$

$$=< 0|\{A_{j-}(x)a_{l'+}(\vec{k}') + [a_{l'+}(\vec{k}'), A_{j-}(x)]\}|0 >= [a_{l'+}(\vec{k}'), A_{j-}(x)]$$

$$= \frac{1}{(2\pi)^{3/2}} \int \frac{1}{\sqrt{2\omega}} e_j^l [a_{l'+}(\vec{k}'), a_{l-}(\vec{k})] e^{-ikx} \, d\vec{k} = \frac{e_j^{l'} e^{-ik'x}}{(2\pi)^{3/2}\sqrt{2\omega'}},$$

since $[a_{l'+}(\vec{k}'), a_{l-}(\vec{k})] = \delta_{l'l}\delta(\vec{k}' - \vec{k})$. Similarly, we have

$$< 0|A_{j+}(x)\, a_{l'-}(\vec{k}')|0 >= \frac{e_j^{l'} e^{ik'x}}{(2\pi)^{3/2}\sqrt{2\omega'}}.$$

We can now write the correspondence between the operators of the electromagnetic field and their matrix components, which are taken between the one-photon state of momentum $\vec{k}$ and polarization l, and the vacuum:

GRAPH TABLE

Feynman-Dyson diagrams and their correspondence in the
configuration and momentum representation

No	Configuration representation	Significance	Graph	Momentum representation
1	$\psi_+(x)$	electron annihilated at x		$(2\pi)^{-3/2}\, \overline{v}_+^{(s)}(\vec{p})$
2	$\overline{\psi}_-(x)$	electron created at x		$(2\pi)^{-3/2}\, \overline{v}_-^{(s)}(\vec{p})$
3	$\overline{\psi}_+(x)$	positron annihilated at x		$(2\pi)^{-3/2}\, \overline{v}_+^{(s)}(\vec{p})$
4	$\psi_-(x)$	positron created at x		$(2\pi)^{-3/2}\, \overline{v}_+^{(s)}(\vec{p})$
5	$A_\mp(x)$	photon created (annihilated) at x		$\frac{1}{(2\pi)^{3/2}}\frac{1}{\sqrt{2\omega}}(e_j)^l$
6	$\overbrace{\psi(x)\overline{\psi}(y)}$ $\overbrace{\overline{\psi}(x)\psi(y)}$	electronic (positronic) propagator		$\frac{1}{(2\pi)^4}\int \frac{i\hat{p}-m}{p^2+m^2-i\alpha}$ $\times(.)\,dp$
7	$\overbrace{A_j(x)A_l(y)}$	photonic propagator		$-\frac{1}{(2\pi)^4}\int \frac{\delta_{jl}}{k^2-i\alpha}$ $\times(.)\,dk$
8	$-e\,\gamma_j\int(.)\,dx$	vertex		$(2\pi)^4\, e\,\gamma_j$ $\times\delta(p_2-p_1\mp k)$

$$\psi_\pm(x) \to \tfrac{1}{(2\pi)^{3/2}} v_\pm^{(s)}(\vec{p})\, e^{\pm ipx};$$

$$\overline{\psi}_\pm(x) \to \tfrac{1}{(2\pi)^{3/2}} \overline{v}_\pm^{(s)}(\vec{p})\, e^{\pm ipx}. \qquad (6.4.2)$$

- In the last column of rows 6 and 7 is shown the momentum representation of the propagators. The dot before the differential element substitutes the functions that depend on the integration variable. Here the exponentials are omitted, but they appear on row 8, labeled "vertex". Indeed, any vertex involves the integral (see (A.44))

$$\int e^{i(p_1 - p_2 \pm k)x}\, dx = (2\pi)^4\, i\, \delta(p_2 - p_1 \mp k). \qquad (6.4.3)$$

This expression, together with the factor ie, yields the formula given in the GRAPH TABLE.

6.5. Examples of Feynman-Dyson diagrams

To exemplify the above concepts, in this section we consider the cases of first and second order diagrams, corresponding to the approximations $S^{(1)}$ and $S^{(2)}$ for the scattering matrix S. According to (6.3.6), these terms are

$$S^{(1)} = ie \int T\{N[\overline{\psi}(x)\,\hat{A}(x)\,\psi(x)]\}\, dx. \qquad (6.5.1)$$

and

$$S^{(2)} = -\frac{e^2}{2} \int \int T\{N[\overline{\psi}(x_1)\,\hat{A}(x_1)\,\psi(x_1)]\, N[\overline{\psi}(x_2)\,\hat{A}(x_2)\,\psi(x_2)]\}$$

$$\times\, dx_1\, dx_2. \qquad (6.5.2)$$

To obtain all the terms containing $S^{(1)}$ and $S^{(2)}$, one must apply the third Wick's theorem and split the non-contracted operators into their positive and negative frequency parts. The operators under T contraction do not split, since they correspond to the causal function (see (3.6.11), (3.6.13), as well as the fourth column of the GRAPH TABLE).

a) First order processes

According to Wick's theorem, we can omit the sign T of the time-ordered product in $S^{(1)}$, so that

$$N[(\overline{\psi}_- + \overline{\psi}_+)(\hat{A}_- + \hat{A}_+)(\psi_- + \psi_+)]$$

$$= N[\overline{\psi}_- \hat{A}_- \psi_+] + N[\overline{\psi}_- \hat{A}_+ \psi_-] + N[\overline{\psi}_- \hat{A}_+ \psi_+]$$

$$+N[\overline{\psi}_+ \hat{A}_- \psi_-] + N[\overline{\psi}_+ \hat{A}_- \psi_+] + N[\overline{\psi}_+ \hat{A}_+ \psi_-], \qquad (6.5.3)$$

where the zero-terms have been left out. The resulting six terms correspond to the following Feynman graphs

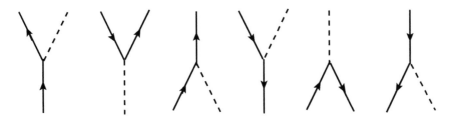

These graphs can be read as follows:
1) An electron emits a photon.
2) A photon transforms into an electron-positron pair.
3) An electron absorbs a photon.
4) A positron emits a photon.
5) An electron-positron pair transforms into a photon.
6) A positron absorbs a photon.

If we write down the corresponding matrix elements in momentum representation, we note the emergence of the factor $\delta(p_1 - p_2 \mp k)$, where p_1 and p_2 are the momenta of the fermions (electron or positron), while k is the momentum of the photon. In order for the corresponding matrix element to be non-zero, we must have

$$p_1 = p_2 \pm k. \qquad (6.5.4)$$

Squaring (6.5.4), we have

$$p_1^2 = p_2^2 + 2(p_2 k) + k^2,$$

or, according to the energy-momentum relation ($p_1^2 = p_2^2 = -m^2$; $k^2 = 0$), we have

$$(p_2 k) = \vec{p}_2 \cdot \vec{k} - E_{p_2}\omega = 0. \qquad (6.5.5)$$

Denoting by θ the angle between $\vec{p}_2$ and $\vec{k}$ and observing that $|\vec{k}| = \omega$, the result is

$$(p_2 k) = (|\vec{p}_2| \cos\theta - \sqrt{|\vec{p}_2|^2 + m^2})\,\omega < 0, \qquad (6.5.6)$$

because $|\vec{p}_2| < \sqrt{|\vec{p}_2|^2 + m^2}$, and $\cos\theta \le 1$. Consequently, relation (6.5.5) cannot be satisfied, therefore $S^{(1)} = 0$. This means that *no process of the first order can exist.* (Exception is met when an external field is present, but this case is not considered here).

It follows that *a free electron (positron) cannot emit (absorb) a photon.* (N.B. The atomic spontaneous emission happens in the presence of the Coulomb field of the nucleus). It also follows that *an electron-positron pair cannot disintegrate into a single photon, and a single photon cannot give birth to an electron-positron pair.*

b) Second order processes

To simplify the calculations, we shall denote by indices 1 and 2 the dependence on x_1 and x_2. Applying the third Wick's theorem to the time-ordered product in (6.5.2), we have:

$$T\{N[(\overline{\psi}\,\hat{A}\,\psi)_1]\,N[(\overline{\psi}\,\hat{A}\,\psi)_2]\}$$

$$= N[(\overline{\psi}\,\hat{A}\,\psi)_1\,(\overline{\psi}\,\hat{A}\,\psi)_2] + N[(\overline{\psi}\,\hat{A}\,\overbrace{\psi)_1\,(\overline{\psi}}\,\hat{A}\,\psi)_2]$$

$$+N[(\overbrace{\overline{\psi}\,\hat{A}\,\psi)_1\,(\overline{\psi}}\,\hat{A}\,\psi)_2] + N[(\overline{\psi}\,\overbrace{\hat{A}\,\psi)_1\,(\overline{\psi}\,\hat{A}}\,\psi)_2]$$

$$+N[(\overline{\psi}\,\overbrace{\hat{A}\,\psi)_1\,(\overline{\psi}\,\hat{A}}\,\psi)_2] + N[(\overline{\psi}\,\overbrace{\hat{A}\psi)_1\,(\overline{\psi}\hat{A}}\,\psi)_2]$$

$$+N[(\overbrace{\overline{\psi}\hat{A}\,\psi)_1\,(\overline{\psi}}\,\hat{A}\psi)_2] + N[(\overline{\psi}\,\hat{A}\,\overbrace{\psi)_1\,(\overline{\psi}}\,\hat{A}\,\psi)_2]. \qquad (6.5.7)$$

Each of these terms is also split, by separating the positive and negative frequency parts.

Let us study the eight terms on the r.h.s. of (6.3.15), by dividing them into six groups, as follows:

1) The term $N[(\overline{\psi}\,\hat{A}\,\psi)_1\,(\overline{\psi}\,\hat{A}\,\psi)_2]$ describes superposition of two first-order processes, one of them taking place at the event x_1, and the other at the event x_2. These processes are independent and, as we have seen, impossible. Therefore, this term brings no contribution to the S-matrix.

2) Next group is composed of two terms

$$N[(\overline{\psi}\,\hat{A}\,\overbrace{\psi)_1\,(\overline{\psi}}\,\hat{A}\psi)_2] + N[(\overbrace{\overline{\psi}\,\hat{A}\psi)_1\,(\overline{\psi}}\,\hat{A}\psi)_2]. \qquad (6.5.8)$$

Before splitting the free operators into positive and negative frequency parts, note that the only terms whose contribution is non-zero are those corresponding to two particles in the initial state, and two in the final state. Indeed, let us assume that in the initial state we have an electron with four-momentum p_1, while in the final state we have one electron with four-momentum p_2 and two photons with four-momenta k_1 and k_2, respectively. The four-momentum conservation law

$$p_1 = p_2 + k_1 + k_2$$

can also be written as

$$p_1 - k_1 = p_2 + k_2. \qquad (6.5.9)$$

Since $p_1^2 = p_2^2 = -m^2$; $\quad k_1^2 = k_2^2 = 0$, then squaring (6.3.17) we obtain

$$(p_1 k_1) + (p_2 k_2) = 0. \qquad (6.5.10)$$

According to (6.5.6), this relation cannot be satisfied. The same conclusion is drawn in any other case when there is a single particle in the initial/final state. In view of these observations, the sum (6.5.8) yields

$$N[(\overline{\psi}\,\hat{A}\,\psi)_1\,(\overline{\psi}\,\hat{A}\,\psi)_2] + N[(\overline{\psi}\,\hat{A}\,\psi)_1\,(\overline{\psi}\,\hat{A}\,\psi)_2]$$

$$= N[(\overline{\psi}_-\,\hat{A}_-\,\psi)_1\,(\overline{\psi}\,\hat{A}_+\,\psi_+)_2] + N[(\overline{\psi}_+\,\hat{A}_+\,\psi)_1\,(\overline{\psi}\,\hat{A}_-\,\psi_-)_2]$$

$$+ N[(\overline{\psi}_-\,\hat{A}_+\,\psi)_1\,(\overline{\psi}\,\hat{A}_-\,\psi_+)_2] + N[(\overline{\psi}_-\,\hat{A}_+\,\psi)_1\,(\overline{\psi}\,\hat{A}_+\,\psi_-)_2]$$

$$+ N[(\overline{\psi}_+\,\hat{A}_-\,\psi)_1\,(\overline{\psi}\,\hat{A}_-\,\psi_+)_2] + N[(\overline{\psi}_+\,\hat{A}_-\,\psi)_1\,(\overline{\psi}\,\hat{A}_+\,\psi_-)_2] \quad (6.5.11)$$

$$+ N[(\overline{\psi}\,\hat{A}_-\,\psi_-)_1\,(\overline{\psi}_+\,\hat{A}_+\,\psi)_2] + N[(\overline{\psi}\,\hat{A}_+\,\psi_+)_1\,(\overline{\psi}_-\,\hat{A}_-\,\psi)_2]$$

$$+ N[(\overline{\psi}\,\hat{A}_-\,\psi_+)_1\,(\overline{\psi}_-\,\hat{A}_+\,\psi)_2] + N[(\overline{\psi}\,\hat{A}_-\,\psi_+)_1\,(\overline{\psi}_+\,\hat{A}_-\,\psi)_2]$$

$$+ N[(\overline{\psi}\,\hat{A}_+\,\psi_-)_1\,(\overline{\psi}_-\,\hat{A}_+\,\psi)_2] + N[(\overline{\psi}\,\hat{A}_+\,\psi_-)_1\,(\overline{\psi}_+\,\hat{A}_-\,\psi)_2].$$

It is convenient to study this expression by groups of terms.

I) First, we consider those terms which describe processes characterized by an electron and a photon in the initial state, implying that the corresponding operators ψ_+ and A_+ are free. Numbering the terms on the r.h.s. of (6.5.11), the terms under discussion have numbers 1, 3, 8, and 9. We rewrite these terms, and find their corresponding Feynman graphs, as shown in the GRAPH TABLE:

$$N[(\overline{\psi}_-\,\hat{A}_-\,\psi)_1\,(\overline{\psi}\,\hat{A}_+\,\psi_+)_2] + N[(\overline{\psi}_-\,\hat{A}_+\,\psi)_1\,(\overline{\psi}\,\hat{A}_-\,\psi_+)_2]$$

$$+N[(\overline{\psi}\,\hat{A}_+\,\psi_+)_1\,(\overline{\psi}_-\,\hat{A}_-\,\psi)_2] + N[(\overline{\psi}\,\hat{A}_-\,\psi_+)_1\,(\overline{\psi}_-\,\hat{A}_+\,\psi)_2] \longrightarrow$$

$$(6.5.12)$$

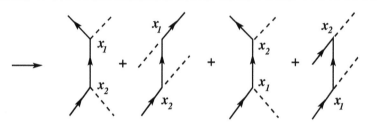

The process described by these graphs is *the Compton effect,* and it will be studied in detail later. It can be seen that graphs 3 and 4 are the same as 1 and 2 if we interchange notations 1 and 2 for the two events. Therefore, the Compton effect is represented by

$$(6.5.13)$$

Taking into account the observation made at the beginning of this section, we remind the reader that the contraction $\overset{\frown}{\psi\overline{\psi}}$ can be represented either by $\overset{\frown}{\psi_+\overline{\psi}_-}$, as shown in the above graphs, or by $\overset{\frown}{\psi_-\overline{\psi}_+}$. This last case is represented by

$$(6.5.14)$$

Graphs (6.5.13) and (6.5.14) describe the same effect. Due to chronological order, they cannot occur simultaneously.

II) The terms 2, 6, 7, and 12 in (6.5.11) describe a similar phenomenon, i.e. the Compton effect for the case when the electron is replaced by a *positron*.

III) Consider now the terms 5 and 10 in (6.5.11). They give

$$N[(\overline{\psi}_+ \hat{A}_- \overbrace{\psi})_1 (\overline{\psi} \hat{A}_- \psi_+)_2] + N[(\overbrace{\overline{\psi} \hat{A}_- \psi_+})_1 (\overline{\psi}_+ \hat{A}_- \psi)_2] \longrightarrow$$

$$(6.5.15)$$

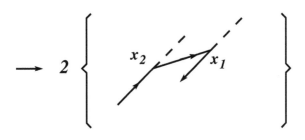

which describes the *annihilation of an electron-positron pair* and the *creation of two photons.*

IV) The remaining two terms in (6.5.11), namely 4 and 11, stand for the inverse effect

$$(6.5.16)$$

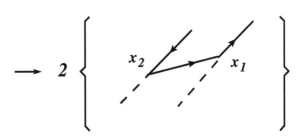

i.e. the *creation of an electron-positron pair, out of two photons.*

3) Let us now go back to (6.5.7) and consider the term

$$N[(\overline{\psi} \hat{A} \overbrace{\psi})_1 (\overline{\psi} \hat{A} \psi)_2]$$

$$= N[(\overline{\psi}_- \overbrace{\hat{A} \psi_+})_1 (\overline{\psi}_- \hat{A} \psi_+)_2] + N[(\overline{\psi}_- \overbrace{\hat{A} \psi_+})_1 (\overline{\psi}_+ \hat{A} \psi_-)_2] \quad (6.5.17)$$

$$+N[(\overline{\psi}_+ \ \overbrace{\hat{A}\,\psi_-)_1 \ (\overline{\psi}_- \ \hat{A}} \ \psi_+)_2] + N[(\overline{\psi}_+ \ \overbrace{\hat{A}\,\psi_-)_1 \ (\overline{\psi}_+ \ \hat{A}} \ \psi_-)_2]$$

$$+N[(\overline{\psi}_- \ \overbrace{\hat{A}\,\psi_-)_1 \ (\overline{\psi}_+ \ \hat{A}} \ \psi_+)_2] + N[(\overline{\psi}_+ \ \overbrace{\hat{A}\,\psi_+)_1 \ (\overline{\psi}_- \ \hat{A}} \ \psi_-)_2].$$

Here we have omitted the terms corresponding to one particle in the initial state, and three particles in the final state (or vice-versa), which would contradict the momentum conservation law, by analogy with the reasoning developed at point (2).

Expression (6.5.17) can also be organized into three groups of terms:

(i) The first term reads

$$N[(\overline{\psi}_- \ \overbrace{\hat{A}\,\psi_+)_1 \ (\overline{\psi}_- \ \hat{A}} \ \psi_+)_2] \ \longrightarrow$$

$$(6.5.18)$$

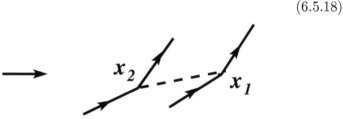

meaning that two electrons interact by exchanging a photon. This phenomenon is called *Möller scattering*.

(ii) The fourth term yields the same phenomenon as (i) for a positron.

(iii) The remaining terms are

$$N[(\overline{\psi}_+ \ \overbrace{\hat{A}\,\psi_-)_1 \ (\overline{\psi}_- \ \hat{A}} \ \psi_+)_2] + N[(\overline{\psi}_- \ \overbrace{\hat{A}\,\psi_-)_1 \ (\overline{\psi}_+ \ \hat{A}} \ \psi_+)_2]$$

$$+N[(\overline{\psi}_- \ \overbrace{\hat{A}\,\psi_+)_1 \ (\overline{\psi}_+ \ \hat{A}} \ \psi_-)_2] + N[(\overline{\psi}_+ \ \overbrace{\hat{A}\,\psi_+)_1 \ (\overline{\psi}_- \ \hat{A}} \ \psi_-)_2] \ \longrightarrow$$

$$(6.5.19)$$

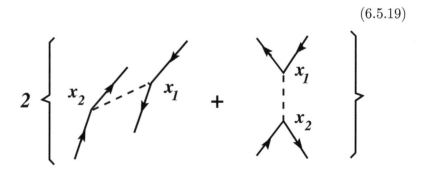

Examples of Feynman-Dyson diagrams

Here the electron-positron pair interact through the exchange of a photon. This is *Bhabba scattering*.

4) Another group of terms in (6.5.7) is

$$N[(\overline{\psi} \, \hat{A} \, \psi)_1 \, (\overline{\psi} \, \hat{A} \, \psi)_2] + N[(\overline{\psi} \, \hat{A}\psi)_1 \, (\overline{\psi}\hat{A} \, \psi)_2]$$

$$= N[(\overline{\psi}_- \, \hat{A} \, \psi)_1 \, (\overline{\psi} \, \hat{A} \, \psi_+)_2] + N[(\overline{\psi} \, \hat{A}\psi_-)_1 \, (\overline{\psi}_+ \, \hat{A} \, \psi)_2]$$

$$+N[(\overline{\psi}_+ \, \hat{A} \, \psi)_1 \, (\overline{\psi} \, \hat{A} \, \psi_-)_2] + N[(\overline{\psi} \, \hat{A}\psi_+)_1 \, (\overline{\psi}_- \, \hat{A} \, \psi)_2] \longrightarrow$$

$$(6.5.20)$$

First diagram is *the electron self-energy graph*, while the second is *the positron self-energy graph*. They will be discussed later, in another context.

5) The fifth group in (6.5.7) contains only one term

$$N[(\overline{\psi} \, \hat{A} \, \psi)_1 \, (\overline{\psi} \, \hat{A}\psi)_2]$$

$$= N[(\overline{\psi}\hat{A}_- \, \psi)_1 \, (\overline{\psi} \, \hat{A}_+ \, \psi)_2] + N[(\overline{\psi}\hat{A}_+ \, \psi)_1 \, (\overline{\psi} \, \hat{A}_- \, \psi)_2] \longrightarrow$$

$$(6.5.21)$$

and expresses *the photon self-energy.*

6) The last term of (6.5.7), namely

$$N[(\overline{\psi} \; \hat{A} \; \psi)_1 \; (\overline{\psi} \; \hat{A} \; \psi)_2] \longrightarrow$$

(6.5.22)

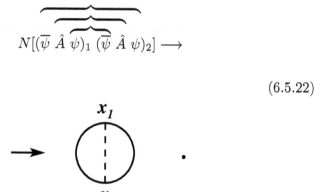

shows that no particle is present, in both initial and final states. This corresponds to a non-observable process, which can take place in the electron-positron vacuum. It is called *non-observable vacuum process.*

6.6. Transition probability

We shall first show that (see (6.2.18) and (6.2.24)), for any $n \geq 1$,

$$< f|S^{(n)}|i > = \delta\left(\sum p_f - \sum p_i\right) < f|M^{(n)}|i >, \qquad (6.6.1)$$

where $\sum p_f$ and $\sum p_i$ are the sums of four-momenta for the particles in the final and initial states, respectively. Relation (6.6.1) constitutes the definition of $M^{(n)}$.

To prove the validity of (6.6.1) it is sufficient to show that after the integration of a propagator with respect to p', the four-momentum p' vanishes, insuring the conservation of the four-momentum entering the first vertex. Indeed, let us consider a second-order process (which could be a segment of a more complex graph) represented by the diagram below, where the four-momenta of the particles are as indicated.

According to the GRAPH TABLE, in the presence of vertices, the S-matrix contains the factor

$$\delta(p_j + k_f - p')\delta(p' - p_i - k_i).$$

If the integration with respect to p' is performed (as the propagator demands - see the same GRAPH TABLE), we are left with

$$\delta[(p_f + k_f) - (p_i + k_i)].$$

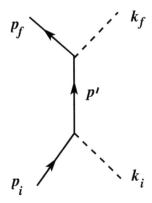

If one line of the graph enters another vertex, we have a new integral and a new delta function, whose argument is the difference between the final and initial momenta. Adding an arbitrary number of vertices, relation (6.6.1) follows easily. Therefore, the factor $\delta(\sum p_f - \sum p_i)$ insures the conservation of the four-momentum.

In view of (6.2.25), the probability density for the system to be in the final state $|f>$, if initially in state $|i>$, should be

$$w(i \to f) = \left[\delta\left(\sum p_f - \sum p_i\right)\right]^2 \, |<f|M^{(n)}|i>|^2. \qquad (6.6.2)$$

But this expression is always divergent. Indeed, integrating $\delta^2(p) = \delta(p)\delta(p)$ with respect to p will take out one variable p, but requires seting $p = 0$ in the other $\delta(p)$, and $\delta(0) = \infty$.

The aforementioned divergence comes from the fact that all integrals are extended over the whole space, and over the entire duration from $-\infty$ to $+\infty$. But, in order to evaluate the probability of any process one must know the transition probability per unit space volume and unit time interval. The appearance of the Dirac delta function in (6.6.1) is due to the presence of exponentials. Using (A.44), the quadruple integral (6.6.1) writes

$$<f|S^{(n)}|i> = \frac{1}{(2\pi)^4 i} \int e^{i(P_f - P_i)x} \, dx \times <f|M^{(n)}|i>, \qquad (6.6.3)$$

where

$$P_f = \sum p_f \; ; \quad P_i = \sum p_i, \qquad (6.6.4)$$

while the integral extends over the whole four-dimensional space. If the integral is extended over a volume V and a time interval T only, instead of (6.6.3) we have

$$<f|S^{(n)}|i>_{VT} = \frac{1}{(2\pi)^4 i} \int_{VT} e^{i(P_f - P_i)x} \, dx \times <f|M^{(n)}|i>, \qquad (6.6.5)$$

meaning that the matrix element $< f|S^{(n)}|i >_{VT}$ describes the transition between the states $|i >$ and $|f >$ in volume V and time interval T. The transition probability per unit volume and unit time is then

$$\frac{\left| < f|S^{(n)}|i >_{VT} \right|^2}{VT}.$$

To calculate the elements of S-matrix, we take the limit of this fraction for V extending over the whole space, and T going from $-\infty$ to $+\infty$. We are therefore interested in calculating

$$w(i \to f) = \lim_{V,T \to \infty} \frac{\left| < f|S^{(n)}|i >_{VT} \right|^2}{VT}, \tag{6.6.6}$$

which is the transition probability from state $|i >$ to state $|f >$, per unit volume and unit time. Using (6.6.5) and (6.6.6), we obtain

$$w(i \to f) = \frac{| < f|M^{(n)}|i > |^2}{(2\pi)^8} \lim_{V,T \to \infty} \frac{1}{VT} \left| \int_{VT} e^{i(P_f - P_i)x} \, dx \right|^2. \tag{6.6.7}$$

The limit in (6.6.7) can be split into four limits, one for each dimension. Let us calculate one of them, e.g. limit for the coordinate z. Denoting $(P_f - P_i)_z = P$, we have:

$$\lim_{Z \to \infty} \frac{1}{2Z} \left| \int_{-Z}^{+Z} e^{iPz} \, dz \right|^2 = \lim_{Z \to \infty} \frac{1}{2Z} \left| \frac{1}{iP} (e^{iPZ} - e^{-iPZ}) \right|^2$$

$$= \lim_{Z \to \infty} \frac{1}{2Z} \left(\frac{2 \sin PZ}{P} \right)^2 = 2\pi \lim_{Z \to \infty} \frac{1}{\pi} \frac{\sin PZ}{P} \times \lim_{Z \to \infty} \frac{\sin PZ}{PZ}.$$

The first limit is $\delta(P)$ (see (A.16)). If $P \neq 0$, then $\delta(P) = 0$ and, since $|\sin PZ/PZ| \leq 1$, the entire expression vanishes. If $P = 0$, we can write

$$\lim_{Z \to \infty} \frac{\sin PZ}{PZ}\bigg|_{P=0} = \lim_{Z \to \infty} \left(\lim_{P \to 0} \frac{\sin PZ}{PZ} \right) = 1.$$

Therefore

$$\lim_{Z \to \infty} \frac{1}{2Z} \left| \int_{-Z}^{+Z} e^{iPz} \, dz \right|^2 = 2\pi \delta(P)$$

which yields

$$\lim_{V,T \to \infty} \frac{1}{VT} \left| \int_{VT} e^{i(P_f - P_i)x} \, dx \right|^2 = (2\pi)^4 \, \delta(P_f - P_i). \tag{6.6.8}$$

166

By means of (6.6.7) and (6.6.8), we finally obtain the transition probability density between the states $|i>$ and $|f>$, per unit volume and unit time

$$w(i \to f) = \frac{1}{(2\pi)^4} |<f|M^{(n)}|i>|^2 \delta\left(\sum p_f - \sum p_i\right). \qquad (6.6.9)$$

6.7. Scattering cross section

Consider an electron with momentum $\vec{p}$ and polarization (spin) s. Its state is described by the vector $a_{s-}^+(\vec{p})|0>$ (see Sections 3.6 and 5.5). Any single-particle state vector can be then written as

$$|1> = \int \psi_s(\vec{p}) \, a_{s-}^+(\vec{p})|0> \, d\vec{p}. \qquad (6.7.1)$$

The normalization condition for the vector $|1>$ demands

$$< 1|1> = \int \psi_{s'}^+(\vec{p}')\psi_s(\vec{p}) < 0|a_{s'+}(\vec{p}')a_{s-}^+(\vec{p})|0> \, d\vec{p} \, d\vec{p}' = 1,$$

however

$$< 0|a_{s'+}(\vec{p}')a_{s-}^+(\vec{p})|0> = \delta_{s's}\delta(\vec{p}' - \vec{p}). \qquad (6.7.2)$$

Indeed, the vacuum state definition yields

$$< 0|a_{s-}^+(\vec{p})a_{s'+}(\vec{p}')|0> = 0,$$

while the sum of the last two relations leads to the well-known anti-commuttion relation

$$\{a_{s'+}(\vec{p}'), \, a_{s-}^+(\vec{p})\} = \delta_{s's}\delta(\vec{p}' - \vec{p}),$$

together with the normalization condition $< 0|0> = 1$. The normalization condition of the single-particle vector $|1>$, together with (6.7.2), then lead to

$$\int \psi_s^+(\vec{p}) \, \psi_s(\vec{p}) \, d\vec{p} = 1, \qquad (6.7.3)$$

where summation over the index s is assumed. It then follows that $|\psi_s(\vec{p})|^2 \, d\vec{p}$ is the probability that in the state described by the single-electron state function $|1>$ there is one electron of momentum $\vec{p}$ and spin s, in the vicinity $d\vec{p}$ of $\vec{p}$.

General problems of field interactions

The Fourier transform of $\psi_s(\vec{p})$, namely

$$\psi_s(\vec{x}) = \frac{1}{(2\pi)^{3/2}} \int e^{i\vec{p}\cdot\vec{x}} \psi_s(\vec{p})\, d\vec{p} \qquad (6.7.4)$$

gives the probability density in configuration space. More precisely, $|\psi_s(\vec{x})|^2\, d\vec{x}$ is the probability that the electron has spin s and is in vicinity $d\vec{x}$ of $\vec{x}$. Using (A.42), we find

$$\int |\psi_s(\vec{x})|^2\, d\vec{x} = \int |\psi_s(\vec{p})|^2\, d\vec{p}. \qquad (6.7.5)$$

Let us now drop the condition $< 1|1 >= 1$, and assume that there is one particle in each volume element. Then,

$$\int_{V=1} |\psi_s(\vec{x})|^2\, d\vec{x} = 1.$$

Since each particle is free, it is naturally to take

$$\psi_s(\vec{x}) = e^{i\vec{p}_0\cdot\vec{x}} \longrightarrow |\psi_s(\vec{x})| = 1, \qquad (6.7.6)$$

where the momentum $\vec{p}_0$ is considered fixed.

Passing now to the momentum representation, we have

$$\psi_s(\vec{p}) = \frac{1}{(2\pi)^{3/2}} \int \psi_s(\vec{x}) e^{-i\vec{p}\cdot\vec{x}}\, d\vec{x} = \frac{1}{(2\pi)^{3/2}} \int e^{i(\vec{p}_0-\vec{p})\cdot\vec{x}}\, d\vec{x},$$

which yields

$$\psi_s(\vec{p}) = (2\pi)^{3/2}\delta(\vec{p}-\vec{p}_0). \qquad (6.7.7)$$

The "single-particle" state vector (here we use quotation marks because this is not a true single-particle state, but the state with one particle per unit volume), which we also denote by $|1>$, by analogy with (6.7.1), then writes

$$|1>= \int \psi_s(\vec{p})\, a_{s-}^+(\vec{p})\, |0> d\vec{p} = (2\pi)^{3/2} a_{s-}^+(\vec{p}_0)|0>. \qquad (6.7.8)$$

Suppose now that we have N types of particles in the initial state (two particles are of different types if their creation operators are different, which means that the particles are either different, or identical, but with different momenta or polarizations), so that we have one particle of each type per unit volume. Repeating the reasoning above, the corresponding state vector is

$$|1_1, 1_2, ...1_N >= (2\pi)^{3N/2} c_{s_1-}^{(1)}(\vec{p}_1)\, c_{s_2-}^{(2)}(\vec{p}_2)...c_{s_N-}^{(N)}(\vec{p}_N)|0>, \qquad (6.7.9)$$

where $c_{s_1-}^{(1)}(\vec{p}_1)$ creates a particle of type 1, with polarization s_1 and momentum $\vec{p}_1$, etc.

According to (6.6.9), the probability density associated with this case is

$$w = (2\pi)^{-4}| < f|M^{(n)}|1_1, 1_2, ..., 1_N > |^2 \delta(\sum p_f - \sum p_i). \quad (6.7.10)$$

One usually denotes by $|i>$ the initial state in the Fock representation, defined as

$$|i >= c_{s_1-}^{(1)}(\vec{p}_1) c_{s_2-}^{(2)}(\vec{p}_2)...c_{s_N-}^{(N)}(\vec{p}_N)|0 >, \quad (6.7.11)$$

and (6.7.10) finally writes

$$w = (2\pi)^{3N-4}| < f|M^{(n)}|i > |^2 \delta(\sum p_f - \sum p_i). \quad (6.7.12)$$

$$* \quad * \quad *$$

Let us now apply this result to the case of two particles in both the initial and final states $(N = 2)$. The probability density of finding the initial particles (with given momenta p_1 and p_2, respectively) in the final state, with momenta contained in the interval $p_1 \to p_1 + dp_1$, $p_2 \to p_2 + dp_2$, is

$$w_1(i \to f) = (2\pi^2|M_{fi}^{(n)}|^2 \delta(p_1' + p_2' - p_1 - p_2) \, d\vec{p}_1' \, d\vec{p}_2', \quad (6.7.13)$$

where

$$M_{fi}^{(n)} =< f|M^{(n)}|i > . \quad (6.7.14)$$

In (6.7.13) we have the product $d\vec{p}_1' \, d\vec{p}_2'$ because (6.7.12) is a probability density, and, as the fourth components of p_1', p_2' are coupled with the space-components through the energy-momentum relation, they are not present in the differential element.

We usually encounter the following situation: the particles are known in the initial state, but in the final state only one type of particles can be detected, say those with momentum $\vec{p}_1$. Therefore, the probability we have to determine is of the form

$$w_2 = \int w_1 \, d\vec{p}_2'.$$

But, since

$$\delta(p_1' + p_2' - p_1 - p_2) = \delta(\vec{p}_1' + \vec{p}_2' - \vec{p}_1 - \vec{p}_2) \, \delta(E_1' + E_2' - E_1 - E_2),$$

we have

$$w_2(i \to f) = (2\pi)^2 |M_{fi}^{(n)}|^2 \delta(E_1' + E_2' - E_1 - E_2) \, d\vec{p}_1', \qquad (6.7.15)$$

together with condition

$$\vec{p}_1' + \vec{p}_2' - \vec{p}_1 - \vec{p}_2 = 0. \qquad (6.7.16)$$

Let N_1 and N_2 be the two types of particles in the initial state. From the definition of the probability W_2, the number dN_1' of particles of type 1 in the final state, with momenta in the interval $\vec{p}_1' \to \vec{p}_1' + d\vec{p}_1'$, is

$$dN' = (2\pi)^2 \, N_1 \, N_2 |M_{fi}^{(n)}|^2 \delta(E_1' + E_2' - E_1 - E_2) \, d\vec{p}_1'. \qquad (6.7.17)$$

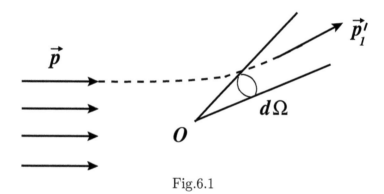

Fig.6.1

In the Figure 6.1 is depicted a very simple scattering process. The particle of type 2 is initially at point O. For a single diffusion centre, we have

$$dN_1' = j_1 \, d\sigma',$$

where j_1 is the current density of the incident particles of the first type, and $d\sigma'$ is the *differential cross section*. For N_2 diffusion centres per unit volume we then have

$$dN_1' = N_2 j_1 \, d\sigma', \qquad (6.7.18)$$

and this is, in fact, the definition of $d\sigma'$.

Since $j_1 = N_1 v_1$, equating (6.7.17) with (6.7.18) we obtain

$$d\sigma' = \frac{(2\pi)^2}{v_1} |M_{fi}^{(n)}|^2 \delta(E_1' + E_2' - E_1 - E_2) \, d\vec{p}_1'. \qquad (6.7.19)$$

It is convenient at this point to introduce spherical coordinates. In this case,

$$d\vec{p}_1' = \vec{p}_1'^2 \sin\theta \, d|\vec{p}_1'| \, d\theta \, d\varphi = \vec{p}_1'^2 \, d|\vec{p}_1'| \, d\Omega',$$

where $d\Omega'$ is the solid angle element. Since $E_1'^2 = \vec{p}_1'^2 + m_1^2$, one can write

$$E_1' \, dE_1' = |\vec{p}_1'| \, d|\vec{p}_1'|,$$

which means

$$d\vec{p}_1' = |\vec{p}_1'| \, E_1' \, dE_1' \, d\Omega', \tag{6.7.20}$$

and (6.7.19) reads

$$d\sigma' = \frac{(2\pi)^2}{v_1} |M_{fi}^{(n)}|^2 \delta(E_1' + E_2' - E_1 - E_2) |\vec{p}_1'| \, E_1' \, dE_1' \, d\Omega'. \tag{6.7.21}$$

The quantity usually determined experimentally is the number of particles scattered through a solid angle $d\omega'$. For this case, we can calculate the differential cross section $d\sigma'$ by integrating with respect to E_1', and finally the differential cross section $d\sigma$. Here are the detailed steps.

To integrate with respect to E_1', let us express the energies as a function of E_1', and denote

$$f(E_1') = E_1' + E_2' - E_1 - E_2. \tag{6.7.22}$$

Assuming that $f(E_1')$ has a single simple root E_{10}', according to (A.8) we can write

$$\delta(E_1' + E_2' - E_1 - E_2) = \delta[f(E_1')] = \frac{\delta(E_1')}{|f'(E_{10}')|}. \tag{6.7.23}$$

Substituting (6.7.23) in (6.7.21), and integrating with respect to E_1', one obtains

$$d\sigma = \frac{(2\pi)^2 |\vec{p}_1'| \, E_1'}{|f'(E_1')| \, v_1} |M_{fi}^{(n)}|^2 \, d\Omega' \Big|_{p_1' + p_2' - p_1 - p_2 = 0}. \tag{6.7.24}$$

Here the index 0 of E_{10}' has been omitted, and we took into account the fact that the existence of the root of $f(E_1')$ implies

$$E_1' + E_2' - E_1 - E_2 = 0. \tag{6.7.25}$$

The last relation, together with (6.7.16), demands the conservation of the four-momentum, and we emphasize this fact in (6.7.24).

In the above calculations the initial particles were monoenergetic and polarized. If the initial particle beam is non-polarized, one must average over the entire initial beam polarization. In this case, (6.7.24) gives only the differential cross section for a given polarization, because the integration is performed only with respect to energy. If, finally, we want to determine the number of particles, irrespective of polarization, one must sum (not average!) over the final polarization, because all particles with all their possible polarizations are taken into account.

CHAPTER VII

NON-DIVERGENT SECOND-ORDER PROCESSES

> *Every particle in Nature has an amplitude
> to move backwards in time, and therefore
> has an anti-particle.*
>
> Richard P. Feynman

In Section 6.5 (b), all second-order processes have been enumerated. It is worthwhile mentioning that only some of them are connected with non-divergent elements of S-matrix, as described in sections (2) and (3). The present chapter will focus on these processes.

7.1. Transition probability for Compton scattering

a) S-matrix element in momentum representation

As we already know, the Compton effect is described, in its lower approximation, by the graphs (see (7.5.13))

$$S^{(2c)} = 2 \left\{ \quad + \quad \right\} = 2\,(S^{(2c)}_I + S^{(2c)}_{II})$$

(7.1.1)

where the letter c stands for "Compton". Using the GRAPH TABLE, let us write the S-matrix elements in (7.1.1) in chronological order, i.e. from bottom to top. Beginning with the graph $S^{(2c)}_I$, we have:

- electron incident at $x_2 \longrightarrow (2\pi)^{-3/2}\, v^{(s)}_+(\vec{p}_1)$

- photon incident at $x_2 \longrightarrow (2\pi)^{-3/2} \frac{1}{\sqrt{2\omega_1}} (e_j)^1$
- vertex $x_2 \longrightarrow (2\pi)^4 e\,\gamma_j\,\delta(p' - p_1 - k_1)$
- propagator $(x_2 \to x_1) \longrightarrow (2\pi)^{-4} \int \frac{i\hat{p}' - m}{p'^2 + m^2 - i\alpha}\, dp'$
- vertex $x_1 \longrightarrow (2\pi)^4 e\,\gamma_l\,\delta(p_2 + k_2 - p')$
- electron emergent from $x_1 \longrightarrow (2\pi)^{-3/2} \overline{v}_-^{(t)}(\vec{p}_2)$
- photon emergent from $x_1 \longrightarrow (2\pi)^{-3/2} \frac{1}{\sqrt{2\omega_2}} (e_l)^2$

Let us now take the product of these factors in bottom-up order, taking care to place the creation operators on the l.h.s., and the annihilation ones on the r.h.s., and multiply the result by $1/2$ according to the factor $(-1)^n/n!$ appearing in (6.2.22). We then have

$$S_I^{(2c)} = \frac{1}{2}\frac{e^2}{(2\pi)^2}\frac{1}{\sqrt{4\omega_1\omega_2}}\, \overline{v}_-^{(t)}(\vec{p}_2)\mathcal{M}_I\, v_+^{(s)}(\vec{p}_1), \qquad (7.1.2)$$

where

$$\mathcal{M}_I = \int \hat{e}_2 \frac{i\hat{p}' - m}{p'^2 + m^2}\, \hat{e}_1\, \delta(p_2 + k_2 - p')\, \delta(p' - p_1 - k_1)\, dp'. \qquad (7.1.3)$$

Similarly, the graph $S_{II}^{(2c)}$ is composed of

- electron incident at $x_2 \longrightarrow (2\pi)^{-3/2} v_+^{(s)}(\vec{p}_1)$
- photon emergent from $x_2 \longrightarrow (2\pi)^{-3/2} \frac{1}{\sqrt{2\omega_2}} (e_j)^2$
- vertex $x_2 \longrightarrow (2\pi)^4 e\,\gamma_j\,\delta(p'' - p_1 + k_2)$
- propagator $(x_2 \to x_1) \longrightarrow (2\pi)^{-4} \int \frac{i\hat{p}'' - m}{p''^2 + m^2 - i\alpha}\, dp''$
- vertex $x_1 \longrightarrow (2\pi)^4 e\,\gamma_l\,\delta(p_2 - k_1 - p'')$
- photon incident at $x_1 \longrightarrow (2\pi)^{-3/2} \frac{1}{\sqrt{2\omega_1}} (e_l)^1$
- electron emergent from $x_1 \longrightarrow (2\pi)^{-3/2} \overline{v}_-^{(t)}(\vec{p}_2)$

Following the same procedure, one finds

$$S_{II}^{(2c)} = \frac{1}{2}\frac{e^2}{(2\pi)^2}\frac{1}{\sqrt{4\omega_1\omega_2}}\, \overline{v}_-^{(t)}(\vec{p}_2)\mathcal{M}_{II}\, v_+^{(s)}(\vec{p}_1), \qquad (7.1.4)$$

with

$$\mathcal{M}_{II} = \int \hat{e}_1 \frac{i\hat{p}'' - m}{p''^2 + m^2}\, \hat{e}_2\, \delta(p_2 - p'' - k_1)\, \delta(p'' + k_2 - p_1)\, dp''. \qquad (7.1.5)$$

In both $\mathcal{M}_I$ and $\mathcal{M}_{II}$ we have omitted the term $i\alpha$, since we have to take the limit $\alpha \to 0$ and, in our case, the integrals are uniformly convergent, so that we are allowed to take the limit under the integral sign. Recalling that $dp' = i \, d\vec{p}' \, dE_{p'}$; $dp'' = i \, d\vec{p}'' \, dE_{p''}$ and working out the integrals, we are left with

$$
\begin{aligned}
\mathcal{M}_I &= i\hat{e}_2 \, \frac{i(\hat{p}_1+\hat{k}_1)-m}{(p_1+k_1)^2+m^2} \, \hat{e}_1 \, \delta(p_2 + k_2 - p_1 - k_1) \; ; \\
\mathcal{M}_{II} &= i\hat{e}_1 \, \frac{i(\hat{p}_1-\hat{k}_2)-m}{(p_1-k_2)^2+m^2} \, \hat{e}_2 \, \delta(p_2 + k_2 - p_1 - k_1) \; ;
\end{aligned}
\tag{7.1.6}
$$

Putting together all these results, we finally obtain

$$
S^{(2c)} = \frac{ie^2}{(2\pi)^2} \frac{1}{\sqrt{4\omega_1\omega_2}} \, \bar{v}_-^{(t)}(\vec{p}_2) \mathcal{M} \, v_+^{(s)}(\vec{p}_1) \, \delta(p_2+k_2-p_1-k_1), \tag{7.1.7}
$$

where

$$
\mathcal{M} = \hat{e}_2 \, \frac{i(\hat{p}_1 + \hat{k}_1) - m}{(p_1 + k_1)^2 + m^2} \, \hat{e}_1 + \hat{e}_1 \, \frac{i(\hat{p}_1 - \hat{k}_2) - m}{(p_1 - k_2)^2 + m^2} \, \hat{e}_2. \tag{7.1.8}
$$

b) The simplified form of $\mathcal{M}$

It is convenient to choose a reference frame in which the initial electron is at rest: $\vec{p}_1 = 0$ and $E_{p_1} = \sqrt{\vec{p}_1^2 + m^2} = m$. In other words,

$$
p_1(\vec{p}_1, \, iE_{p_1}) = p_1(\vec{0}, \, im). \tag{7.1.9}
$$

Since the rest mass of the photon is zero, the components of four-vectors k_1 and k_2 are

$$
k_1(\vec{k}_1, \, i\omega_1), \quad \text{and} \quad k_2(\vec{k}_2, \, i\omega_2), \quad \text{with } \omega_1 = |\vec{k}_1|, \; \omega_2 = |\vec{k}_2|. \tag{7.1.10}
$$

Using (7.1.9) and (7.1.10), the two denominators in (7.1.8) write

$$
\begin{aligned}
(p_1 + k_1)^2 + m^2 &= -m^2 + 2(p_1 k_1) + m^2 = -2 \, m \, \omega_1, \\
(p_1 - k_2)^2 + m^2 &= -m^2 - 2(p_1 k_2) + m^2 = 2 \, m \, \omega_2
\end{aligned}
\tag{7.1.11}
$$

since $k_1^2 = k_2^2 = 0$.

Next, we shall consider the fact that both the initial and final photons are real. To this end, one must realize that the polarization vectors e_1 and e_2 cannot have temporal components, while the spatial part has to be orthogonal to the wave vector. Therefore,

$$
e_1 = e_1(\vec{e}_1, 0) \; ; \quad e_2 = e_2(\vec{e}_2, 0) \; ; \quad \vec{e}_1 \cdot \vec{k}_1 = 0 \; ; \quad \vec{e}_2 \cdot \vec{k}_2 = 0. \tag{7.1.12}
$$

Let a and b be two arbitrary four-vectors. Then,

$$
\hat{a}\hat{b} = (a_i\gamma_i)(b_k\gamma_k) = a_ib_k \, (2\delta_{ik} - \gamma_k\gamma_i)
$$

Non-divergent second-order processes

$$= 2a_i b_i - (b_k \gamma_k)(a_i \gamma_i) = 2(ab) - \hat{b}\hat{a}$$

which means that

$$\hat{a}\hat{b} + \hat{b}\hat{a} = 2(ab). \tag{7.1.13}$$

In this case, (7.1.9) and (7.1.12) give

$$(e_1 p_1) = (e_2 p_1) = 0, \tag{7.1.14}$$

and (7.1.13) leads to

$$\hat{e}_1 \hat{p}_1 + \hat{p}_1 \hat{e}_1 = 0 \; ; \quad \hat{e}_2 \hat{p}_1 + \hat{p}_1 \hat{e}_2 = 0. \tag{7.1.15}$$

Using these relations, let us modify the denominators in (7.1.8). We can write

$$\hat{e}_2[i(\hat{p}_1 + \hat{k}_1) - m]\hat{e}_1 = i\hat{e}_2\hat{k}_1\hat{e}_1 - i\hat{e}_2\hat{e}_1\hat{p}_1 - m\hat{e}_2\hat{e}_1$$
$$= i\hat{e}_2\hat{k}_1\hat{e}_1 - i\hat{e}_2\hat{e}_1(\hat{p}_1 - im);$$
$$\hat{e}_1[i(\hat{p}_1 - \hat{k}_2) - m]\hat{e}_2 = -i\hat{e}_1\hat{k}_2\hat{e}_2 - i\hat{e}_1\hat{e}_2\hat{p}_1 - m\hat{e}_1\hat{e}_2$$
$$= -i\hat{e}_1\hat{k}_2\hat{e}_2 - i\hat{e}_1\hat{e}_2(\hat{p}_1 - im). \tag{7.1.16}$$

In view of (7.1.11) and (7.1.16), (7.1.8) becomes

$$\mathcal{M} = -\frac{i\hat{e}_2\hat{k}_1\hat{e}_1}{2m\omega_1} - \frac{i\hat{e}_1\hat{k}_2\hat{e}_2}{2m\omega_2} + \left(\frac{i\hat{e}_2\hat{e}_1}{2m\omega_1} - \frac{i\hat{e}_1\hat{e}_2}{2m\omega_2}\right)(\hat{p}_1 - im). \tag{7.1.17}$$

As can be seen from (7.1.7), the operator $\mathcal{M}$ acts upon $v_+^{(s)}(\vec{p}_1)$ and, according to Dirac equation (5.1.7),

$$(\hat{p}_1 - im)v_+^{(s)}(\vec{p}_1) = 0,$$

which means that $\mathcal{M}$ reduces to

$$\mathcal{M} = -\frac{i}{2m\omega_1\omega_2}(\omega_2\hat{e}_2\hat{k}_1\hat{e}_1 + \omega_1\hat{e}_1\hat{k}_2\hat{e}_2). \tag{7.1.18}$$

Consequently, the form of $S^{(2c)}$ we are looking for is

$$S^{(2c)} = M_{ts}^{(2c)}\delta(p_2 + k_2 - p_1 - k_1), \tag{7.1.19}$$

where

$$M_{ts}^{(2c)} = A\,\bar{v}_-^{(t)}(\vec{p}_2)\,\mathcal{M}'\,v_+^{(s)}(\vec{p}_1), \tag{7.1.20}$$

176

with

$$A = -\frac{e^2}{(2\pi)^2}\frac{1}{4m(\omega_1\omega_2)^{3/2}} \; ; \quad \mathcal{M}' = \omega_2\hat{e}_2\hat{k}_1\hat{e}_1 + \omega_1\hat{e}_1\hat{k}_2\hat{e}_2. \quad (7.1.21)$$

c) Transition probability

As can be observed, in expression (7.6.9) for the transition probability, appears the squared magnitude of the quantity $M_{ts}^{(2c)}$ given by (7.1.20). This can be obtained by means of the Hermitian conjugate

$$M_{ts}^{(2c)+} = A[v_+^{(s)}(\vec{p}_1)]^+\mathcal{M}'^+[\overline{v}_-^{(t)}(\vec{p}_2)]^+. \quad (7.1.22)$$

But, in view of (5.3.33) and (5.3.41),

$$[v_+^{(s)}]^+ = v_-^{(s)+} = v_-^{(s)+}\gamma_4\gamma_4 = \overline{v}_-^{(s)}\gamma_4 \; ; \\ [\overline{v}_-^{(t)}]^+ = [v_-^{(t)+}\gamma_4]^+ = \gamma_4[v_-^{(t)+}]^+ = \gamma_4 v_+^{(t)}, \quad (7.1.23)$$

and (7.1.22) becomes

$$M_{ts}^{(2c)+} = A\,\overline{v}_-^{(s)}(\vec{p}_1)\gamma_4\mathcal{M}'^+\gamma_4\,v_+^{(t)}(\vec{p}_2),$$

or

$$M_{ts}^{(2c)+} = A\,\overline{v}_-^{(s)}(\vec{p}_1)\overline{\mathcal{M}'}v_+^{(t)}(\vec{p}_2), \quad (7.1.24)$$

where

$$\overline{\mathcal{M}'} = \gamma_4\mathcal{M}'^+\gamma_4. \quad (7.1.25)$$

The desired result is now obtained by means of (7.1.20) and (7.1.24):

$$|M_{ts}^{(2c)}|^2 = A^2\overline{v}_-^{(s)}(\vec{p}_1)\overline{\mathcal{M}'}\,v_+^{(t)}(\vec{p}_2)\,\overline{v}_-^{(t)}(\vec{p}_2)\mathcal{M}'\,v_+^{(s)}(\vec{p}_1). \quad (7.1.26)$$

As mentioned in the previous chapter, we now have to average over all possible spins in the initial state and sum over the spins in the final state. To this end, one has to calculate

$$|M^{(2c)}|^2 = \frac{1}{2}\sum_{s,t}|M_{ts}^{(2c)}|^2, \quad (7.1.27)$$

which is equal to half of the trace of matrix (7.1.26). Note that under the trace sign one can perform a circular permutation, which gives

$$|M^{(2c)}|^2 = \frac{1}{2}A^2\,Tr\{v_+^{(s)}(\vec{p}_1)\overline{v}_-^{(s)}(\vec{p}_1)\overline{\mathcal{M}'}\,v_+^{(t)}(\vec{p}_2)\overline{v}_-^{(t)}(\vec{p}_2)\}. \quad (7.1.28)$$

Recalling that (see (5.3.53))

$$\sum_s v_+^{(s)}(\vec{p})\bar{v}_-^{(s)}(\vec{p}) = \frac{-i\hat{p} + m}{2E_p},$$

we have

$$\left|M^{(2c)}\right|^2 = \frac{A^2}{8E_{p_1}E_{p_2}}Tr[(i\hat{p}_1 - m)\overline{\mathcal{M}'}(i\hat{p}_2 - m)\mathcal{M}']. \qquad (7.1.29)$$

Next, we need to explicitly write $\overline{\mathcal{M}'}$. Using (7.1.21) and (7.1.25), we have:

$$\overline{\mathcal{M}'} = \gamma_4(\omega_2\hat{e}_2\hat{k}_1\hat{e}_1 + \omega_1\hat{e}_1\hat{k}_2\hat{e}_2)^+\gamma_4$$

$$= \gamma_4(\omega_2\hat{e}_1^+\hat{k}_1^+\hat{e}_2^+ + \omega_1\hat{e}_2^+\hat{k}_2^+\hat{e}_1^+)\gamma_4. \qquad (7.1.30)$$

On the other hand, if a is an arbitrary four-vector, we can write:

$$\gamma_4\hat{a}^+ = \gamma_4(a_1\gamma_1 + a_2\gamma_2 + a_3\gamma_3 + ia_0\gamma_4)^+$$

$$= \gamma_4(a_1\gamma_1 + a_2\gamma_2 + a_3\gamma_3 - ia_0\gamma_4)$$

$$= -(a_1\gamma_1 + a_2\gamma_2 + a_3\gamma_3 + ia_0\gamma_4)\gamma_4,$$

therefore

$$\gamma_4\hat{a}^+ = -\hat{a}\,\gamma_4, \qquad (7.1.31)$$

and (7.1.30) becomes

$$\overline{\mathcal{M}'} = -(\omega_2\hat{e}_1\hat{k}_1\hat{e}_2 + \omega_1\hat{e}_2\hat{k}_2\hat{e}_1) \equiv -\mathcal{M}''. \qquad (7.1.32)$$

With this result, (7.1.29) writes

$$\left|M^{(2c)}\right|^2 = \frac{A^2}{8E_{p_1}E_{p_2}}Tr[(\hat{p}_2 + im)\mathcal{M}'(\hat{p}_1 + im)\mathcal{M}'']. \qquad (7.1.33)$$

Calculation of traces

To put the trace (7.1.33) into a more explicit form, we make the substitution $\hat{p}_2 = \hat{p}_1 + (\hat{p}_2 - \hat{p}_1)$ and split the trace into three terms, as follows:

$$Tr[(\hat{p}_2 + im)\mathcal{M}'(\hat{p}_1 + im)\mathcal{M}'']$$

$$= Tr[(\hat{p}_1 + im)\mathcal{M}'(\hat{p}_1 + im)\mathcal{M}'']$$

$$+ Tr[(\hat{p}_2 - \hat{p}_1)\mathcal{M}'(\hat{p}_1 + im)\mathcal{M}''].$$

Transition probability for Compton scattering

This can be written as

$$Tr[(\hat{p}_2 + im)\mathcal{M}'(\hat{p}_1 + im)\mathcal{M}''] = A_1 + A_2 + A_3, \qquad (7.1.34)$$

where

$$A_1 = Tr[(\hat{p}_1 + im)\mathcal{M}'(\hat{p}_1 + im)\mathcal{M}''] \; ;$$
$$A_2 = Tr[(\hat{p}_2 - \hat{p}_1)\mathcal{M}'\hat{p}_1\mathcal{M}''] \; ; \qquad (7.1.35)$$
$$A_3 = im\, Tr[(\hat{p}_2 - \hat{p}_1)\mathcal{M}'\mathcal{M}''].$$

As we know, the trace of an odd number of gamma matrices is zero, which means that $A_3 = 0$, and (7.1.33) writes

$$|M^{(2c)}|^2 = \frac{A^2}{8E_{p_1}E_{p_2}}(A_1 + A_2). \qquad (7.1.36)$$

Calculation of A_1

Making use of (7.1.21) and (7.1.32), (8.1.35)$_1$ writes

$$A_1 = Tr[(\hat{p}_1 + im)(\omega_2\hat{e}_2\hat{k}_1\hat{e}_1 + \omega_1\hat{e}_1\hat{k}_2\hat{e}_2)$$

$$\times(\hat{p}_1 + im)(\omega_2\hat{e}_1\hat{k}_1\hat{e}_2 + \omega_1\hat{e}_2\hat{k}_2\hat{e}_1). \qquad (7.1.37)$$

Next, we reverse the order of $(\hat{p}_1 + im)$ and $\mathcal{M}'$. Using (7.1.13) and (7.1.15), we have

$$\hat{p}_1\hat{e}_2\hat{k}_1\hat{e}_1 = -\hat{e}_2\hat{p}_1\hat{k}_1\hat{e}_1$$
$$= -\hat{e}_2[-\hat{k}_1\hat{p}_1 + 2(k_1p_1)]\hat{e}_1$$
$$= -2(k_1p_1)\hat{e}_2\hat{e}_1 + \hat{e}_2\hat{k}_1\hat{p}_1\hat{e}_1,$$

or

$$\hat{p}_1\hat{e}_2\hat{k}_1\hat{e}_1 = -2(k_1p_1)\hat{e}_2\hat{e}_1 - \hat{e}_2\hat{k}_1\hat{e}_1\hat{p}_1. \qquad (7.1.38)$$

A similar procedure gives

$$\hat{p}_1\hat{e}_1\hat{k}_2\hat{e}_2 = -2(k_2p_1)\hat{e}_1\hat{e}_2 - \hat{e}_1\hat{k}_2\hat{e}_2\hat{p}_1. \qquad (7.1.39)$$

Summarizing these results, one obtains

$$A_1 = -2\,Tr\{[\omega_2(k_1p_1)\hat{e}_2\hat{e}_1 + \omega_1(k_2p_1)\hat{e}_1\hat{e}_2$$

$$\times(\hat{p}_1 + im)(\omega_2\hat{e}_1\hat{k}_1\hat{e}_2 + \omega_1\hat{e}_2\hat{k}_2\hat{e}_1)\} + A_1', \qquad (7.1.40)$$

179

with
$$A'_1 = Tr[\mathcal{M}'(-\hat{p}_1 + im)(\hat{p}_1 + im)\mathcal{M}'']. \tag{7.1.41}$$

But $A'_1 = 0$, according to energy-momentum relation (5.1.1). By means of (7.1.11), we have

$$A_1 = 2m\,\omega_1\omega_2 Tr[(\hat{e}_2\hat{e}_1 + \hat{e}_1\hat{e}_2)(\hat{p}_1 + im)(\omega_2\hat{e}_1\hat{k}_1\hat{e}_2 + \omega_1\hat{e}_2\hat{k}_2\hat{e}_1)]. \tag{7.1.42}$$

Using again (7.1.13), and the property that the trace of the product of an odd number of gamma matrices is zero as well, we have

$$A_1 = 4\,m\,\omega_1\omega_2(e_1e_2)Tr(\omega_2\hat{p}_1\hat{e}_1\hat{k}_1\hat{e}_2 + \omega_1\hat{p}_1\hat{e}_2\hat{k}_2\hat{e}_1). \tag{7.1.43}$$

But, according to (5.1.34),

$$Tr(\hat{a}\hat{b}\hat{c}\hat{d}) = a_ib_jc_kd_l\,Tr(\gamma_i\gamma_j\gamma_k\gamma_l)$$

$$= 4[(ab)(cd) - (ac)(bd) + (ad)(bc)], \tag{7.1.44}$$

and therefore, making use of (7.1.11) and (7.1.13), one obtains

$$\begin{aligned}
Tr(\hat{p}_1\hat{e}_1\hat{k}_1\hat{e}_2) &= -4(p_1k_1)(e_1e_2) = 4m\omega_1(e_1e_2)\;; \\
Tr(\hat{p}_1\hat{e}_2\hat{k}_2\hat{e}_1) &= -4(p_1k_2)(e_2e_1) = 4m\omega_2(e_1e_2)
\end{aligned} \tag{7.1.45}$$

and (7.1.43) finally writes

$$A_1 = 32m^2(\omega_1\omega_2)^2(e_2e_2)^2. \tag{7.1.46}$$

Calculation of A_2

Using the four-momentum conservation law

$$p_1 + k_1 = p_2 + k_2, \tag{7.1.47}$$

relation $(7.1.35)_2$ can be written as

$$A_2 = Tr(BC), \tag{7.1.48}$$

where
$$\begin{aligned}
B &= (\hat{k}_1 - \hat{k}_2)(\omega_2\hat{e}_2\hat{k}_1\hat{e}_1 + \omega_1\hat{e}_1\hat{k}_2\hat{e}_2)\;; \\
C &= \hat{p}_1(\omega_2\hat{e}_1\hat{k}_1\hat{e}_2 + \omega_1\hat{e}_2\hat{k}_2\hat{e}_1)
\end{aligned} \tag{7.1.49}$$

To calculate B, let us first use (7.1.13). We can write:

$$B = 2\omega_2(k_1e_2)\hat{k}_1\hat{e}_1 - \omega_2\hat{e}_2\hat{k}_1\hat{k}_1\hat{e}_1 + 2\omega_1(k_1e_1)\hat{k}_2\hat{e}_2 - \omega_1\hat{e}_1\hat{k}_1\hat{k}_2\hat{e}_2$$

180

$$-2\omega_2(k_2e_2)\hat{k}_1\hat{e}_1 + \omega_2\hat{e}_2\hat{k}_2\hat{k}_1\hat{e}_1 - 2\omega_1(k_2e_1)\hat{k}_2\hat{e}_2 + \omega_1\hat{e}_1\hat{k}_2\hat{k}_2\hat{e}_2.$$

But, using again (7.1.11), we obtain

$$\begin{aligned}
\hat{k}_1\hat{k}_1 &= (k_1k_1) = 0 \; ; \quad (k_1e_1) = 0 \; ; \quad (p_1k_1) = -m\omega_1 \; ; \\
\hat{k}_2\hat{k}_2 &= (k_2k_2) = 0 \; ; \quad (k_2e_2) = 0 \; ; \quad (p_1k_2) = -m\omega_2.
\end{aligned} \tag{7.1.50}$$

Let us write

$$B = B_1 + B_2 + B_3 + B_4, \tag{7.1.51}$$

with

$$\begin{aligned}
B_1 &= \omega_2\hat{e}_2\hat{k}_2\hat{k}_1\hat{e}_1 \; ; \quad B_3 = 2\omega_2(k_1e_2)\hat{k}_1\hat{e}_1 \; ; \\
B_2 &= -\omega_1\hat{e}_1\hat{k}_1\hat{k}_2\hat{e}_2 \; ; \quad B_4 = -2\omega_1(k_2e_1)\hat{k}_2\hat{e}_2.
\end{aligned} \tag{7.1.52}$$

In the same manner, we get

$$\begin{aligned}
\hat{p}_1\hat{e}_1\hat{k}_1\hat{e}_2 &= 2m\omega_1\hat{e}_1\hat{e}_2 - \hat{e}_1\hat{k}_1\hat{e}_2\hat{p}_1 \; ; \\
\hat{p}_1\hat{e}_2\hat{k}_2\hat{e}_1 &= 2m\omega_2\hat{e}_2\hat{e}_1 - \hat{e}_2\hat{k}_2\hat{e}_1\hat{p}_1,
\end{aligned} \tag{7.1.53}$$

which allows to write C as

$$C = C_1 + C_2 + C_3, \tag{7.1.54}$$

where

$$C_1 = 4m\omega_1\omega_2(e_1e_2) \; ; \quad C_2 = -\omega_2\hat{e}_1\hat{k}_1\hat{e}_2\hat{p}_1 \; ; \quad C_3 = -\omega_1\hat{e}_2\hat{k}_2\hat{e}_1\hat{p}_1. \tag{7.1.55}$$

In view of (7.1.52) and (6.1.55), one finds

$$Tr\, B_1C_1 = 4m\omega_1\omega_2^2(e_1e_2)\, Tr\, \hat{e}_2\hat{k}_2\hat{k}_1\hat{e}_1$$

$$= 16m\omega_1\omega_2^2(e_1e_2)[(e_2k_1)(e_1k_2) - (e_1e_2)(k_1k_2)],$$

as well as

$$Tr\, B_2C_1 = -16m\omega_1^2\omega_2(e_1e_2)[(e_1k_2)(e_2k_1) - (e_1e_2)(k_1k_2)];$$

$$Tr\, B_3C_1 = 8m\omega_1\omega_2^2(e_1e_2)(k_1e_2).4(e_1k_1) = 0;$$

$$Tr\, B_4C_1 = 0.$$

Adding together the last four relations, we have

$$Tr\, BC_1 = 16m\omega_1\omega_2(\omega_1 - \omega_2)(e_1e_2)[(e_1k_2)(k_1e_2) - (e_1e_2)(k_1k_2)]. \tag{7.1.56}$$

Non-divergent second-order processes

Furthermore, we get:

$$Tr\, B_1 C_2 = -\omega_2^2\, Tr(\hat{e}_2 \hat{k}_2 \hat{k}_1 \hat{e}_1 \hat{e}_1 \hat{k}_1 \hat{e}_2 \hat{p}_1)$$

$$= -4\omega_2^2 (e_1 e_1) Tr(\hat{e}_2 \hat{k}_2 \hat{k}_1 \hat{k}_1 \hat{e}_2 \hat{p}_1) \qquad (7.1.57)$$

$$= -16\omega_2^2 (e_1 e_2)(k_1^2)\, Tr(\hat{e}_2 \hat{k}_2 \hat{e}_2 \hat{p}_1) = 0,$$

and

$$Tr\, B_3 C_2 = 0. \qquad (7.1.58)$$

In the following, we shall consider the obvious relations

$$\begin{aligned}\hat{e}_1 \hat{e}_1 = (e_1 e_1) = \vec{e}_1 \cdot \vec{e}_1 = 1\ ; \\ \hat{e}_2 \hat{e}_2 = (e_2 e_2) = \vec{e}_2 \cdot \vec{e}_2 = 1\ ,\end{aligned} \qquad (7.1.59)$$

and underline those pairs of neighboring quantities whose order we want to invert, so as to apply (7.1.13). We seriatim have:

$$Tr\, B_2 C_2 = \omega_1 \omega_2\, Tr(\hat{e}_1 \hat{k}_1 \hat{k}_2 \hat{e}_2 \underline{\hat{e}_1 \hat{k}_1} \hat{e}_2 \hat{p}_1)$$

$$= -\omega_1 \omega_2\, Tr(\hat{e}_1 \hat{k}_1 \hat{k}_2 \underline{\hat{e}_2 \hat{k}_1} \hat{e}_1 \hat{e}_2 \hat{p}_1)$$

$$= -2\omega_1 \omega_2 (e_2 k_1) Tr(\hat{e}_1 \hat{k}_1 \underline{\hat{k}_2 \hat{e}_1} \hat{e}_2 \hat{p}_1) + \omega_1 \omega_2\, Tr(\hat{e}_1 \hat{k}_1 \underline{\hat{k}_2 \hat{k}_1} \hat{e}_2 \hat{e}_1 \hat{e}_2 \hat{p}_1)$$

$$= -4\omega_1 \omega_2 (e_2 k_1)(e_1 k_2)\, Tr(\hat{e}_1 \hat{k}_1 \hat{e}_2 \hat{p}_1) + 2\omega_1 \omega_2 (e_2 k_1)\, Tr(\hat{e}_1 \underline{\hat{k}_1 \hat{e}_1} \hat{k}_2 \hat{e}_2 \hat{p}_1)$$

$$+ 2\omega_1 \omega_2 (k_1 k_2)\, Tr(\hat{e}_1 \hat{k}_1 \hat{e}_2 \underline{\hat{e}_1} \hat{e}_2 \hat{p}_1)$$

$$= -4\omega_1 \omega_2 (e_1 k_2)(e_2 k_1)\, Tr(\hat{e}_1 \hat{k}_1 \hat{e}_2 \hat{p}_1)$$

$$- 2\omega_1 \omega_2 (e_2 k_1)\, Tr(\hat{k}_1 \hat{k}_2 \hat{e}_2 \hat{p}_1)$$

$$+ 4\omega_1 \omega_2 (k_1 k_2)(e_1 e_2)\, Tr(\hat{e}_1 \hat{k}_1 \hat{e}_2 \hat{p}_1)$$

$$- 2\omega_1 \omega_2 (k_1 k_2)\, Tr(\hat{e}_1 \hat{k}_1 \hat{e}_1 \hat{p}_1),$$

which yields

$$\begin{aligned}Tr\, B_2 C_2 = -8m\omega_1 \omega_2 [2\omega_1 (e_1 k_2)(e_2 k_1)(e_1 e_2) \\ + \omega_2 (e_2 k_1)^2 - 2\omega_1 (k_1 k_2)(e_1 e_2)^2 + \omega_1 (k_1 k_2)].\end{aligned} \qquad (7.1.60)$$

Using a similar procedure, we also find:

$$Tr\, B_4 C_2 = 2\omega_1 \omega_2 (k_2 e_1)\, Tr(\hat{k}_2 \hat{e}_2 \hat{e}_1 \underline{\hat{k}_1 \hat{e}_2} \hat{p}_1$$

$$= 4\omega_1 \omega_2 (e_1 k_2)(e_2 k_1)\, Tr(\hat{k}_2 \hat{e}_2 \hat{e}_1 \hat{p}_1)$$

$$-2\omega_1\omega_2(e_1k_2)\,Tr(\hat{k}_2\hat{e}_2\hat{e}_1\hat{e}_2\hat{k}_1\hat{p}_1)$$

$$= 4\omega_1\omega_2(e_1k_2)(e_2k_1)\,Tr(\hat{k}_2\hat{e}_2\hat{e}_1\hat{p}_1)$$

$$-4\omega_1\omega_2(k_2e_1)(e_1e_2)\,Tr(\hat{k}_2\hat{e}_2\hat{k}_1\hat{p}_1)$$

$$+2\omega_1\omega_2(e_1k_2)\,Tr(\hat{k}_2\hat{e}_1\hat{k}_1\hat{p}_1),$$

or, if the traces are expanded

$$Tr\,B_4C_2 = -8m\omega_1\omega_2[\omega_1(e_1k_2)^2]. \tag{7.1.61}$$

Adding now (7.1.57), (7.1.58), (7.1.60) and (7.1.61), we are left with

$$Tr\,BC_2 = -8m\omega_1\omega_2[2\omega_1(e_1k_2)(e_2k_1)(e_1e_2)$$
$$+\omega_2(k_1e_2)^2 + \omega_1(k_2e_1)^2 - 2\omega_1(k_1k_2)(e_1e_2)^2 + \omega_1(k_1k_2)]. \tag{7.1.62}$$

By looking at (7.1.55), we realize that C_3 is obtained from C_2 if the indices of ω, k, and e are interchanged.Since this permutation also transforms B into $-B$, we can then obtain the quantity $Tr\,BC_3$ by interchanging the indices and reversing the sign in (7.1.62). The result is

$$Tr\,BC_3 = 8m\omega_1\omega_2[2\omega_2(e_1k_2)(e_2k_1)(e_1e_2)$$
$$+\omega_1(k_2e_1)^2 + \omega_2(k_1e_2)^2 - 2\omega_2(k_1k_2)(e_1e_2)^2 + \omega_2(k_1k_2)]. \tag{7.1.63}$$

Adding (7.1.56), (7.1.62), and (7.1.63), we finally obtain A_2 as

$$A_2 = 8m\omega_1\omega_2(\omega_2 - \omega_1)(k_1k_2). \tag{7.1.64}$$

$$* \quad * \quad *$$

In view of (7.1.46) and (7.1.64), the trace (7.34) is therefore

$$A_1 + A_2 = 8m\omega_1\omega_2[4m\omega_1\omega_2(e_1e_2)^2 + (\omega_2 - \omega_1)(k_1k_2)]. \tag{7.1.65}$$

But, according to (7.1.47),

$$(p_1 - p_2)^2 = (k_1 - k_2)^2,$$

that is

$$p_1^2 + p_2^2 - 2(p_1p_2) = -2(k_1k_2),$$

which gives

$$(k_1k_2) = m^2 + (p_1p_2) = m^2 - E_{p_1}E_{p_2} = m(E_{p_1} - E_{p_2}) = m(\omega_2 - \omega_1). \tag{7.1.66}$$

This yields

$$A_1 + A_2 = 8m^2\omega_1^2\omega_2^2\left[4(e_1e_2)^2 + \frac{(\omega_2 - \omega_1)^2}{\omega_1\omega_2}\right],$$

and the quantity (7.1.36) finally writes

$$|M^{(2c)}|^2 = \frac{e^4}{(2\pi)^4.16m\omega_1\omega_2 E_{p_2}}\left[4(e_1e_2)^2 + \frac{(\omega_2 - \omega_1)^2}{\omega_1\omega_2}\right]. \qquad (7.1.67)$$

7.2. Differential cross section for Compton scattering

To find the scattering cross section, we make use of the relation (6.7.24). For the Compton effect, $\vec{p}_1'$ and E_1' characterize the emerging photon, while v_1 is the velocity of the incident photon, therefore

$$|\vec{p}_1'| = |\vec{k}_2| = \omega_2 \; ; \quad E_1' = \omega_2 \; ; \quad v_1 = c = 1 \qquad (7.2.1)$$

and, taking into account (6.7.24), (7.1.67) and (7.2.1), we obtain

$$d\sigma = \frac{e^4}{(8\pi)^2 m E_{p_2}}\frac{2}{\omega_1|f'(\omega_2)|}[4(e_1e_2)^2 + \frac{(\omega_2 - \omega_1)^2}{\omega_1\omega_2}]\,d\Omega'. \qquad (7.2.2)$$

Let us now consider $f'(\omega_2)$. Using (6.7.22), we can write

$$f(\omega_2) = E_{p_2} + \omega_2 - E_{p_1} - \omega_1 = E_{p_2} + \omega_2 - m - \omega_1 = 0. \qquad (7.2.3)$$

But $\vec{p}_2 = \vec{k}_1 - \vec{k}_2$, because $\vec{p}_1 = 0$, so that

$$\vec{p}_2^2 = \omega_1^2 + \omega_2^2 - 2\omega_1\omega_2 \cos\theta, \qquad (7.2.4)$$

where θ is the scattering angle (i.e. the angle between $\vec{k}_1$ and $\vec{k}_2$). We then find

$$E_{p_2} = \sqrt{\vec{p}_2^2 + m^2} = \sqrt{m^2 + \omega_1^2 + \omega_2^2 - 2\omega_1\omega_2 \cos\theta}, \qquad (7.2.5)$$

and (7.2.3) becomes

$$f(\omega_2) = \omega_2 - m - \omega_1 + \sqrt{m^2 + \omega_1^2 + \omega_2^2 - 2\omega_1\omega_2 \cos\theta}. \qquad (7.2.6)$$

Let us now take the derivative of (7.2.6) with respect to ω_2. The result is

$$f'(\omega_2) = 1 + \frac{\omega_2 - \omega_1 \cos\theta}{E_{p_2}} = \frac{E_{p_2} + \omega_2 - \omega_1 \cos\theta}{E_{p_2}},$$

or

$$f'(\omega_2) = \frac{m + \omega_1(1 - \cos\theta)}{E_{p_2}}.$$ (7.2.7)

To simplify (7.2.7), we shall use *Compton's formula*. This can be deduced by squaring (7.1.47)

$$(p_1 k_1) = (p_2 k_2),$$ (7.2.8)

then multiplying (7.1.47) by k_2

$$(p_2 k_2) = (p_1 k_2) + (k_1 k_2),$$ (7.2.9)

and finally combining the last two relations as

$$(p_1 k_1) = (p_1 k_2) + (k_1 k_2),$$

or

$$-m\omega_1 = -m\omega_2 + \omega_1\omega_2 \cos\theta - \omega_1\omega_2,$$

which then yields Compton's formula

$$\omega_2 = \frac{m\,\omega_1}{m + \omega_1\,(1 - \cos\theta)}.$$ (7.2.10)

By means of (7.2.7) and (7.2.10) one finds $f'(\omega_2) = m\omega_1/\omega_2 E_{p_2}$, and the differential cross section (7.2.2) writes

$$d\sigma = \frac{r_o^2}{4}\left(\frac{\omega_2}{\omega_1}\right)^2\left[4(e_1 e_2)^2 + \frac{(\omega_1 - \omega_2)^2}{\omega_1\omega_2}\right]d\Omega',$$ (7.2.11)

where

$$r_o = \frac{e^2}{4\pi m}$$ (7.2.12)

is the classical electron radius.

Next we shall discuss the *photon polarization*. One can distinguish two cases:

a) If both initial and final photons are polarized, then

$$(e_1 e_2) = \vec{e}_1 \cdot \vec{e}_2 = \cos\phi,$$ (7.2.13)

where ϕ is the angle between the polarizations of the initial and final photons. In this case (7.2.11) becomes

$$d\sigma = \frac{r_o^2}{4}\left(\frac{\omega_2}{\omega_1}\right)^2\left(4\,\cos^2\phi + \frac{\omega_1}{\omega_2} + \frac{\omega_2}{\omega_1} - 2\right)d\Omega',$$ (7.2.14)

Non-divergent second-order processes

which is known as the *Klein-Nishina formula*[1].

b) If the photons are not polarized, one must average over the polarizations in the initial state and sum over the final polarizations. The differential cross section (7.2.11) then is

$$\overline{d\sigma} = \frac{1}{2}\sum_{e_1,e_2} d\sigma = \frac{r_o^2}{2}\left(\frac{\omega_2}{\omega_1}\right)^2\left[\sum_{e_1,e_2}(e_1e_2) + \frac{1}{4}\sum_{e_1,e_2}\frac{(\omega_1-\omega_2)^2}{\omega_1\omega_2}\right]d\Omega'.$$

(7.2.15)

Since both initial and final photons are real, there are two independent initial polarizations, and two independent final polarizations, which yields

$$\frac{1}{4}\sum_{e_1,e_2}\frac{(\omega_1-\omega_2)^2}{\omega_1\omega_2} = \frac{1}{4}\frac{(\omega_1-\omega_2)^2}{\omega_1\omega_2}\sum_{e_1,e_2}1 = \frac{(\omega_1-\omega_2)^2}{\omega_1\omega_2}.$$

(7.2.16)

Evaluation of the sum over (e_1e_2) is a little more difficult. Consider three vectors $\vec{v}_1$, $\vec{v}_2$, $\vec{v}_3 = \vec{k}_1/\omega_1$, which form an orthogonal trihedron. Naturally, we have

$$(\vec{v}_1\cdot\vec{e}_2)^2 + (\vec{v}_2\cdot\vec{e}_2)^2 + (\vec{v}_2\cdot\vec{e}_2)^2 = 1.$$

Since $\vec{e}_1$ has to lie in the plane determined by $\vec{v}_1$ and $\vec{v}_2$, we have

$$\sum_{e_1}(e_1e_2)^2 = (\vec{v}_1\cdot\vec{e}_2)^2 + (\vec{v}_2\cdot\vec{e}_2)^2,$$

therefore

$$\sum_{e_1}(e_1e_2)^2 = 1 - \frac{(\vec{k}_1\cdot\vec{e}_2)^2}{\omega_1^2}.$$

Let us now take a second orthogonal trihedron of unit vectors $\vec{w}_1$, $\vec{w}_2$, $\vec{w}_3 = \vec{k}_2/\omega_2$. It can be easily shown that

$$\left[1-\frac{(\vec{k}_1\cdot\vec{w}_1)^2}{\omega_1^2}\right] + \left[1-\frac{(\vec{k}_1\cdot\vec{w}_2)^2}{\omega_1^2}\right] + \left[1-\frac{(\vec{k}_1\cdot\vec{w}_3)^2}{\omega_1^2}\right] = 3 - \frac{\vec{k}_1^2}{\omega_1^2} = 2.$$

Since $\vec{e}_2$ lies in the plane $(\vec{w}_1,\vec{w}_2)$, we also have

$$\sum_{e_1,e_2}(e_1e_2)^2 = \left[1-\frac{(\vec{k}_1\cdot\vec{w}_1)^2}{\omega_1^2}\right] + \left[1-\frac{(\vec{k}_1\cdot\vec{w}_2)^2}{\omega_1^2}\right],$$

---

[1] O.Klein, S.Nishina, Z. Physik, **82**, 1929, p.853.

Differential cross section for Compton scattering

which means

$$\sum_{e_1,e_2}(e_1e_2)^2 = 2 - \left[1 - \frac{(\vec{k}_1\cdot\vec{w}_3)^2}{w_1^2}\right] = 1 + \frac{(\vec{k}_1\cdot\vec{k}_2)^2}{w_1^2 w_2^2},$$

or

$$\sum_{e_1,e_2}(e_1e_2)^2 = 1 + \cos^2\theta. \qquad (7.2.17)$$

Alternatively, this formula can be obtained by means of (4.2.17) and (4.2.19). Indeed,

$$\sum_{e_1,e_2}(e_1e_2)^2 = \sum_{a,b=1}^{2} e_{1\alpha}^a e_{2\alpha}^b e_{1\beta}^a e_{2\beta}^b$$

$$= \left(\delta_{\alpha\beta} - \frac{k_{1\alpha}k_{1\beta}}{w_1^2}\right)\left(\delta_{\alpha\beta} - \frac{k_{2\alpha}k_{2\beta}}{w_2^2}\right)$$

$$= 1 + \left(\frac{\vec{k}_1\cdot\vec{k}_2}{w_1 w_2}\right)^2 = 1 + \cos^2\theta.$$

Substituting (7.2.16) and (7.2.17) into (7.2.14), one finds

$$\overline{d\sigma} = \frac{r_0^2}{2}\left(\frac{w_2}{w_1}\right)^2\left[\frac{w_1}{w_2} + \frac{w_2}{w_1} - \sin^2\theta\right]d\Omega', \qquad (7.2.18)$$

which is the *Klein-Nishina-Tamm*[1] *formula.*
Let us calculate the total cross section, which is

$$\overline{\sigma} = \frac{r_0^2}{2}\int_0^\pi\left(\frac{w_2}{w_1}\right)^2\left[\frac{w_1}{w_2} + \frac{w_2}{w_1} - \sin^2\theta\right]\sin\theta\,d\theta\int_0^{2\pi}d\varphi. \qquad (7.2.19)$$

To integrate, we make the substitution (see (7.2.10))

$$\frac{w_1}{w_2} = 1 + a(1 - \cos\theta) = x\,; \qquad a = \frac{w_1}{m} \qquad (7.2.20)$$

which yields

$$\overline{\sigma} = \frac{\pi r_0^2}{a}\left[\frac{4}{a} + \frac{a^2 - 2a - 2}{a^2}\ln(1 + 2a) + \frac{2a(a+1)}{(1+2a)^2}\right]. \qquad (7.2.11)$$

---

[1] I.E.Tamm, Z. Physik, **62**, 1930, p.545.

187

Non-divergent second-order processes

* * *

Consider the following two extreme cases:

i) *The non-relativistic approximation*, characterized by

$$a = \frac{\omega_1}{m} << 1. \tag{7.2.12}$$

Then, it follows from (7.2.10) that $\omega_1 \approx \omega_2$, and (7.2.14) becomes

$$\overline{d\sigma} = \frac{r_0^2}{2}(1 + \cos^2 \theta)\, d\Omega', \tag{7.2.13}$$

which is known as *Thomson's formula*.

A series expansion of $\ln(1 + 2a)$ and $(1 + 2a)^{-2}$ in (7.2.11) leads to

$$\overline{\sigma} = \frac{\pi r_0^2}{a}\left[\frac{4}{a} + \left(1 - \frac{2}{a} + \frac{2}{a^2}\right)\left(2a - 2a^2 + \frac{8a^3}{3} - 4a^4 + ...\right)\right.$$

$$\left. + (2a^2 + 2a)(1 - 4a + 12\,a^2 + ...)\right],$$

which gives

$$\overline{\sigma} \simeq \frac{8\pi r_0^2}{3}(1 - 2a). \tag{7.2.14}$$

This means that in the case of small energies $\overline{\sigma}$ varies linearly with a, and, if a is small enough, $\overline{\sigma}$ can be considered constant.

ii) *The ultra-relativistic approximation*, defined by

$$a = \frac{\omega_1}{m} >> 1. \tag{7.2.15}$$

Here we distinguish, again, two cases, depending on whether the scattering angle θ is small or large. In the first situation we have $1 - \cos\theta \simeq 0$, that is $\omega_1 \approx \omega_2$, and we retrieve (7.2.13) and (7.2.14), while in the second case we can write

$$\frac{\omega_1}{m}(1 - \cos\theta) \simeq \frac{\omega_1}{m}\frac{e^2}{m} >> 1 ; \quad e >> \sqrt{\frac{2m}{\omega_1}}, \tag{7.2.16}$$

therefore

$$\frac{\omega_2}{\omega_1} = \frac{1}{1 + a(1 - \cos\theta)} \simeq \frac{1}{a(1 - \cos\theta)} << 1. \tag{7.2.17}$$

In this approximation, we may keep only the term ω_1/ω_2 inside the parentheses in (7.2.14), which gives

$$d\sigma = \frac{r_0^2}{4}\frac{m}{\omega_1(1-\cos\theta)}\,d\Omega'. \tag{7.2.18}$$

We also have, instead of (7.2.18),

$$\overline{d\sigma} = \frac{r_0^2}{2}\frac{m}{\omega_1(1-\cos\theta)}\,d\Omega' = 2\,d\sigma. \tag{7.2.18}$$

Inside the square brackets in (7.2.11) we can take

$$\frac{4}{a}\approx 0\ ;\quad \frac{a^2-2aq-2}{a^2}\approx 1\ ;\quad 1+2a\approx 2a\ ;\quad \frac{2a^2+2a}{4a^4+a+1}\approx\frac{1}{2},$$

which gives

$$\bar{\sigma} = \frac{\pi r_0^2}{a}\left(\ln 2a+\frac{1}{2}\right). \tag{7.2.19}$$

Therefore $\bar{\sigma}$ is a descending function of a, i.e. the photon energy.

7.3. Electron-positron annihilation

Electron-positron annihilation occurs when an electron and a positron collide. The result of the collision is the annihilation of the electron-positron pair, and the creation of gamma ray photons or, at higher energies, other particles. There are three types of electron-positron annihilation:

a) A free electron may annihilate with a free positron.

b) Positronium-type annihilation. Positronium is a system consisting of an electron and a positron, bound together into an "exotic atom". This system being unstable, the two particles annihilate each other to produce photons.

c) Electron-positron annihilation in an external field, such as the Coulomb field of the nucleus.

We devote this section to the study of type (a) electron-positron annihilation . Our final goal is to calculate the cross section of the process of annihilation of a free electron with a free positron. As we shall see, the corresponding diagrams are topologically related to the diagrams for the Compton effect, which means that we can successfully use the substitution rule.

As we have seen (e.g. (7.5.15)), the process of annihilation of an electron-positron pair with the creation of two photons in lowest order is described by the graphs

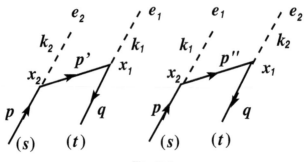

Fig.7.1

where $p = (\vec{p}, iE_-)$ and $q = (\vec{q}, iE_+)$ are the momentum four-vectors of the electron and positron, respectively, $k_1 = (\vec{k}_1, i\omega_1)$, e_1, and $k_2 = (\vec{k}_2, i\omega_2)$, e_2 are the momenta and polarizations of the emergent photons, while p' and p'' are the four-momenta of the virtual electrons, respectively.

To write the matrix element corresponding to the process described by these graphs we shall use the GRAPH TABLE. The components of the first graph, taken in chronological order, write:

- electron incident at $x_2 \longrightarrow (2\pi)^{-3/2} v_+^{(s)}(\vec{p})$
- vertex $x_2 \longrightarrow (2\pi)^4 e\gamma_l \, \delta(p' + k_2 - p)$
- photon emergent from $x_2 \longrightarrow (2\pi)^{-3/2} \frac{1}{\sqrt{2\omega_2}} (e_l)_2$
- propagator $(x_2 \to x_1) \longrightarrow (2\pi)^{-4} \int \frac{i\hat{p}' - m}{p'^2 + m^2 - i\alpha} \, dp'$
- positron incident at $x_1 \longrightarrow (2\pi)^{-3/2} \bar{v}_+^{(t)}(\vec{q})$
- vertex $x_1 \longrightarrow (2\pi)^4 e\gamma_j \, \delta(k_1 - q - p')$
- photon emergent from $x_1 \longrightarrow (2\pi)^{-3/2} \frac{1}{\sqrt{2\omega_1}} (e_j)_1$

Let us now take the product of these factors in down-up order, taking care to place the creation operators on the l.h.s. and the annihilation operators on the r.h.s., and then multiply the result by $1/2$, as required by the factor $(-1)^n/n!$ appearing in (6.2.22). The result is

$$S_I^{(2)} = \frac{1}{2} \frac{e^2}{(2\pi)^2} \frac{1}{\sqrt{4\omega_1\omega_2}} \int \bar{v}_+^{(t)}(\vec{q}) \, hat e_1 \, \frac{i\hat{p}' - m}{p'^2 + m^2} \, \hat{e}_2 \, v_+^{(s)}(\vec{p})$$

$$\times \delta(k_1 - q - p') \, \delta(p' + k_2 - p) \, dp'. \qquad (7.3.1)$$

Working in the same manner, the second graph gives:

- electron incident at $x_2 \longrightarrow (2\pi)^{-3/2} v_+^{(s)}(\vec{p})$
- vertex $x_2 \longrightarrow (2\pi)^4 e\gamma_l \, \delta(p'' + k_1 - p)$
- photon emergent from $x_2 \longrightarrow (2\pi)^{-3/2} \frac{1}{\sqrt{2\omega_1}} (e_l)_1$

190

- propagator $(x_2 \to x_1) \longrightarrow (2\pi)^{-4} \int \frac{i p'' - m}{p''^2 + m^2 - i\alpha} \, dp''$
- positron incident at $x_1 \longrightarrow (2\pi)^{-3/2} \overline{v}_+^{(t)}(\vec{q})$
- vertex $x_1 \longrightarrow (2\pi)^4 \, e \, \gamma_j \, \delta(k_2 - q - p'')$
- photon emergent from $x_1 \longrightarrow (2\pi)^{-3/2} \frac{1}{\sqrt{2\omega_2}} (e_j)_2$

This gives

$$S_{II}^{(2)} = \frac{1}{2} \frac{e^2}{(2\pi)^2} \frac{1}{\sqrt{4\omega_1\omega_2}} \int \overline{v}_+^{(t)}(\vec{q}) \, \hat{e}_2 \frac{i\hat{p}'' - m}{p''^2 + m^2} \hat{e}_1 \, v_+^{(s)}(\vec{p})$$

$$\times \delta(k_2 - q - p'') \, \delta(p'' + k_1 - p) \, dp''. \qquad (7.3.2)$$

As one can see, in both (7.3.1) and (7.3.2) the term $i\alpha$ in the denominator has been omitted. Recalling that the matrix element corresponding to the two graphs is

$$S^{(2)} = 2 \, (S_I^{(2)} + S_{II}^{(2)}), \qquad (7.3.3)$$

one obtains

$$S^{(2)} = M^{(2)} \, \delta(k_1 + k_2 - p - q), \qquad (7.3.4)$$

where

$$M^{(2)} = \frac{i \, e^2}{(2\pi)^2} \frac{1}{\sqrt{4\omega_1\omega_2}} \overline{v}_+^{(t)}(\vec{q}) \Big[\hat{e}_2 \frac{i(\hat{p} - \hat{k}_1) - m}{(p - k_1)^2 + m^2} \hat{e}_1$$

$$+ \hat{e}_1 \frac{i(\hat{p} - \hat{k}_2) - m}{(p - k_2)^2 + m^2} \hat{e}_2 \Big] v_+^{(s)}(\vec{p}), \qquad (7.3.5)$$

or, if the propagators are expressed in terms of q,

$$M^{(2)} = \frac{i \, e^2}{(2\pi)^2} \frac{1}{\sqrt{4\omega_1\omega_2}} \overline{v}_+^{(t)}(\vec{q}) \Big[\hat{e}_2 \frac{i(\hat{k}_2 - \hat{q}) - m}{(k_2 - q)^2 + m^2} \hat{e}_1$$

$$+ \hat{e}_1 \frac{i(\hat{k}_1 - \hat{q}) - m}{(k_1 - q)^2 + m^2} \hat{e}_2 \Big] v_+^{(s)}(\vec{p}), \qquad (7.3.5')$$

If we take a look at the Compton's effect, as shown by (7.1.7) and (7.1.8), we observe that it can also be expressed as

$$S^{(2c)} = M^{(2c)} \, \delta(p_2 + k_2 - p_1 - p_1), \qquad (7.3.6)$$

with

$$M^{(2c)} = \frac{i \, e^2}{(2\pi)^2} \frac{1}{\sqrt{4\omega_1\omega_2}} \overline{v}_-^{(t)}(\vec{p}_2) \Big[\hat{e}_2 \frac{i(\hat{p}_1 + \hat{k}_1) - m}{(p_1 + k_1)^2 + m^2} \hat{e}_1$$

$$+\hat{e}_1 \frac{i(\hat{p}_1 - \hat{k}_2) - m}{(p_1 - k_2)^2 + m^2} \hat{e}_2\Big] v_+^{(s)}(\vec{p}_1), \tag{7.3.7}$$

or, if the propagators are expressed in terms of p_2,

$$M^{(2c)} = \frac{i\,e^2}{(2\pi)^2} \frac{1}{\sqrt{4\omega_1\omega_2}} \overline{v}_-^{(t)}(\vec{p}_2)\Big[\hat{e}_2 \frac{i(\hat{p}_2 + \hat{k}_2) - m}{(p_2 + k_2)^2 + m^2} \hat{e}_1$$

$$+\hat{e}_1 \frac{i(\hat{p}_2 - \hat{k}_1) - m}{(p_2 - k_1)^2 + m^2} \hat{e}_2\Big] v_+^{(s)}(\vec{p}_1), \tag{7.3.7'}$$

Comparing (7.3.4), (7.3.5), and (7.3.5'), with (7.3.6), (7.3.7), and (7.3.7'), one observes that $S^{(2c)} \longrightarrow \frac{1}{i}S^{(2)}$ following the substitutions:

$$p_1 \to p \; ; \; p_2 \to -q \; ; \; k_1 \to -k_1 \; ; \; k_2 \to k_2 \; ; \; e_1 \to e_1 \; ; \; e_2 \to e_2 \; ;$$
$$v_+^{(s)}(\vec{p}_1) \to v_+^{(s)}(\vec{p}) \; ; \; \overline{v}_-^{(t)}(\vec{p}_2) \to \overline{v}_+^{(t)}(\vec{q}). \tag{7.3.8}$$

We have just applied the *substitution law*, proposed by *Jauch and Rohrlich*[1] for quantum electrodynamics. This law postulates that if the elementary particle process

$$a + b + c + ... \longrightarrow \alpha + \beta + \gamma + ... \tag{i}$$

is possible, then another process is also possible, namely

$$b + c + ... \longrightarrow \overline{a} + \alpha + \beta + \gamma + ..., \tag{ii}$$

where $\overline{a}$ is the anti-particle of a. This law allows us to write a quantitative formula for processes (i) and (ii). Let the S-matrix elements for these two processes be $M(k_\alpha, k_\beta, ...; k_a, k_b, ...)$, and $M(k_{\overline{a}}, k_\alpha, ...; k_b, k_c...)$, respectively, where $k_a, k_b, ...$ denote the four-momenta of the corresponding particles. Then the substitution law writes

$$M(k_{\overline{a}}, k_\alpha, ...; k_b, k_c...) = M(k_\alpha, k_\beta, ...; -k_a, k_b, ...),$$

where $-k_\alpha$ denotes the quantity k_α with the sign reversed for all four components.

Going now back to the Compton effect, we have obtained (see (7.1.67))

$$|M^{(2c)}|^2 = \frac{e^4}{(4\pi)^4 \omega_1\omega_2 E_{p_1} E_{p_2}}\Big[4(e_1e_2)^2 + \frac{(\omega_2 - \omega_1)^2}{\omega_1\omega_2}\Big]$$

[1] J.M.Jauch, and F.Rohrlich, Theory of photons and electrons, Addison-Wesley, Cambridge, 1955.

in the reference frame in which the initial electron was at rest (laboratory reference frame, LF), that is $p_1 = (\vec{0}, im)$, while the photons were supposed to be real (transversal), which means $e_1 = (\vec{e}_1, 0)$, $e_2 = (\vec{e}_2, 0)$, and $\vec{e}_1 \cdot \vec{k}_1 = \vec{e}_2 \cdot \vec{k}_2 = 0$.

Using the substitution law (7.3.8), we obtain for the conversion of the electron-positron pair into two photons

$$|M^{(2)}|^2 = -\frac{e^4}{2(4\pi)^4 \omega_1 \omega_2 m E_+}\left[4(e_1 e_2)^2 - \frac{(\omega_1 + \omega_2)^2}{\omega_1 \omega_2}\right],$$

where the factor $1/2$ appears due to averaging over the initial spins of the positron beam (N.B. averaging over the spins of the electron beam has already been done). As one observes, the substitution law introduces a minus sign for each electron becoming a positron, and vice-versa.

To calculate the differential cross section we use relation (6.7.24). In our case $\vec{p}_1'$ and E_1' characterize the first emergent photon, and we replace v_1 by v_r, the relative velocity between the positron and the electron (the second one being at rest), therefore

$$|\vec{p}_1'| = E_1' = \omega_1 ; \quad v_r = \beta ; \quad E_+ = \gamma m ; \quad \gamma = \frac{1}{\sqrt{1 - \beta^2}}, \quad (7.3.9)$$

and the differential cross section (6.7.24) writes

$$d\sigma = \frac{e^4}{8(4\pi)^2 m^2 \beta \gamma} \frac{\omega_1}{\omega_2} \frac{X}{|f'(\omega_1)|} d\Omega_1, \quad (7.3.10)$$

with

$$X = \frac{\omega_1}{\omega_2} + \frac{\omega_2}{\omega_1} + 2 - 4(e_1 e_2)^2. \quad (7.3.11)$$

In order to determine the cross section, one must first calculate $f'(\omega_1)$. This task is accomplished by means of the appropriate conservation laws. Using (6.7.22), we can write

$$f(\omega_1) = \omega_1 + \omega_2 - E_- - E_+, \quad (7.3.12)$$

and therefore

$$f'(\omega_1) = \frac{df(\omega_1)}{d\omega_1} = \frac{d(\omega_1 + \omega_2)}{d\omega_1} = 1 + \frac{d\omega_2}{d\omega_1}. \quad (7.3.13)$$

Let us now the square of the momentum conservation law $\vec{k}_2 = \vec{p} + \vec{q} - \vec{k}_1$. In view of (7.1.10), we have:

$$\omega_2^2 = p^2 + q^2 + \omega_1^2 + 2[|\vec{p}||\vec{q}|\cos\alpha - \omega_1(|\vec{p}|\cos\theta_1' + |\vec{q}|\cos\theta_1)], \quad (7.3.14)$$

193

where θ_1 and θ_1' are the angles between $\vec{k}_1$, $\vec{q}$, and $\vec{k}_1$, $\vec{p}$, respectively, while α is the angle between $\vec{p}$ and $\vec{q}$.

Next, we project the momentum conservation equation on two directions, parallel and perpendicular to $\vec{k}_1$, respectively

$$\omega_1 + \omega_2 \cos\theta = |\vec{q}| \cos\theta_1 + |\vec{p}| \cos\theta_1'$$
$$\omega_2 \sin\theta = |\vec{q}| \sin\theta_1 + |\vec{p}| \sin\theta_1', \qquad (7.3.15)$$

where θ is the angle between $\vec{k}_1$ and $\vec{k}_2$ (see Fig.7.2).

Further, we take the derivative of (7.3.14) with respect to ω_1, where θ_1, θ_1' as well as the initial state $(\vec{p}, \vec{q}, \alpha)$ are maintained constant. The result is

$$2\omega_2 \frac{d\omega_2}{d\omega_1} = 2\omega_1 - 2(|\vec{p}| \cos\theta_1' + |\vec{q}| \cos\theta_1),$$

or, in view of $(7.3.15)_1$

$$\frac{d\omega_2}{d\omega_1} = -\cos\theta, \qquad (7.3.16)$$

and (7.3.13) finally yields

$$f'(\omega_1) = 1 - \cos\theta. \qquad (7.3.17)$$

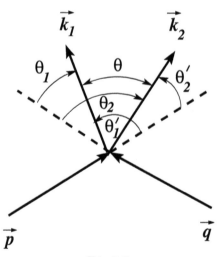

Fig.7.2

Resuming the calculation of (7.3.10), one must sum over the emergent photons. Using the formula (see (7.2.17))

$$\sum_{e_1,e_2} (e_1 e_2)^2 = 1 + \cos^2\theta, \qquad (7.3.18)$$

as well as (7.2.16), since a factor of 4 appears due to the terms independent of e_1 and e_2, one obtains

$$\overline{X} = \sum_{e_1,e_2} X = 4\left[\frac{\omega_1}{\omega_2} + \frac{\omega_2}{\omega_1} + 2 - (1 + \cos^2\theta)\right]. \tag{7.3.19}$$

Let us now apply the substitution law (7.3.8) to Compton's formula (7.2.10)

$$\frac{1}{\omega_1} + \frac{1}{\omega_2} = \frac{1 - \cos\theta}{m}, \tag{7.3.20}$$

which easily yields

$$\frac{\omega_1}{\omega_2} + \frac{\omega_2}{\omega_1} = \frac{1 - \cos\theta}{m}(\omega_1 + \omega_2) - 2.$$

In view of the energy conservation law $\omega_1 + \omega_2 = E_- + E_+ = m(1+\gamma)$, the last relation becomes

$$\frac{\omega_1}{\omega_2} + \frac{\omega_2}{\omega_1} + 2 = (1 + \gamma)(1 - \cos\theta). \tag{7.3.21}$$

Substituting (7.3.21) into (7.3.19), and the result into (7.3.10), we have

$$d\sigma = \frac{1}{2}r_0^2\frac{\omega_1}{\beta\gamma\omega_2}\left(1 + \gamma - \frac{1 + \cos^2\theta}{1 - \cos\theta}\right)d\Omega_1, \tag{7.3.22}$$

where $r_0 = e^2/4\pi m$ is the *electron classical radius* (also known as *the Lorentz radius* or *the Thomson scattering length*).

This is the differential cross section for the annihilation of two photons, in the laboratory reference frame (LF), in which the initial electron is at rest ($\vec{p} = 0$). Relation (7.3.22) gives the angular distribution of the first photon, for any arbitrary direction of the other photon.

Let us write $d\sigma$ in terms of θ_1. To this end, we start with the general relations

$$|\vec{p}| = \beta_- E_- \ ; \quad |\vec{q}| = \beta_+ E_+ \ ; \quad E_\pm = \gamma_\pm m$$
$$\gamma_\pm = \frac{1}{\sqrt{1-\beta_\pm^2}} \ ; \quad |\vec{k_1}| = \omega_1 \ ; \quad |\vec{k_2}| = \omega_2. \tag{7.3.23}$$

Then, we square the conservation law $p - k_2 = k_1 - q$, which yields

$$p.k_2 = q.k_1, \tag{7.3.24}$$

or

$$\beta_- E_- \omega_2 \cos \theta_2' - E_- \omega_2 = \beta_+ E_+ \omega_1 \cos \theta_1 - E_+ \omega_1, \qquad (7.3.25)$$

which results in

$$\frac{\omega_1}{\omega_2} = \frac{\gamma_-(1 - \beta_- \cos \theta_2')}{\gamma_+(1 - \beta_+ \cos \theta_1)}. \qquad (7.3.26)$$

In our case the electron is at rest, therefore

$$\beta_- = 0; \quad \gamma_- = 1; \quad |\vec{q}| = \beta \, E_+; \quad E_- = m; \quad E_+ = \gamma \, m, \quad (7.3.27)$$

and (7.3.26) becomes

$$\frac{\omega_1}{\omega_2} = \frac{1}{\gamma(1 - \beta \cos \theta_1)}. \qquad (7.3.28)$$

Combining this relation with the energy conservation law used in (7.3.21), one obtains

$$\frac{\omega_1}{m} = \frac{1 + \gamma}{\gamma(1 - \beta \, \cos \theta_1)} \; ; \qquad \frac{\omega_2}{m} = \frac{\gamma(1 + \gamma)(1 - \beta \, \cos \theta_1)}{1 + \gamma(1 - \beta \, \cos \theta_1)}. \qquad (7.3.29)$$

Then we have:

$$1 - \cos \theta = \frac{[1 + \gamma(1 - \beta \, cos\theta_1)]^2}{\gamma(1 + \gamma)(1 - \beta \, \cos \theta_1)} \qquad (7.3.30)$$

and

$$1 + \cos^2 \theta = 2 - 2\frac{[1 + \gamma(1 - \beta \, cos\theta_1)]^2}{\gamma(1 + \gamma)(1 - \beta \, \cos \theta_1)} + \frac{[1 + \gamma(1 - \beta \, cos\theta_1)]^4}{\gamma^2(1 + \gamma)^2(1 - \beta \, \cos \theta_1)^2}, \qquad (7.3.31)$$

as well as

$$d\Omega_1 = 2\pi \, \sin \theta_1 \, d\theta_1. \qquad (7.3.32)$$

Next, we denote $x = \cos \theta_1$ and use (7.3.28), (7.3.30), (7.3.31) and (7.3.32). The differential cross section then writes (the velocity β of the positron being given):

$$d\sigma = \frac{\pi \, r_0^2 \, dx}{\beta \gamma^2 (\beta \, x - 1)} \left\{ \gamma + 3 + \frac{[1 + \gamma(1 - \beta \, x)]^2}{\gamma(1 + \gamma)(\beta \, x - 1)} \right.$$

$$\left. + \frac{2\gamma(\gamma + 1)(\beta \, x - 1)}{[1 + \gamma(1 - \beta \, x)]^2} \right\}, \qquad (7.3.33)$$

which shows the angular distribution for the annihilation of a free electron-positron pair into two photons, in (LF).

To calculate the total cross section, which is independent of the reference system, one must integrate (7.3.33). To do this one makes the substitution $y = \gamma(1 - \beta x)$. Since θ_1 varies between 0 and π, x goes from -1 to 1, while y takes values between $\gamma(1-\beta)$ and $\gamma(1+\beta)$. Hence,

$$\sigma = \frac{\pi r_0^2}{\beta^2 \gamma^2} \int_{\gamma(1-\beta)}^{\gamma(1+\beta)} \left[\gamma + 3 - \frac{(1+y)^2}{(1+\gamma)y} - 2\frac{(1+\gamma)y}{(1+y)^2} \right] \frac{dy}{y}, \qquad (7.3.34)$$

or, if we work out the integral,

$$\sigma = \frac{\pi r_0^2}{\beta^2 \gamma^2} \left\{ \frac{\gamma^2 + 4\gamma + 1}{1+\gamma} \ln\frac{1+\beta}{1-\beta} - 2\frac{\beta}{1+\gamma}\left[\gamma + \frac{1}{\gamma(1-\beta^2)} \right] \right.$$

$$\left. - \frac{4\beta\gamma(1+\gamma)}{(1+\gamma)^2 - \beta^2\gamma^2} \right\}. \qquad (7.3.35)$$

According to (7.3.9)$_4$, one obtains $1 - \beta^2 = 1/\gamma^2$; $\beta^2\gamma^2 = \gamma^2 - 1$; $\beta = \sqrt{\gamma^2 - 1}/\gamma$, so that

$$\frac{1+\beta}{1-\beta} = \frac{(1+\beta)^2}{1-\beta^2} = \frac{\left(1 + \frac{\sqrt{\gamma^2-1}}{\gamma}\right)^2}{\frac{1}{\gamma^2}} = \left(\gamma + \sqrt{\gamma^2 - 1}\right)^2$$

and (7.3.35) becomes

$$\sigma = \frac{\pi r_0^2}{\beta^2 \gamma(\gamma+1)} \left[\left(\gamma + 4 + \frac{1}{\gamma}\right) \ln(\gamma + \sqrt{\gamma^2 - 1}) - \beta(\gamma + 3) \right]. \quad (7.3.36)$$

This result was first obtained by *Dirac*[1]. Here we have introduced the factor $1/2$, because the integrating over all directions of propagation of one of the photons doubles the result for the final state.

Let us consider, as before, the following two particular cases:

i) *The non-relativistic approximation.* If the momentum $|\vec{q}|$ of the incident positron is small, in other words if $\beta << 1$, we get $\gamma \approx 1$. The series expansion

$$\ln(\gamma + \sqrt{\gamma^2 - 1}) = \ln[\gamma(1 + \beta)] = \ln\gamma + \ln(1 + \beta) \approx \beta$$

---

[1] P.A.M.Dirac, Proc. Camb. Phil. Soc. **26**, 1930, pag.261

then yields

$$\sigma = \frac{\pi \, e_0^2}{\beta}. \tag{7.3.37}$$

This expression goes to infinity, when β goes to zero, but the number of annihilation processes per unit time remains finite, because the current $e\beta$ of the incident positrons goes to zero in the same limit. In other words, the number of annihilation processes per unit time goes to a constant value

$$P = \frac{1}{\tau_2} = \pi \, r_0^2 \rho, \tag{7.3.38}$$

where ρ is the electron density in the medium. For example, in a solid body the lifetime τ_2 is about 10^{-10} sec.

ii) *The ultra-relativistic approximation.* In this limit $\beta \to 1$, and γ becomes very large. Then $\beta^2 = 1 - 1/\gamma^2 \approx 1$ and, neglecting the rest of terms containing γ, we get

$$\sigma = \frac{\pi \, r_0^2 \, \ln(2\gamma)}{\gamma}. \tag{7.3.39}$$

This expression has a maximum for very small velocities of the incident positrons, and goes to zero for large energies of these particles.

7.4. Transition probability for Möller scattering

Non-relativistic quantum mechanics is able to investigate the scattering of two particles interacting via an electromagnetic field, but only if the velocities of these particles are small enough. The case of high-energy particle interaction can only be correctly studied within the framework of quantum electrodynamics.

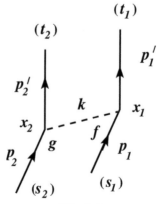

Fig.7.3

To study the scattering of two charged particles (electrons) of spin 1/2 we shall use again the Feynman-Dyson formalism. Let the four-momenta of the electrons in the initial and final states be p_1, p_2, and p'_1, p'_2, respectively, while their corresponding polarizations are (e_1), (e_2) and (t_1), (t_2). The graph associated with this process is represented in Fig.7.3.

Using the GRAPH TABLE, here is the correspondence between the graphic signs and the S-matrix elements, in momentum representation, following the graph in chronological order:

- electron incident at $x_2 \longrightarrow (2\pi)^{-3/2} v_+^{(s_2)}(\vec{p}_2)$
- vertex $x_2 \longrightarrow (2\pi)^4 e \gamma_g \delta(p'_2 - p_2 + k)$
- electron emergent from $x_2 \longrightarrow (2\pi)^{-3/2} \overline{v}_-^{(t_2)}(\vec{p}'_2)$
- photonic propagator $(x_2 \to x_1) \longrightarrow -(2\pi)^{-4} \int \frac{\delta_{fg}}{k^2 - i\alpha}\, dk$
- electron incident at $x_1 \longrightarrow (2\pi)^{-3/2} v_+^{(s_1)}(\vec{p}_1)$
- vertex $x_1 \longrightarrow (2\pi)^4 e \gamma_f \delta(p'_1 - p_1 - k)$
- electron emergent from $x_1 \longrightarrow (2\pi)^{-3/2} \overline{v}_-^{(t_1)}(\vec{p}'_1)$

Applying the known method, the S-matrix element associated with the graph given in Fig.7.3, in which the factor 1/2! corresponding to $S^{(n)}$ expansion is also taken into account, writes (here index "m" stands for "Möller")

$$S_1^{(2m)} = -\frac{e^2}{2(2\pi)^2} \int \overline{v}_-^{(t_1)}(\vec{p}'_1)\gamma_f\, v_+^{(s_1)}(\vec{p}_1)\delta(p'_1 - p_1 - k)$$

$$\times \frac{\delta_{fg}}{k^2} \overline{v}_-^{(t_2)}(\vec{p}'_2)\gamma_g\, v_+^{(s_2)}(\vec{p}_2)\delta(p'_2 - p_2 + k)\, dk, \qquad (7.4.1)$$

or, if we work out the integral,

$$S_1^{(2m)} = -\frac{i\,e^2}{2(2\pi)^2}\, \overline{v}_-^{(t_1)}(\vec{p}'_1)\gamma_f\, v_+^{(s_1)}(\vec{p}_1)\frac{1}{(p'_1 - p_1)^2}\overline{v}_-^{(t_2)}(\vec{p}'_2)$$

$$\times \gamma_f\, v_+^{(s_2)}(\vec{p}_2)\, \delta(p'_1 + p'_2 - p_1 - p_2). \qquad (7.4.2)$$

But in the case of identical particles the *indiscernibility principle* holds, which allows the permutation of particles without changing the state of the system. This means that to each process we can associate three more graphs, by permuting both $p'_1 \leftrightarrow p'_2$ and $p_1 \leftrightarrow p_2$, or $p'_1 \leftrightarrow p'_2$ only, or, finally, $p'_1 \leftrightarrow p'_2$ only. In other words, the graphs shown in Fig.7.4 also contribute to the matrix element.

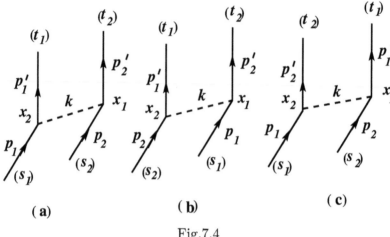

Fig.7.4

However, only the graph given in Fig.7.4(b) is different from the one shown in Fig.7.3, while the graph in Fig.7.4(a) reduces to the graph given in Fig.7.3 by changing the vertex notation, and Fig.7.4(c) reduces to Fig.7.4(b) by the same procedure.

The matrix element associated with the graph drawn in Fig.7.4(b) is calculated using the same procedure as for the graph shown in Fig.7.3, the only difference being the permutation $p'_1 \leftrightarrow p'_2$ in (7.4.2). The result is

$$S_{2b}^{(2m)} = -\frac{ie^2}{2(2\pi)^2}\bar{v}_-^{(t_2)}(\vec{p}'_2)\gamma_f\, v_+^{(s_1)}(\vec{p}_1)\frac{1}{(p'_2 - p_1)^2}\bar{v}_-^{(t_1)}(\vec{p}'_1)$$

$$\times\gamma_f\, v_+^{(s_2)}(\vec{p}_2)\delta(p'_1 + p'_2 - p_1 - p_2). \qquad (7.4.3)$$

Formulas (7.4.2) and (7.4.3) describe the same process. Since in the case of fermionic particles the state vectors have to be antisymmetric, the total matrix element equals the difference of these two expressions (divided by 2, to obtain the value for a single graph, and multiplied by 4, because there are 4 graphs associated with this process), that is

$$S^{(2m)} = 2(S_1^{(2m)} - S_{2b}^{(2m)}), \qquad (7.4.4)$$

or, in view of (7.4.2) and (7.4.3)

$$S^{(2m)} = M^{(2m)}\delta(p'_1 + p'_2 - p_1 - p_2), \qquad (7.4.5)$$

with

$$M_{t_1 t_2 s_1 s_2}^{(2m)} = -\frac{ie^2}{2(2\pi)^2}\left\{\frac{1}{(p'_1 - p_1)^2}\left[\bar{v}_{-a}^{(t_1)}(\vec{p}'_1)\gamma_{ab}^f v_{+b}^{(s_1)}(\vec{p}_1)\right.\right.$$

200

$$\times \overline{v}_{-c}^{(t_2)}(\vec{p}\,'_2)\gamma_{cd}^f v_{+d}^{(s_2)}(\vec{p}_2)\Big]$$

$$-\frac{1}{(p'_2 - p_1)^2}\Big[\overline{v}_{-h}^{(t_2)}(\vec{p}\,'_2)\gamma_{hi}^f v_{+i}^{(s_1)}(\vec{p}_1)$$

$$\times \overline{v}_{-j}^{(t_1)}(\vec{p}\,'_1)\gamma_{jk}^f v_{+k}^{(s_2)}(\vec{p}_2)\Big]\Big\}, \tag{7.4.6}$$

where the matrix elements have been highlighted. The propagators can also be written alternately, and we obtain $M^{(2m)}$ as

$$M^{(2m)} = -\frac{ie^2}{2(2\pi)^2}\Big\{\frac{1}{(p_2 - p'_2)^2}\Big[\overline{v}_{-}^{(t_1)}(\vec{p}\,'_1)\gamma_f v_{+}^{(s_1)}(\vec{p}_1)\Big]$$

$$\times \Big[\overline{v}_{-}^{(t_2)}(\vec{p}\,'_2)\gamma_f v_{+}^{(s_2)}(\vec{p}_2)\Big]$$

$$-\frac{1}{(p_2 - p'_1)^2}\Big[v_{-}^{(t_2)}(\vec{p}\,'_2)\gamma_f v_{+}^{(s_1)}(\vec{p}_1)\Big]$$

$$\times \Big[\overline{v}_{-}^{(t_1)}(\vec{p}\,'_1)\gamma_f v_{+}^{(s_2)}(\vec{p}_2)\Big]\Big\}. \tag{7.4.7}$$

To calculate the square of this matrix element we need to derive $M_{t_1 t_2 s_1 s_2}^{(2m)+}$ first. The adjoint of (7.4.7) is obtained by inverting the order of the operators and applying the Hermitian conjugation rule for operator products. We also use the relations (7.1.23) and $\gamma_4\gamma_f\gamma_4 = \pm\gamma_f$ (since $f = 4$, or $f \neq 4$). Taking into account all these observations, the first term inside the curly brackets yields:

$$\Big[\overline{v}_{-a}^{(t_1)}(\vec{p}\,'_1)\gamma_{ab}^f v_{+b}^{(s_1)}(\vec{p}_1)\overline{v}_{-c}^{(t_2)}(\vec{p}\,'_2)\gamma_{cd}^f v_{+d}^{(s_2)}(\vec{p}_2)\Big]^+$$

$$= \overline{v}_{-d}^{(s_2)}(\vec{p}_2)\gamma_4\gamma_{dc}^{f+}\gamma_4 v_{+c}^{(t_2)}(\vec{p}\,'_2)\overline{v}_{-b}^{(s_1)}(\vec{p}_1)\gamma_4\gamma_{ba}^{f+}\gamma_4 v_{+a}^{(t_1)}(\vec{p}\,'_1)$$

$$= \overline{v}_{-d}^{(s_2)}(\vec{p}_2)\gamma_{dc}^f v_{+c}^{(t_2)}(\vec{p}\,'_2)\overline{v}_{-b}^{(s_1)}(\vec{p}_1)\gamma_{ba}^f v_{+a}^{(t_1)}(\vec{p}\,'_1). \tag{7.4.8}$$

Proceeding in the same manner with the second term in (7.4.6), and making a convenient change of summation indices, we finally obtain:

$$M_{t_1 t_2 s_1 s_2}^{(2m)+} = \frac{ie^2}{(2\pi)^2}\Big\{\frac{1}{(p'_1 - p_1)^2}$$

$$\times \Big[\overline{v}_{-l}^{(s_2)}(\vec{p}_2)\gamma_{lm}^g v_{+m}^{(t_2)}(\vec{p}\,'_2)\Big]\Big[\overline{v}_{-n}^{(s_1)}(\vec{p}_1)\gamma_{no}^g v_{+c}^{(t_1)}(\vec{p}\,'_1)\Big]$$

$$-\frac{1}{(p'_2 - p_1)^2}\Big[\overline{v}_{-p}^{(s_2)}(\vec{p}_2)\gamma_{pq}^g v_{+q}^{(t_1)}(\vec{p}\,'_1)\Big]\Big[\overline{v}_{-r}^{(s_1)}(\vec{p}_1)\gamma_{ru}^g v_{+u}^{(t_2)}(\vec{p}\,'_2)\Big]\Big\}. \tag{7.4.9}$$

Non-divergent second-order processes

So we have

$$M^{(2m)+}_{t_1 t_2 s_1 s_2} M^{(2m)}_{t_1 t_2 s_1 s_2}$$

$$= \frac{e^4}{(2\pi)^4} \left\{ \frac{N}{(p_1' - p_1)^4} + \frac{O}{(p_2' - p_1)^4} - \frac{P + R}{(p_1' - p_1)^2 (p_2' - p_1)^2} \right\}, \quad (7.4.10)$$

where the following notations have been introduced:

$$N = \left[\overline{v}^{(s_2)}_{-l}(\vec{p}_2) \gamma^g_{lm} v^{(t_2)}_{+m}(\vec{p}_2') \right] \left[\overline{v}^{(s_1)}_{-n}(\vec{p}_1) \gamma^g_{no} v^{(t_1)}_{+o}(\vec{p}_1') \right]$$

$$\times \left[\overline{v}^{(t_1)}_{-a}(\vec{p}_1') \gamma^f_{ab} v^{(s_1)}_{+b}(\vec{p}_1) \right] \left[\overline{v}^{(t_2)}_{-c}(\vec{p}_2') \gamma^f_{cd} v^{(s_2)}_{+d}(\vec{p}_2) \right] ; \quad (7.4.11)$$

$$O = \left[\overline{v}^{(s_2)}_{-p}(\vec{p}_2) \gamma^g_{pq} v^{(t_1)}_{+q}(\vec{p}_1') \right] \left[\overline{v}^{(s_1)}_{-r}(\vec{p}_1) \gamma^g_{ru} v^{(t_2)}_{+u}(\vec{p}_2') \right]$$

$$\times \left[\overline{v}^{(t_2)}_{-h}(\vec{p}_2') \gamma^f_{hi} v^{(s_1)}_{+i}(\vec{p}_1) \right] \left[\overline{v}^{(t_1)}_{-j}(\vec{p}_1') \gamma^f_{jk} v^{(s_2)}_{+k}(\vec{p}_2) \right] ; \quad (7.4.12)$$

$$P = \left[\overline{v}^{(s_2)}_{-l}(\vec{p}_2) \gamma^g_{lm} v^{(t_2)}_{+m}(\vec{p}_2') \right] \left[\overline{v}^{(s_1)}_{-n}(\vec{p}_1) \gamma^g_{no} v^{(t_1)}_{+o}(\vec{p}_1') \right]$$

$$\times \left[\overline{v}^{(t_2)}_{-h}(\vec{p}_2') \gamma^f_{hi} v^{(s_1)}_{+i}(\vec{p}_1) \right] \left[\overline{v}^{(t_1)}_{-j}(\vec{p}_1') \gamma^f_{jk} v^{(s_2)}_{+k}(\vec{p}_2) \right] ; \quad (7.4.13)$$

$$R = \left[\overline{v}^{(s_2)}_{-p}(\vec{p}_2) \gamma^g_{pq} v^{(t_1)}_{+q}(\vec{p}_1') \right] \left[\overline{v}^{(s_1)}_{-r}(\vec{p}_1) \gamma^g_{ru} v^{(t_2)}_{+u}(\vec{p}_2') \right]$$

$$\times \left[\overline{v}^{(t_1)}_{-a}(\vec{p}_1') \gamma^f_{ab} v^{(s_1)}_{+b}(\vec{p}_1) \right] \left[\overline{v}^{(t_2)}_{-c}(\vec{p}_2') \gamma^f_{cd} v^{(s_2)}_{+d}(\vec{p}_2) \right] . \quad (7.4.14)$$

To find the differential cross section one must sum over the initial and final spins in (7.4.10) and average over the initial spins. Since in (7.4.11)-(7.4.14) we have matrix elements, and the order of the factors can be interchanged, it can be shown that the sums over the spinorial indices are in fact traces. Indeed, we can write:

$$N = \left[\overline{v}^{(s_1)}_{-n}(\vec{p}_1) \gamma^g_{no} v^{(t_1)}_{+o}(\vec{p}_1') \overline{v}^{(t_1)}_{-a}(\vec{p}_1') \gamma^f_{ab} v^{(s_1)}_{+b}(\vec{p}_1) \right]$$

$$\times \left[\overline{v}^{(s_2)}_{-l}(\vec{p}_2) \gamma^g_{lm} v^{(t_2)}_{+m}(\vec{p}_2') \overline{v}^{(t_2)}_{-c}(\vec{p}_2') \gamma^f_{cd} v^{(s_2)}_{+d}(\vec{p}_2) \right] . \quad (7.4.15)$$

Using now relation (5.3.53), which is

$$v^{(r)}_{+a}(\vec{p}) \, v^{(r)}_{-b}(\vec{p}) = -\frac{i}{2E_p} (\hat{p} + im)_{ab} \quad (7.4.16)$$

and noting that the sums over s_1, t_1 and s_2, t_2, respectively, are independent, the result will be the product of two traces. To prove this statement, let us first perform the summation over t_1 and t_2 in the factors underlined in (7.4.15). The result is

$$N = \left(-\frac{i}{2E_{p_1'}}\right)\left(-\frac{i}{2E_{p_2'}}\right)\left[\overline{v}_{-n}^{(s_1)}(\vec{p}_1)\gamma_{no}^g(\hat{p}_1' + im)_{oa}\gamma_{ab}^f\, v_{+b}^{(s_1)}(\vec{p}_1)\right]$$

$$\times \left[\overline{v}_{-l}^{(s_2)}(\vec{p}_2)\gamma_{lm}^g\,(p_2' + im)_{mc}\gamma_{cd}^f\, v_{+d}^{(s_2)}(\vec{p}_2)\right],$$

Performing the permutations $v_{+b}^{(s_1)}(\vec{p}_1) \leftrightarrow \overline{v}_{-n}^{(s_1)}(\vec{p}_1);\ v_{+d}^{(s_2)}(\vec{p}_2) \leftrightarrow \overline{v}_{-l}^{(s_2)}(\vec{p}_2)$, we have

$$N = -\frac{1}{4E_{p_1'}E_{p_2'}}\left[\underline{v_{+b}^{(s_1)}(\vec{p}_1)\overline{v}_{-n}^{(s_1)}(\vec{p}_1)}\gamma_{no}^g(\hat{p}_1' + im)_{oa}\gamma_{ab}^f\right]$$

$$\times \left[\underline{v_{+d}^{(s_2)}(\vec{p}_2)\overline{v}_{-l}^{(s_2)}(\vec{p}_2)}\gamma_{lm}^g\,(p_2' + im)_{mo}\gamma_{od}^f\right].$$

Next, we sum over s_1 and s_2, use (7.4.16) again, and perform the permutations indicated by the underlined products. Combining the result with the last relation, we finally get

$$N = \frac{1}{16E_{p_1}E_{p_2}E_{p_1'}E_{p_2'}}\left[\gamma_{ab}^f(\hat{p}_1 + im)_{bn}\gamma_{no}^g(\hat{p}_1' + im)_{oa}\right]$$

$$\left[\gamma_{od}^f(\hat{p}_2 + im)_{dl}\gamma_{lm}^g(\hat{p}_2' + im)_{mc}\right]. \qquad (7.4.17)$$

Using the same procedure, we obtain for the next term:

$$O = \left[\overline{v}_{-r}^{(s_1)}(\vec{p}_1)\gamma_{ru}^g\underline{v_{+u}^{(t_2)}(\vec{p}_2')\overline{v}_{-h}^{(t_2)}(\vec{p}_2')}\gamma_{hi}^f\, v_{+i}^{(s_1)}(\vec{p}_1)\right]$$

$$\times \left[\overline{v}_{-p}^{(s_2)}(\vec{p}_2)\gamma_{pq}^g\underline{v_{+q}^{(t_1)}(\vec{p}_1')\overline{v}_{-j}^{(t_1)}(\vec{p}_1')}\gamma_{jk}^f\, v_{+k}^{(s_2)}(\vec{p}_2)\right]$$

$$= -\frac{1}{4E_{p_1'}E_{p_2'}}\left[\overline{v}_{-r}^{(s_1)}(\vec{p}_1)\gamma_{ru}^g(\hat{p}_2' + im)_{uh}\gamma_{hi}^f\, v_{+i}^{(s_1)}(\vec{p}_1)\right]$$

$$\times \left[\overline{v}_{-p}^{(s_2)}(\vec{p}_2)\gamma_{pq}^g(\hat{p}_1' + im)_{qj}\gamma_{jk}^f\, v_{+k}^{(s_2)}(\vec{p}_2)\right] \qquad (7.4.18)$$

$$= -\frac{1}{4E_{p_1'}E_{p_2'}}\left[\underline{v_{+i}^{(s_1)}(\vec{p}_1)\overline{v}_{-r}^{(s_1)}(\vec{p}_1)}\gamma_{ru}^g(\hat{p}_2' + im)_{uh}\gamma_{hi}^f\right]$$

$$\times \left[\underline{v_{+k}^{(s_2)}(\vec{p}_2)\overline{v}_{-p}^{(s_2)}(\vec{p}_2)}\gamma_{pq}^g(\hat{p}_1' + im)_{qj}\gamma_{jk}^f\right]$$

Non-divergent second-order processes

$$= \frac{1}{16 E_{p_1} E_{p_2} E_{p_1'} E_{p_2'}} \left[\gamma_{hi}^f (\hat{p}_1 + im)_{ir} \gamma_{ru}^g (\hat{p}_2' + im)_{uh} \right]$$

$$\times \left[\gamma_{jk}^f (\hat{p}_2 + im)_{kp} \gamma_{pq}^g (\hat{p}_1' + im)_{qj} \right].$$

In the last two terms (P and R), the sums over spinorial indices are not independent anymore, and this will result in a single trace. Indeed, a similar procedure yields:

$$P = \left[\overline{v}_{-n}^{(s_1)}(\vec{p}_1) \gamma_{no}^g v_{+o}^{(t_1)}(\vec{p}_1') \overline{v}_{-j}^{(t_1)}(\vec{p}_1') \gamma_{jk}^f \right.$$

$$\times v_{+k}^{(s_2)}(\vec{p}_2) \overline{v}_{-l}^{(s_2)}(\vec{p}_2) \gamma_{lm}^g v_{+m}^{(t_2)}(\vec{p}_2') \overline{v}_{-h}^{(t_2)}(\vec{p}_2') \gamma_{hi}^f v_{+i}^{(s_1)}(\vec{p}_1) \left. \right]$$

$$= \left(- \frac{i}{2E_{p_1'}} \right) \left(- \frac{i}{2E_{p_2'}} \right) \left(- \frac{i}{2E_{p_2}} \right) \left[\overline{v}_{-n}^{(s_1)}(\vec{p}_1) \gamma_{no}^g \right.$$

$$\times (\hat{p}_1' + im)_{oj} \gamma_{jk}^f (\hat{p}_2 + im)_{kl} \gamma_{lm}^g (\hat{p}_2' + im)_{mh} \gamma_{hi}^f v_{+i}^{(s_1)}(\vec{p}_1) \left. \right] \quad (7.4.19)$$

$$= \frac{1}{8 E_{p_2} E_{p_1'} E_{p_2'}} \left[\underline{v_{+i}^{(s_1)}(\vec{p}_1) \overline{v}_{-n}^{(s_1)}(\vec{p}_1)} \gamma_{no}^g (\hat{p}_1' + im)_{oj} \gamma_{jk}^f \right.$$

$$\times (\hat{p}_2 + im)_{kl} \gamma_{lm}^g (\hat{p}_2' + im)_{mh} \gamma_{hi}^f \left. \right]$$

$$= \frac{1}{16 E_{p_1} E_{p_2} E_{p_1'} E_{p_2'}} \left[\gamma_{hi}^f (\hat{p}_1 + im)_{in} \gamma_{no}^g (\hat{p}_1' + im)_{oj} \gamma_{jk}^f \right.$$

$$\times (\hat{p}_2 + im)_{kl} \gamma_{lm}^g (\hat{p}_2' + im)_{mh} \left. \right],$$

as well as

$$R = \left[\overline{v}_{-r}^{(s_1)}(\vec{p}_1) \gamma_{ru}^g v_{+u}^{(t_2)}(\vec{p}_2') \overline{v}_{-o}^{(t_2)}(\vec{p}_2') \gamma_{cd}^f \right.$$

$$\times v_{+d}^{(s_2)}(\vec{p}_2) \overline{v}_{-p}^{(s_2)}(\vec{p}_2) \gamma_{pq}^g v_{+q}^{(t_1)}(\vec{p}_1') \overline{v}_{-a}^{(t_1)}(\vec{p}_1') \gamma_{ab}^f v_{+b}^{(s_1)}(\vec{p}_1) \left. \right]$$

$$= \frac{1}{8 E_{p_2} E_{p_1'} E_{p_2'}} \left[v_{-r}^{(s_1)}(\vec{p}_1) \gamma_{ru}^g (\hat{p}_2' + im)_{uc} \gamma_{cd}^f \right.$$

$$\times (\hat{p}_2 + im)_{dp} \gamma_{pq}^g (\hat{p}_1' + im)_{qa} \gamma_{ab}^f v_{+b}^{(s_1)}(\vec{p}_1) \left. \right] \quad (7.4.20)$$

$$= \frac{1}{8 E_{p_2} E_{p_1'} E_{p_2'}} \left[\underline{v_{+b}^{(s_1)}(\vec{p}_1) \overline{v}_{-r}^{(s_1)}(\vec{p}_1)} \gamma_{ru}^g (\hat{p}_2' + im)_{uc} \gamma_{cd}^f \right.$$

$$\times (\hat{p}_2 + im)_{dp} \gamma_{pq}^g (\hat{p}_1' + im)_{qa} \gamma_{ab}^f \left. \right]$$

204

Transition probability for Möller scattering

$$= \frac{1}{16E_{p_1}E_{p_2}E_{p'_1}E_{p'_2}}\Big[\gamma^f_{ab}(\hat{p}_1+im)_{br}\gamma^g_{ru}(\hat{p}'_2+im)_{uc}\gamma^f_{cd}$$

$$\times(\hat{p}_2+im)_{dp}\gamma^g_{pq}(\hat{p}'_1+im)_{qa}\Big].$$

Substituting (7.4.17)-(7.4.20) into (7.4.10) and averaging over the initial spins (the sum over final spins has already been done), one obtains

$$\left|M^{(2m)}\right|^2 = \frac{1}{4}M^{(2)+}_{t_1t_2s_1s_2}M^{(2)}_{t_1t_2s_1s_2}. \tag{7.4.21}$$

Since in both the initial and final states the electrons are not polarized, each of the two incident electron beams introduces a factor of $1/2$ (there are only two possible spin orientations). Therefore,

$$\left|M^{(2m)}\right|^2 = \frac{e^4}{64(2\pi)^4E_1E_2E'_1E'_2}\Big[\frac{A}{(p'_1-p_1)^4}+\frac{B}{(p'_2-p_1)^4}$$

$$-\frac{C+D}{(p'_1-p_1)^2(p'_2-p_1)^2}\Big], \tag{7.4.22}$$

where $E_{p_{1,2}} \equiv E_{1,2}$, $E_{p'_{1,2}} \equiv E'_{1,2}$, and

$$A = Tr[\gamma_f(\hat{p}_1+im)\gamma_g(\hat{p}'_1+im)]\,Tr[\gamma_f(\hat{p}_2+im)\gamma_g(\hat{p}'_2+im)]$$
$$B = Tr[\gamma_f(\hat{p}_1+im)\gamma_g(\hat{p}'_2+im)]\,Tr[\gamma_f(\hat{p}_2+im)\gamma_g(\hat{p}'_1+im)]$$
$$C = Tr[\gamma_f(\hat{p}_1+im)\gamma_g(\hat{p}'_1+im)\gamma_f(\hat{p}_2+im)\gamma_g(\hat{p}'_2+im)]$$
$$D = Tr[\gamma_f(\hat{p}_1+im)\gamma_g(\hat{p}'_2+im)\gamma_f(\hat{p}_2+im)\gamma_g(\hat{p}'_1+im)]$$
$$\tag{7.4.23}$$

As can be seen, by the interchange $p'_1 \leftrightarrow p'_2$ A becomes B, B becomes A, while $C = D$, so that we have to calculate only A and B. To this end, we use the relations (see Section 5.1)

$$Tr\,I = 4\,; \quad Tr\,\gamma_i\gamma_k = 4\delta_{ik}\,; \quad Tr\,\gamma_{i_1...i_{2k+1}} = 0\,; \tag{7.4.24}$$
$$Tr\,\gamma_i\gamma_k\gamma_l\gamma_k = 4(\delta_{ik}\delta_{lm}+\delta_{im}\delta_{kl}-\delta_{km}\delta_{il}).$$

Let us denote by $A = A_{1fg}A_{2fg}$ the product of traces in $(7.4.23)_1$. This gives

$$A_{1fg} = Tr(\gamma_f\hat{p}_1\gamma_g\hat{p}'_1)-m^2\,Tr(\gamma_f\gamma_g)+im\,Tr(\gamma_f\hat{p}_1\gamma_g)+im\,Tr(\gamma_f\gamma_g\hat{p}'_1).$$

Since $\hat{p}_1 = p_{1k}\gamma_k$, and $\hat{p}'_1 = p'_{1l}\gamma_l$, in view of (7.4.24) we have

$$A_{1fg} = 4p_{1k}p'_{1l}(\delta_{fk}\delta_{gl}+\delta_{fl}\delta_{gk}-\delta_{fg}\delta_{kl}) - 4m^2\delta_{fg}$$

$$= 4(p_{1f}p'_{1g}+p'_{1f}p_{1g}-p_{1k}p'_{1k}\delta_{fg}-m^2\delta_{fg}).$$

205

Non-divergent second-order processes

Changing $p_1 \to p_2$ and $p'_1 \to p'_2$, we have

$$A_{2fg} = 4(p_{2f}p'_{2g} + p'_{2f}p_{2g} - p_{2k}p'_{2k}\delta_{fg} - m^2\delta_{fg}).$$

This leads to

$$A = 16[(p_1p_2)(p'_1p'_2) + (p_1p'_2)(p'_1p_2) - (p_1p'_1)(p_2p'_2) - m^2(p_1p'_1)$$

$$+(p'_1p_2)(p_1p'_2) + (p_1p_2)(p'_1p'_2) - (p_1p'_1)((p_2p'_2) - m^2(p_1p'_1) - (p_1p'_1)(p_2p'_2)$$

$$-(p_1p'_1)(p_2p'_2) + 4(p_1p'_1)(p_2p'_2) + 4m^2(p_1p'_1)$$

$$-m^2(p_2p'_2) - m^2(p_2p'_2) + 4m^2(p_2p'_2) + 4m^4],$$

which finally yields

$$A = 32[2m^4 + m^2p_1p'_1 + m^2p_2p'_2 + (p_1p_2)(p'_1p'_2) + (p_1p'_2)(p'_1p_2)] \quad (7.4.25)$$

Interchanging $p'_1 \leftrightarrow p'_2$, we also have

$$B = 32[2m^4 + m^2p_1p'_1 + m^2p_2p'_2 + (p_1p_2)(p'_1p'_2) + (p_1p'_2)(p'_1p_2)] \quad (7.4.26)$$

Moving on to $Tr\,C$, in view of $(7.4.23)_3$ one can write

$$C = Tr[(\gamma_f\hat{p}_1\gamma_g\hat{p}'_1 + im\gamma_f\hat{p}_1\gamma_g + im\gamma_f\gamma_g\hat{p}'_1 - m^2\gamma_f\gamma_g)$$

$$\times(\gamma_f\hat{p}_2\gamma_g\hat{p}'_2 + im\gamma_f\hat{p}_2\gamma_g + im\gamma_f\gamma_g\hat{p}'_2 - m^2\gamma_f\gamma_g)].$$

First note that, in agreement with $(7.4.24)_3$, the terms containing m and m^3 vanish, and C can be written as

$$C = \sum_{i=1}^{8} C_i, \quad (7.4.27)$$

with

$$C_1 = Tr(\gamma_f\hat{p}_1\gamma_g\hat{p}'_1\gamma_f\hat{p}_2\gamma_g\hat{p}'_2) \; ;$$

$$C_2 = -m^2\,Tr(\gamma_f\hat{p}_1\gamma_g\hat{p}'_1\gamma_f\gamma_g) \; ;$$

$$C_3 = -m^2\,Tr(\gamma_f\hat{p}_1\gamma_g\gamma_f\hat{p}_2\gamma_g) \; ;$$

$$C_4 = -m^2\,Tr(\gamma_f\hat{p}_1\gamma_g\gamma_f\gamma_g\hat{p}'_2) \; ;$$

$$C_5 = -m^2\,Tr(\gamma_f\gamma_g\hat{p}'_1\gamma_f\hat{p}_2\gamma_g) \; ;$$

$$C_6 = -m^2\,Tr(\gamma_f\gamma_g\hat{p}'_1\gamma_f\gamma_g\hat{p}'_2) \; ;$$

$$C_7 = -m^2\,Tr(\gamma_f\gamma_g\gamma_f\hat{p}_2\gamma_g\hat{p}'_2) \; ;$$

$$C_8 = m^4 \, Tr(\gamma_f \gamma_g \gamma_f \gamma_g) \ .$$

Applying now $(7.4.24)_4$, one obtains

$$C_8 = 4m^4(\delta_{fg}\delta_{fg} + \delta_{fg}\delta_{gf} - \delta_{ff}\delta_{gg})$$

$$= 4m^4(4 + 4 - 16) = -32\,m^4. \tag{7.4.28}$$

Writing $\hat{p} = p_i p_i$, we also have for $C_2, ..., C_7$:

$$C_2 = -m^2 p_{1k} p'_{1j} \, Tr(\gamma_f \gamma_k \gamma_g \gamma_j \gamma_f \gamma_g) \ ;$$

$$C_3 = -m^2 p_{1k} p_{2j} \, Tr(\gamma_f \gamma_k \gamma_g \gamma_f \gamma_j \gamma_g) \ ;$$

$$C_4 = -m^2 p_{1k} p'_{2j} \, Tr(\gamma_f \gamma_k \gamma_g \gamma_f \gamma_g \gamma_j) \ ;$$

$$C_5 = -m^2 p'_{1k} p_{2j} \, Tr(\gamma_f \gamma_g \gamma_k \gamma_f \gamma_j \gamma_g) \ ;$$

$$C_6 = -m^2 p'_{1k} p'_{2j} \, Tr(\gamma_f \gamma_g \gamma_k \gamma_f \gamma_g \gamma_j) \ ;$$

$$C_7 = -m^2 p_{2k} p'_{2j} \, Tr(\gamma_f \gamma_g \gamma_f \gamma_k \gamma_g \gamma_j) \ .$$

The above traces are easily calculated using the Dirac anti-commutation relations, as well as the invariance of the trace of a matrix product under the cyclic permutation of the matrices. Using $(7.4.24)_4$ and the relation $\gamma_f \gamma_f = \gamma_1 \gamma_1 + \gamma_2 \gamma_2 + \gamma_3 \gamma_3 + \gamma_4 \gamma_4 = 4$, the trace appearing in C becomes:

$$Tr\gamma_f \gamma_k \gamma_g \gamma_j \gamma_f \gamma_g = Tr\gamma_f \gamma_g \gamma_f \gamma_k \gamma_g \gamma_j = Tr(2\delta_{fg} - \gamma_g \gamma_f)\gamma_f \gamma_k \gamma_g \gamma_j$$

$$= 2\,Tr\gamma_g \gamma_k \gamma_g \gamma_j - 4\,Tr\gamma_g \gamma_k \gamma_g \gamma_j = -2\,Tr\gamma_g \gamma_k \gamma_g \gamma_j$$

$$= -2 \times 4\,(\delta_{gk}\delta_{gj} - \delta_{gg}\gamma_{kj} + \delta_{gj}\delta_{gk}) = -8(\delta_{kj} - 4\delta_{kj} + \delta_{kj}) = 16\delta_{kj}.$$

A similar procedure yields:

$$Tr\,\gamma_f \gamma_k \gamma_g \gamma_f \gamma_j \gamma_g = Tr(2\delta_{fk} - \gamma_k \gamma_f)\gamma_g \gamma_f \gamma_j \gamma_g$$

$$= 2\,Tr\,\gamma_g \gamma_k \gamma_j \gamma_g - Tr\,\gamma_k \gamma_f \gamma_g \gamma_f \gamma_j \gamma_g = 2\,Tr\,\gamma_g \gamma_k \gamma_j \gamma_g$$

$$-Tr\,\gamma_k(2\delta_{fg} - \gamma_g \gamma_f)\gamma_f \gamma_j \gamma_g = 2\,Tr\,\gamma_g \gamma_k \gamma_j \gamma_g - 2\,Tr\,\gamma_k \gamma_g \gamma_j \gamma_g$$

$$+4\,Tr\,\gamma_k \gamma_g \gamma_j \gamma_g = 2\,Tr\,\gamma_g \gamma_k \gamma_j \gamma_g + 2\,Tr\,\gamma_k \gamma_g \gamma_j \gamma_g = 16\,\delta_{kj} \ ;$$

$$Tr\,\gamma_f \gamma_k \gamma_g \gamma_f \gamma_g \gamma_j = Tr\,\gamma_f \gamma_k(2\delta_{fg} - \gamma_f \gamma_g)\gamma_g \gamma_j$$

$$= 2Tr\,\gamma_f \gamma_k \gamma_f \gamma_j - 4Tr\,\gamma_f \gamma_k \gamma_f \gamma_j = -2Tr\,\gamma_f \gamma_k \gamma_f \gamma_j = 16\,\delta_{kj} \ ;$$

$$Tr\,\gamma_f\gamma_g\gamma_k\gamma_f\gamma_j\gamma_g = Tr\,\gamma_g\gamma_f\gamma_g\gamma_k\gamma_f\gamma_j = Tr\,(2\delta_{fg} - \gamma_f\gamma_g)\gamma_g\gamma_k\gamma_f\gamma_j$$

$$= 2Tr\,\gamma_f\gamma_k\gamma_f\gamma_j - 4Tr\,\gamma_f\gamma_k\gamma_f\gamma_j = -2Tr\,\gamma_f\gamma_k\gamma_f\gamma_j = 16\,\delta_{kj}\;;$$

$$Tr\,\gamma_f\gamma_g\gamma_k\gamma_f\gamma_g\gamma_j = Tr(2\delta_{fg} - \gamma_g\gamma_f)\gamma_k\gamma_f\gamma_g\gamma_j$$

$$= 2Tr\,\gamma_k\gamma_f\gamma_f\gamma_j - Tr\,\gamma_g\gamma_f\gamma_k\gamma_f\gamma_g\gamma_j$$

$$= 8Tr\,\gamma_k\gamma_j - 2Tr\,\gamma_g\gamma_k\gamma_g\gamma_j + 4Tr\,\gamma_g\gamma_k\gamma_g\gamma_j$$

$$= 8Tr\gamma_k\gamma_j + 2Tr\gamma_g\gamma_k\gamma_g\gamma_j = 16\,\delta_{kj}\;;$$

$$Tr\,\gamma_f\gamma_g\gamma_f\gamma_k\gamma_g\gamma_j = Tr\,(2\delta_{fg} - \gamma_g\gamma_f)\gamma_f\gamma_k\gamma_g\gamma_j$$

$$= 2Tr\,\gamma_g\gamma_k\gamma_g\gamma_j - 4Tr\,\gamma_g\gamma_k\gamma_g\gamma_j = -2Tr\,\gamma_g\gamma_k\gamma_g\gamma_j = 16\delta_{kj}\;.$$

Replacing these results for the traces, one obtains

$$C_2 = -16\,m^2\,(p_1 p_1')\;;\quad C_3 = -16\,m^2\,(p_1 p_2)\;;$$

$$C_4 = -16\,m^2\,(p_1 p_2')\;;\quad C_5 = -16\,m^2\,(p_1' p_2)\;;\qquad (7.4.29)$$

$$C_6 = -16\,m^2\,(p_1' p_2')\;;\quad C_7 = -16\,m^2\,(p_2 p_2')\;;$$

We are left with calculating C_1, which is

$$C_1 = p_{1k}p_{1j}'p_{2l}p_{2m}'Tr\,\gamma_f\gamma_k\gamma_g\gamma_j\gamma_f\gamma_l\gamma_g\gamma_m.$$

Using the same procedure, one can write:

$$Tr\,\gamma_f\gamma_k\gamma_g\gamma_j\gamma_f\gamma_l\gamma_g\gamma_m = Tr(2\delta_{fk} - \gamma_k\gamma_f)\gamma_g\gamma_j\gamma_f\gamma_l\gamma_g\gamma_m$$

$$= 2Tr\,\gamma_g\gamma_j\gamma_k\gamma_l\gamma_l\gamma_m - Tr\,\gamma_k\gamma_f\gamma_g\gamma_i\gamma_f\gamma_l\gamma_g\gamma_m$$

$$= 2Tr\,\gamma_g\gamma_m\gamma_g\gamma_j\gamma_k\gamma_l - Tr\,\gamma_k(2\delta_{fg} - \gamma_g\gamma_f)\gamma_j\gamma_f\gamma_l\gamma_g\gamma_m$$

$$= 2Tr\,(2\delta_{gm} - \gamma_m\gamma_g)\gamma_g\gamma_j\gamma_k\gamma_l - 2Tr\,\gamma_k\gamma_j\gamma_g\gamma_l\gamma_g\gamma_m$$

$$+Tr\,\gamma_k\gamma_g\gamma_f\gamma_j\gamma_f\gamma_l\gamma_g\gamma_m = 4Tr\,\gamma_m\gamma_j\gamma_k\gamma_l - 8Tr\,\gamma_m\gamma_j\gamma_k\gamma_l$$

$$-2Tr\,\gamma_k\gamma_j(2\delta_{gl} - \gamma_l\gamma_g)\gamma_g\gamma_m + Tr\gamma_k\gamma_g(2\delta_{fj} - \gamma_j\gamma_f)\gamma_f\gamma_l\gamma_g\gamma_m$$

$$= -4Tr\,\gamma_m\gamma_j\gamma_k\gamma_l - 4Tr\,\gamma_k\gamma_j\gamma_l\gamma_m + 8Tr\gamma_k\gamma_j\gamma_l\gamma_m$$

$$+2Tr\,\gamma_k\gamma_g\gamma_j\gamma_l\gamma_g\gamma_m - 4Tr\gamma_k\gamma_g\gamma_j\gamma_l\gamma_g\gamma_m$$

Transition probability for Möller scattering

$$= -4Tr\,\gamma_m\gamma_j\gamma_k\gamma_l + 4Tr\gamma_k\gamma_j\gamma_l\gamma_m - 2Tr\,\gamma_k(2\delta_{gj} - \gamma_j\gamma_g)\gamma_l\gamma_g\gamma_m$$

$$= -4Tr\,\gamma_m\gamma_j\gamma_k\gamma_l + 4Tr\gamma_k\gamma_j\gamma_l\gamma_m - 4Tr\gamma_k\gamma_l\gamma_j\gamma_m + 2Tr\,\gamma_k\gamma_j\gamma_g\gamma_l\gamma_g\gamma_m$$

$$= -4Tr\,\gamma_m\gamma_j\gamma_k\gamma_l + 4Tr\gamma_k\gamma_j\gamma_l\gamma_m$$

$$-4Tr\gamma_k\gamma_l\gamma_j\gamma_m + 2Tr\gamma_k\gamma_j(2\delta_{gl} - \gamma_l\gamma_g)\gamma_g\gamma_m,$$

that is

$$Tr\,\gamma_f\gamma_k\gamma_g\gamma_j\gamma_f\gamma_l\gamma_g\gamma_m = -4Tr\,\gamma_m\gamma_j\gamma_k\gamma_l + 4Tr\gamma_k\gamma_j\gamma_l\gamma_m$$

$$-4Tr\gamma_k\gamma_l\gamma_j\gamma_m + 4Tr\gamma_k\gamma_j\gamma_l\gamma_m - 8Tr\gamma_k\gamma_j\gamma_l\gamma_m$$

$$= -4Tr\,\gamma_m\gamma_j\gamma_k\gamma_l - 4Tr\gamma_k\gamma_l\gamma_j\gamma_m - 4\times 4(\delta_{mj}\delta_{kl} - \delta_{mk}\delta_{jl} + \delta_{ml}\delta_{kj})$$

$$-4\times 4(\delta_{kl}\delta_{mj} - \delta_{kj}\delta_{ml} + \delta_{km}\delta_{lj}) = -32\,\delta_{kl}\delta_{mj}$$

which finally leads to

$$C_1 = -32(p_1p_2)(p_1'p_2'). \tag{7.4.30}$$

Taking into account (7.4.27) - (7.4.29), the trace C given by (7.4.26) writes

$$C = -16[2m^4 + m^2(p_1p_1') + m^2(p_2p_2') + m^2(p_1p_2)$$

$$+m^2(p_1'p_2') + m^2(p_1p_2') + m^2(p_1'p_2) + 2(p_1p_2)(p_1'p_2')]. \tag{7.4.31}$$

Interchanging $p_1' \leftrightarrow p_2'$ one obtains the trace D:

$$D = -16[2m^4 + m^2(p_1p_2') + m^2(p_1'p_2) + m^2(p_1p_2)$$

$$+m^2(p_1'p_2') + m^2(p_1p_1') + m^2(p_2p_2') + 2(p_1p_2)(p_1'p_2')]. \tag{7.4.32}$$

As one can see, $C = D$.

Before using the traces A, B, and C, we realize that they can be written in a simpler form. To this end, we recall that we are subject to the relation

$$p_1' + p_2' = p_1 + p_2, \tag{7.4.33}$$

due to the presence of the Dirac delta function in (7.4.5). Squaring (7.4.33), we obtain

$$p_1'^2 + p_2'^2 + 2p_1'p_2' = p_1^2 + p_2^2 + 2p_1p_2,$$

or

$$\vec{p}_1'^2 - E_1'^2 + \vec{p}_2'^2 - E_2'^2 + 2p_1'p_2' = \vec{p}_1^2 - E_1^2 + \vec{p}_2^2 - E_2^2 + 2p_1p_2.$$

Using now the energy-momentum relation $E^2 = \vec{p}^2 + m^2$, after simplifications we get

$$p_1 p_2 = p_1' p_2' = -m^2 \kappa. \tag{7.4.34}$$

Next, we can write (7.4.32) as $p_1' - p_2 = p_2' - p_1$ and proceed in the same way. The result is

$$p_1 p_2' = p_1' p_2 = -m^2 \mu. \tag{7.4.35}$$

Finally, since $p_1' - p_1 = p_2' - p_2$, we also obtain

$$p_1 p_1' = p_2 p_2' = -m^2 \lambda. \tag{7.4.36}$$

Here κ, μ, and λ are shorthand notations. By means of (7.4.33)-(7.4.35), the traces (7.4.25), (7.4.26), (7.4.31), and (7.4.32) are

$$\begin{array}{c} A = 32\, m^4 (\kappa^2 + \mu^2 - 2\lambda + 2) \; ; \\ B = 32\, m^4 (\kappa^2 + \lambda^2 - 2\mu + 2) \; ; \\ C = D = -32 m^4 (\kappa^2 - \kappa - \mu - \lambda + 1) \end{array} \quad , \tag{7.4.37}$$

and (7.4.22) becomes

$$\left| M^{2m} \right|^2 = \frac{e^4}{(2\pi)^4 E_1 E_2 E_1' E_2'} X, \tag{7.4.38}$$

where

$$X = \frac{1}{2} m^4 \left[\frac{\kappa^2 + \mu^2 - 2\lambda + 2}{(p_1' - p_1)^4} \right.$$

$$\left. + \frac{\kappa^2 + \lambda^2 - 2\mu + 2}{(p_2' - p_1)^4} + 2 \frac{\kappa^2 - \kappa - \mu - \lambda + 1}{(p_1' - p_1)^2 (p_2' - p_1)^2} \right]. \tag{7.4.39}$$

7.5. Möller scattering cross section

According to (6.7.24), the differential cross section is

$$d\sigma = (2\pi)^4 \frac{|\vec{p}_1'| E_1'}{v_r\, |f'(E_1')|} \left| M^{(2m)} \right|^2 \bigg|_{p_1' + p_2' - p_1 - p_2 = 0} d\Omega_1',$$

or, in view of (7.4.39)

$$d\sigma = \frac{4 r_0^2 m^2 |\vec{p}_1'|\, X}{E_1 E_2 E_2' v_r |f'(E_1')|} \bigg|_{p_1' + p_2' - p_1 - p_2 = 0} d\Omega_1', \tag{7.5.1}$$

where $r_0 = e^2/4\pi m$ is the electron classical radius, and (see (6.7.22))

$$f(E'_1) = E'_1 + E'_2 - E_1 - E_2. \qquad (7.5.2)$$

Let us first calculate $f'(E'_1)$. To this end, we use the well-known relativistic kinematics relations

$$|\vec{p}| = \beta E \quad \longrightarrow \quad E = \gamma m \ ;$$
$$\gamma = \frac{1}{\sqrt{1-\beta^2}} \quad \longrightarrow \quad \beta = v. \qquad (7.5.3)$$

Let us define the four-vector $P = (\vec{P}, iE)$ to describe the motion of the center of mass of the system, and take the polar axis of the coordinate system along $\vec{P}$. Then,

$$\vec{P} = \vec{p}_1 + \vec{p}_2 \ ; \qquad E = E_1 + E_2 \ . \qquad (7.5.4)$$

To calculate the derivative of $f(E'_1)$ with respect to E'_1, one must take into account that E'_2 depends on E'_1. In view of (7.4.33) and (7.5.4), we can write

$$\vec{p}'_2 = \vec{P} - \vec{p}'_1.$$

Using the energy-momentum relation and (7.5.3), we have:

$$E'^2_2 = \vec{p}'^2_2 + m^2 = (\vec{P} - \vec{p}'_1)^2 + m^2$$

$$= \vec{P}^2 - 2\vec{p}'_1 \cdot \vec{P} + \vec{p}'^2_1 + m^2$$

$$= \vec{P}^2 - 2|\vec{p}'_1||\vec{P}| \cos(\vec{p}'_1, \vec{P}) + E'^2_1$$

$$= \vec{P}^2 + E'^2_1 - 2\beta'_1 E'_1 |\vec{P}| \cos(\vec{p}'_1, \vec{P}),$$

or

$$E'_2 = [E'^2_1 - 2\beta'_1 E'_1 |\vec{P}| \cos(\vec{p}'_1, \vec{P}) + \vec{P}^2]^{1/2}. \qquad (7.5.5)$$

By means of (7.5.2), (7.5.4)$_2$, (7.5.5), and the temporal component of (7.4.33), one obtains

$$f'(E'_1) = 1 + \frac{dE'_2}{dE'_1} = 1 + \frac{E'_1 - \beta'_1 |\vec{P}| \cos(\vec{p}'_1, \vec{P})}{E'_2} = \frac{E - \beta'_1 |\vec{P}| \cos\theta'_1}{E'_2}. \qquad (7.5.6)$$

Using now (7.5.3) and (7.5.6), the differential cross section (7.5.1) finally writes

$$d\sigma = \frac{4\, r_0^2 m^2 \beta'_1\, E'_1\, X\, d\Omega'_1}{v_r\, E_1\, E_2 (E - \beta'_1 |\vec{P}| \cos\theta'_1)}, \qquad (7.5.7)$$

where X is given by (7.4.39). This is the most general formula, but we shall particularize it for the centre-of-mass frame (CMF), and for the system in which one of the initial electrons is at rest. A good approximation for the last reference frame is the laboratory frame (LF).

The centre-of-mass frame is characterized by

$$\begin{aligned}
\vec{p}_1 &= -\vec{p}_2 = -\vec{p} \; ; \\
\vec{p}'_1 &= -\vec{p}'_2 = -\vec{p}' \; ; \\
E_1 &= E'_1 = E_2 = E'_2 = \epsilon \; ; \\
|\vec{p}| &= |\vec{p}'| = p,
\end{aligned}$$

(7.5.8)

together with (7.5.3). Using (7.5.3), (7.5.8), as well as Fig.7.5, we have:

$$\begin{aligned}
\vec{p} \cdot \vec{p}' &= |\vec{p}||\vec{p}'| \cos\theta = \beta^2 \epsilon^2 \cos\theta \\
&= \beta^2 \gamma^2 \, m^2 \cos\theta = m^2(\gamma^2 - 1) \cos\theta.
\end{aligned}$$

(7.5.9)

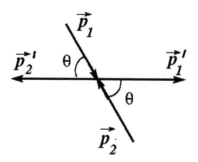

Fig.7.5

To find κ, μ, and λ in CMF, we use (7.4.34)-(7.4.36), together with (7.5.3) and (7.5.8). The results are:

$$\begin{aligned}
\kappa &= -\tfrac{1}{m^2}(\vec{p}_1 \cdot \vec{p}_2 - E_1 E_2) = \tfrac{1}{m^2}\vec{p}^2 + \epsilon^2) \\
&= \tfrac{\epsilon^2}{m^2}(1 + \beta^2) = 2\gamma^2 - 1 \; ;
\end{aligned}$$

(7.5.10)

$$\begin{aligned}
\mu &= -\tfrac{1}{m^2}(\vec{p}_1 \cdot \vec{p}'_2 - E_1 E'_2) = \tfrac{1}{m^2}(\vec{p} \cdot \vec{p}' + \epsilon^2) \\
&= \tfrac{1}{m^2}\left[m^2(\gamma^2 - 1)\cos\theta + m^2\gamma^2\right] = \gamma^2(1 + \beta^2 \cos\theta) \; ;
\end{aligned}$$

(7.5.11)

$$\begin{aligned}
\lambda &= -\tfrac{1}{m^2}(\vec{p}_1 \cdot \vec{p}'_1 - E_1 E'_1) = \tfrac{1}{m^2}(\epsilon^2 - \vec{p} \cdot \vec{p}') \\
&= \gamma^2(1 - \beta^2 \cos\theta).
\end{aligned}$$

(7.5.12)

Transition probability for Möller scattering

Let us now calculate the denominators in (7.4.39). Making use of (7.5.3), (7.5.8), and (7.5.9), we find:

$$(p'_1 - p_1)^2 = p_1^2 + p_1'^2 - 2p_1 p'_1 = -2m^2 - 2(\vec{p}_1 \cdot \vec{p}'_1 - E_1 E'_1)$$

$$= -2m^2 - 2(\vec{p} \cdot \vec{p}' - \epsilon^2) = -2m^2 - 2[m^2(\gamma^2 - 1)\cos\theta - m^2\gamma^2] \quad (7.5.13)$$

$$= 2m^2(\gamma^2 - 1)(1 - \cos\theta) = 4m^2(\gamma^2 - 1)\sin^2\frac{\theta}{2} = m^2(2\gamma\beta)^2 \sin^2\frac{\theta}{2},$$

as well as

$$(p'_2 - p_1)^2 = -2m^2 - 2(\vec{p}_1 \cdot \vec{p}'_2 - E_1 E'_2) = -2m^2 - 2(-\vec{p} \cdot \vec{p}' - \epsilon^2)$$

$$= -2m^2 - 2[-m^2(\gamma^2 - 1)\cos\theta - \gamma^2 m^2] = 2m^2(\gamma^2 - 1)(1 + \cos\theta)$$
$$\tag{7.5.14}$$

$$= 4m^2\gamma^2\beta^2 \cos^2\frac{\theta}{2} = m^2(2\gamma\beta)^2 \cos^2\frac{\theta}{2}.$$

In view of (7.4.22), (7.4.37), (7.4.39), and (7.5.10)-(7.5.14), we have:

$$X = \frac{1}{64m^4(2\beta\gamma)^4}\left[\frac{A}{\sin^4\frac{\theta}{2}} + \frac{B}{\cos^4\frac{\theta}{2}} - 2\frac{C}{\sin^2\frac{\theta}{2}\cos^2\frac{\theta}{2}}\right], \quad (7.5.15)$$

or

$$X = \frac{1}{2(2\beta\gamma)^4 \sin^4\frac{\theta}{2}\cos^4\frac{\theta}{2}}Z, \quad (7.5.16)$$

with

$$Z = \cos^4\frac{\theta}{2}\left[(2\gamma^2 - 1)^2 + \gamma^4(1 + \beta^2\cos\theta)^2 - 2\gamma^2(1 - \beta^2\cos\theta) + 2\right]$$

$$+ \sin^4\frac{\theta}{2}\left[(2\gamma^2 - 1)^2 + \gamma^4(1 - \beta^2\cos\theta)^2 - 2\gamma^2(1 + \beta^2\cos\theta) + 2\right]$$

$$+ 2\sin^2\frac{\theta}{2}\cos^2\frac{\theta}{2}\left[(2\gamma^2 - 1)^2 - (2\gamma^2 - 1)\right.$$

$$\left. - \gamma^2(1 + \beta^2\cos\theta) - \gamma^2(1 - \beta^2\cos\theta) + 1\right].$$

Since $\gamma^2\beta^2 = \gamma^2 - 1$, we have

$$Z = \cos^4\frac{\theta}{2}\left[5\gamma^4 - 6\gamma^2 + 3 + 2\gamma^2\beta^2(\gamma^2 + 1)\cos\theta + (\gamma^2 - 1)^2(1 - \sin^2\theta)\right]$$

$$+\sin^4\frac{\theta}{2}\left[5\gamma^4-6\gamma^2+3-2\gamma^2\beta^2(\gamma^2+1)\cos\theta+(\gamma^2-1)^2(1-sin^2\theta)\right]$$

$$+2\sin^2\frac{\theta}{2}\cos^2\frac{\theta}{2}(4\gamma^4-8\gamma^2+3)$$

$$=\cos^4\frac{\theta}{2}\left[6\gamma^4-8\gamma^2+4+2(\gamma^4-1)\cos\theta-(\gamma^2-1)^2\sin^2\theta\right]$$

$$+\sin^4\frac{\theta}{2}\left[6\gamma^4-8\gamma^2+4+2(\gamma^4-1)\cos\theta-(\gamma^2-1)^2\sin^2\theta\right]$$

$$+2\sin^2\frac{\theta}{2}\cos^2\frac{\theta}{2}(4\gamma^4-8\gamma^2+3)$$

$$=6\gamma^4(\cos^4\frac{\theta}{2}+\sin^4\frac{\theta}{2})-8\gamma^2(\cos^4\frac{\theta}{2}+\sin^4\frac{\theta}{2})+4(\cos^4\frac{\theta}{2}+\sin^4\frac{\theta}{2})$$

$$-(\gamma^2-1)\sin^2\theta(\cos^4\frac{\theta}{2}+\sin^4\frac{\theta}{2})+2(\gamma^4-1)\cos\theta(\cos^4\frac{\theta}{2}-\sin^4\frac{\theta}{2})$$

$$+8\gamma^4\sin^2\frac{\theta}{2}\cos^2\frac{\theta}{2}-16\gamma^2\sin^2\frac{\theta}{2}\cos^2\frac{\theta}{2}+6\sin^2\frac{\theta}{2}\cos^2\frac{\theta}{2}.$$

Using the well-known trigonometric formulas

$$\cos^4\frac{\theta}{2}+\sin^4\frac{\theta}{2}=1-\sin^2\frac{\theta}{2}\cos^2\frac{\theta}{2}\ ;$$

$$\cos^4\frac{\theta}{2}-\sin^4\frac{\theta}{2}=\cos^2\frac{\theta}{2}-\sin^2\frac{\theta}{2}=\cos\theta\ ;\qquad(7.5.17)$$

$$2\sin\frac{\theta}{2}\cos\frac{\theta}{2}=\sin\theta\ ,$$

we have

$$Z=6\gamma^4-8\gamma^2+4-(\gamma^2-1)^2\sin^2\theta-3\gamma^4\sin^2\theta$$

$$+4\gamma^2\sin^2\theta-2\sin^2\theta+\frac{1}{2}(\gamma^2-1)^2\sin^4\theta$$

$$+2(\gamma^4-1)(1-\sin^2\theta)+2\gamma^4\sin^2\theta-4\gamma^2\sin^2\theta+\frac{3}{2}\sin^2\theta\qquad(7.5.18)$$

$$=2\left[(2\gamma^2-1)^2-(2\gamma^4-\gamma^2-\frac{1}{4})\sin^2\theta+\frac{1}{4}(\gamma^2-1)^2\sin^4\theta\right].$$

Introducing this result in (7.5.16), one obtains

$$X=\frac{1}{(\gamma^2-1)^2}\left[\frac{(2\gamma^2-1)^2}{\sin^4\theta}-\frac{2\gamma^4-\gamma^2-\frac{1}{4}}{\sin^2\theta}+\frac{(\gamma^2-1)^2}{4}\right],\qquad(7.5.19)$$

written in CMF.

The differential cross section in CMF is found by means of (7.5.7), and taking into account (7.5.8), where X is given by (7.5.16). Using (7.5.3), (7.5.4), and (7.5.8), one gets: $\gamma = \frac{\epsilon}{m}$, $E = 2\epsilon$, $\vec{P} = 0$, $\beta_1' = v$. We remind the reader that in (7.5.7) v_r is the relative velocity of particle 1 with respect to particle 2. Relations (7.5.8) and Fig.7.5 show that $\vec{v}_2$ and $\vec{v}_1$ have opposite orientations, but the same modulus ($|\vec{v}_2| = |\vec{v}_1| = v$), which means that the relative velocity is $v_r = 2v$. The differential cross section is then (in CMF)

$$d\sigma = \frac{r_0^2 \, X}{\gamma^2} \, d\Omega. \tag{7.5.20}$$

In applications it is more convenient to determine the differential cross section in laboratory reference frame (LF). Denoting by "*" the set of quantities determined with respect to LF, let us find the relations connecting the quantities expressed in CMF and LF. To this end, we remind ourselves that the differential cross section is invariant with respect to the Lorentz transformation, while the transition from CMF to LF takes place via a Lorentz transformation with velocity $-v = -\beta$. Therefore, the components of the momentum four-vector transform as

$$\begin{cases} p_x^* = \gamma(p_x + \beta\epsilon) \; ; \\ p_y^* = p_y \; ; \\ p_z^* = p_z \; ; \\ \epsilon^* = \gamma(\epsilon + \beta p_x) \; . \end{cases} \tag{7.5.21}$$

Since the direction of $\vec{p}$ in CMF has been chosen as polar axis, we have $\vec{p} = (p, 0, 0)$, and $(7.5.21)_4$ writes (see also Fig.7.6)

$$\epsilon^* = \gamma(\epsilon + \beta p). \tag{7.5.22}$$

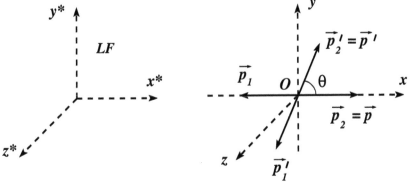

Fig.7.6

Non-divergent second-order processes

Using now (6.5.3) and (7.5.22), one obtains

$$\gamma^* m = \gamma(\gamma m + \beta^2 \gamma m),$$

or, after simplification

$$\gamma^* = \gamma^2 + \beta^2 \gamma^2.$$

Since $\gamma^2 \beta^2 = \gamma^2 - 1$, we can write

$$\gamma^* = 2\gamma^2 - 1. \tag{7.5.23}$$

To get the Einstein-Lorentz transformation relations for p', note that $\vec{p}'$ and $\vec{p}$ have the same magnitude, but their directions make an angle θ (see Fig.7.6). Then (7.5.8) yield

$$\begin{aligned} p'_x &= p'_{2x} = p'_2 \cos\theta = p' \cos\theta = p\cos\theta\ ; \\ p'_y &= p'_{2y} = p'_2 \sin\theta = p' \sin\theta = p\sin\theta\ . \end{aligned} \tag{7.5.24}$$

In view of $(7.5.8)_3$ and (7.5.24), the first two components of $\vec{p}'$ transform according to

$$\begin{aligned} p^* \cos\theta^* &= \gamma(p\cos\theta + \beta\epsilon)\ ; \\ p^* \sin\theta^* &= p\sin\theta\ , \end{aligned} \tag{7.5.25}$$

or, making use of (7.5.3)

$$\begin{aligned} \gamma^* \beta^* \cos\theta^* &= \gamma^2 \beta(\cos\theta + 1)\ ; \\ \gamma^* \beta^* \sin\theta^* &= \gamma\beta \sin\theta\ . \end{aligned} \tag{7.5.26}$$

We then square these two relations, recalling that $\gamma^2 \beta^2 = \gamma^2 - 1$, and divide one by the other. The result is

$$\tan^2 \theta^* = \frac{\sin^2\theta}{\gamma^2(1+\cos\theta)^2} = \frac{4\sin^2\frac{\theta}{2}\cos^2\frac{\theta}{2}}{4\gamma^2 \cos^4\frac{\theta}{2}} = \frac{1}{\gamma^2}\tan^2\frac{\theta}{2}. \tag{7.5.27}$$

On the other hand, (7.5.23) gives

$$\gamma^2 = \frac{1}{2}(\gamma^* + 1), \tag{7.5.28}$$

and we can write

$$\cos\theta = \frac{1-\tan^2\frac{\theta}{2}}{1+\tan^2\frac{\theta}{2}} = \frac{1-\gamma^2\tan^2\theta^*}{1+\gamma^2\tan^2\theta^*}$$

216

Transition probability for Möller scattering

$$= \frac{2 - (\gamma^* + 1)\frac{\sin^2 \theta^*}{1 - \sin^2 \theta^*}}{2 + (\gamma^* + 1)\frac{\sin^2 \theta^*}{1 - \sin^2 \theta^*}} = \frac{2 - (\gamma^* + 3)\sin^2 \theta^*}{2 + (\gamma^* - 1)\sin^2 \theta^*} = x, \qquad (7.5.29)$$

where x is a shorthand notation.

The last step is to find the relation between $d\Omega$ and $d\Omega^*$. Using the definition of the elementary solid angle and (7.5.29), one finds

$$\frac{d\Omega}{d\Omega^*} = \frac{2\pi \sin \theta \, d\theta}{2\pi \sin \theta^* \, d\theta^*} = \frac{d(\cos \theta)}{d(\cos \theta^*)}$$

$$= \frac{d}{d(\cos\theta^*)} \left[\frac{2 - (\gamma^* + 3)(1 - \cos^2 \theta^*)}{2 + (\gamma^* - 1)(1 - \cos^2 \theta^*)} \right],$$

namely

$$\frac{d\Omega}{d\Omega^*} = \frac{d}{d(\cos\theta^*)} \left[\frac{-(\gamma^* + 1) + (\gamma^* + 3)\cos^2 \theta^*}{(\gamma^* + 1) - (\gamma^* - 1)\cos^2 \theta^*} \right]$$

$$= \left[(\gamma^* + 1) - (\gamma^* - 1)\cos^2 \theta^* \right]^{-2} \left\{ 2(\gamma^* + 3)\cos \theta^* \right.$$

$$\times \left[(\gamma^* + 1) - (\gamma^* - 1)\cos^2 \theta^* \right]$$

$$\left. + 2(\gamma^* - 1)\cos \theta^* \left[-(\gamma^* + 1) + (\gamma^* + 3)\cos^2 \theta^* \right] \right\}$$

$$= \left[2 + (\gamma^* - 1)\sin^2 \theta^* \right]^{-2} 2\left\{ \left[(\gamma^* + 3)(\gamma^* + 1) \right.\right.$$

$$\left. -(\gamma^* - 1)(\gamma^* + 1) \right] \cos \theta^*$$

$$\left. - \left[(\gamma^* + 3)(\gamma^* - 1) - (\gamma^* - 1)(\gamma^* + 3) \right] \cos^3 \theta^* \right\},$$

and, after some simplifications

$$d\Omega = \frac{8 \, \cos \theta^* (\gamma^* + 1)}{[2 + (\gamma^* - 1)\sin^2 \theta^*]^2} \, d\Omega^*. \qquad (7.5.30)$$

The differential cross section in LF is then obtained by substituting (7.5.28), (7.5.29), and (7.5.30) into (7.5.20) (with X given by (7.5.19)):

$$d\sigma^* = r_0^2 \frac{2}{\gamma^* + 1} \frac{4}{(\gamma^* - 1)^2} \left[\frac{\gamma^{*2}}{(1 - x^2)^2} - \frac{2(\gamma^* + 1)^2 - 2(\gamma^* + 1) - 1}{4(1 - x^2)} \right.$$

217

$$+\frac{(\gamma^*-1)^2}{16}\Bigg]\frac{8(\gamma^*+1)\cos\theta^*}{[2+(\gamma^*-1)\sin^2\theta^*]^2}\,d\Omega^*,$$

and therefore,

$$d\sigma^* = \frac{16\,r_0^2}{(\gamma^*-1)^2}\left[\frac{4\gamma^{*2}}{(1-x^2)^2}-\frac{2\gamma^{*2}+2\gamma^*-1}{1-x^2}+\frac{\gamma^{*2}-2\gamma^*+1}{4}\right]$$

$$\times\frac{\cos\theta^*}{[2+(\gamma^*-1)\sin^2\theta^*]^2}\,d\Omega^*$$

$$=\frac{r_0^2\cos\theta^*\,d\Omega^*}{[2+\gamma^*-1)\sin^2\theta^*]^2}\left(\frac{4\gamma^*}{\gamma^*-1}\right)^2\left[\frac{4}{(1-x^2)^2}-\frac{2+(2/\gamma^*)-(1/\gamma^{*2})}{1-x^2}\right.$$

$$\left.+\frac{1-(2/\gamma^*)+(1/\gamma^{*2})}{4}\right]=\frac{r_0^2\cos\theta^*\,d\Omega^*}{[2+\gamma^*-1)\sin^2\theta^*]^2}\left[\frac{4\gamma^*(\gamma^*+1)^2}{\gamma^{*2}-1}\right]^2$$

$$\times\left[\frac{4}{(1-x^2)^2}-\frac{3}{1-x^2}+\frac{(1/\gamma^{*2}-(2/\gamma^*)+1}{1-x^2}+\frac{1-((2/\gamma^*)+(1/\gamma^{*2})}{4}\right]$$

$$=\frac{r_0^2\cos\theta^*\,d\Omega^*}{[2+(\gamma^*-1)\sin^2\theta^*]^2}\left[\frac{4\gamma^*(\gamma^*+1)}{\gamma^{*2}\beta^{*2}}\right]^2$$

$$\times\left[\frac{4}{(1-x^2)^2}-\frac{3}{1-x^2}+\frac14\left(\frac{1}{\gamma^{*2}}-\frac{2}{\gamma^*}+1\right)\left(1+\frac{4}{1-x^2}\right)\right],$$

or, finally

$$d\sigma^* = r_0^2\left(4\frac{\gamma^*+1}{\beta^{*2}\gamma^*}\right)^2\frac{\cos\theta^*\,d\Omega^*}{[2+(\gamma^*-1)\sin^2\theta^*]^2}\left[\frac{4}{(1-x^2)^2}\right.$$

$$\left.-\frac{3}{1-x^2}+\left(\frac{\gamma^*-1}{2\gamma^*}\right)^2\left(1+\frac{4}{1-x^2}\right)\right],\quad(LF)\qquad(7.5.31)$$

where x is given by (7.5.29). This is *Möller's formula*. The first two terms correspond to the scattering of two identical spinless particles, satisfying the Pauli exclusion principle. But this is impossible, because this principle is only satisfied by fermions, with spin 1/2. Consequently, the last term in (7.5.31) reflects the effect of the spin.

The total cross section is obtained by integrating over all angles, and dividing the result by 2 (because the particles are identical, and each state is considered twice, if the direction of scattering varies over the entire sphere). In fact, relations (7.5.20) and (7.5.31) cannot be integrated over *all* angles, because the integrals are divergent for $\theta = 0$ and $\theta = \pi$. This divergence corresponds to the scattering of the two electrons, without the emission of photons, which physically is impossible. On the other hand, the emission of a photon with very small energy cannot be neglected when the momentum transfer becomes very small ($\theta \longrightarrow 0$). Since the two electrons are indiscernible, the case $\theta \longrightarrow \pi$ leads to the same difficulty. This type of divergence is called *infrared divergence*.

At the end of this section, let us consider the non-relativistic limit of Möller scattering. This means that β^* is small, and $\gamma^* \longrightarrow 1$, while (7.5.29) becomes

$$x = \cos\theta = \frac{1}{2}(2 - 4\sin^2\theta^*) = 1 - 2\sin^2\theta^* = \cos 2\theta^*. \qquad (7.5.32)$$

As can be seen, in the non-relativistic approximation the scattering angle θ in the CMF is twice as large as the angle θ^* in LF, with θ^* varying from 0 to $\frac{\pi}{2}$.

To find the differential cross section in CMF, in the non-relativistic approximation, first note that $\gamma \longrightarrow 1$ (β is small) and (7.5.10)-(7.5.12) give : $k = 1$, $\mu = \lambda = \gamma^2$, while (7.4.38), (7.5.15) lead to

$$X = \frac{1}{(2\beta)^4}\left(\frac{1}{\sin^4 \frac{\theta}{2}} + \frac{1}{\cos^4 \frac{\theta}{2}} - \frac{1}{\sin^2 \frac{\theta}{2}\cos^2 \frac{\theta}{2}}\right). \qquad (7.5.33)$$

Taking into account (7.5.33), the differential cross section (7.5.20) in CMF becomes

$$d\sigma = \frac{r_0^2}{(2\beta)^4}\left(\frac{1}{\sin^4 \frac{\theta}{2}} + \frac{1}{\cos^4 \frac{\theta}{2}} - \frac{1}{\sin^2 \frac{\theta}{2}\cos^2 \frac{\theta}{2}}\right) d\Omega. \qquad (7.5.34)$$

Let us finally calculate the differential cross section in LS, in the non-relativistic approximation. In this respect, we rewrite $(7.5.21)_1$ as

$$p^* = \gamma(p + \beta\epsilon). \qquad (7.5.35)$$

Using (7.5.3), we have
$$\gamma^*\beta^* = 2\gamma^2\beta, \qquad (7.5.36)$$

or, by means of (7.5.22)

$$\beta^* = \frac{2\gamma^2}{2\gamma^2 - 1}\beta. \qquad (7.5.37)$$

In the non-relativistic approximation this formula becomes

$$\beta = \frac{1}{2}\beta^*, \qquad (7.5.38)$$

and (7.5.32) gives

$$\theta = 2\theta^*. \tag{7.5.39}$$

In the same approximation, (7.5.30) becomes

$$d\Omega = 4\,\cos\theta^*\,d\Omega^*. \tag{7.5.40}$$

In view of (7.5.38)-(7.5.40), the differential cross section (7.5.34) writes

$$d\sigma^* = \frac{4\,r_0^2\cos\theta^*}{\beta^{*4}}\left(\frac{1}{\sin^4\theta^*}+\frac{1}{\cos^4\theta^*}-\frac{1}{\sin^2\theta^*\cos^2\theta^*}\right)d\Omega^*. \tag{7.5.41}$$

This formula was first obtained by *Sir Nevill Francis Mott* (Nobel Prize, 1977). He showed that the last two terms are due to the particle interchange. If the particles are different, the last two terms disappear, and (7.5.41) becomes the well-known *Rutherford formula*. This conclusion is also valid for (7.5.34).

The differential cross sections calculated above have been experimentally verified, and the theory proved to be in very good agreement with the experiment.

7.6. Photon-photon scattering with electron-positron pair production

This process is the opposite to the annihilation of a free electron-positron pair into two photons. In lowest order, this is described (according to (7.5.16)) by the diagrams given in Fig.7.7, where p, q, p', p'', k_1, e_1, k_2, and e_2 have the same significance as for the electron-positron pair annihilation, analyzed in the previous section.

Using the GRAPH TABLE, and writing the S-matrix elements of the first diagram in chronological order, we obtain:
- photon incident at $x_2 \longrightarrow (2\pi)^{-3/2}\frac{1}{\sqrt{2\omega_2}}(e_l)_2$
- vertex $x_2 \longrightarrow (2\pi)^4\,e\,\gamma_l\delta(p'+q-k_2)$
- positron emergent from $x_2 \longrightarrow (2\pi)^{-3/2}v_-^{(s)}(\vec{q})$
- propagator $(x_2 \to x_1) \longrightarrow (2\pi)^{-4}\int \frac{i\hat{p}'-m}{p'^2+m^2-i\alpha}\,dp'$
- photon incident at $x_1 \longrightarrow (2\pi)^{-3/2}\frac{1}{\sqrt{2\omega_1}}(e_j)_1$
- vertex $x_1 \longrightarrow (2\pi)^4\,e\,\gamma_j\delta(p-k_1-p')$
- electron emergent from $x_1 \longrightarrow (2\pi)^{-3/2}\overline{v}_-^{(t)}(\vec{p})$.

We multiply these factors, in down-up order, in such a way that the creation operators appear on the l.h.s, and the annihilation operators on the r.h.s., and take into account the multiplication factor $1/2$.

Photon-photon scattering with electron-positron pair production

The result is

$$S_I^{(2)} = \frac{1}{2} \frac{e^2}{(2\pi)^2} \frac{1}{\sqrt{4\omega_1\omega_2}} \int \overline{v}_-^{(t)}(\vec{p})\hat{e}_1 \frac{i\hat{p}' - m}{p'^2 + m^2}\hat{e}_2 v_-^{(s)}(\vec{q})$$

$$\times \delta(p - k_1 - p')\delta(p' + q - k_2) \, dp'. \qquad (7.6.1)$$

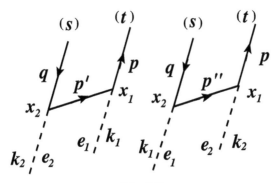

Fig.7.7

The second diagram is composed of

     - photon incident at $x_2 \longrightarrow (2\pi)^{-3/2} \frac{1}{\sqrt{2\omega_1}} (e_l)_1$

     - vertex $x_2 \longrightarrow (2\pi)^4 e \gamma_l \delta(p'' + q - k_1)$

     - positron emergent from $x_2 \longrightarrow (2\pi)^{-3/2} v_-^{(s)}(\vec{q})$

     - propagator $(x_2 \to x_1) \longrightarrow (2\pi)^{-4} \int \frac{i\hat{p}'' - m}{p''^2 + m^2 - i\alpha} \, dp''$

     - photon incident at $x_1 \longrightarrow (2\pi)^{-3/2} \frac{1}{\sqrt{2\omega_2}} (e_j)_2$

     - vertex $x_1 \longrightarrow (2\pi)^4 e \gamma_j \delta(p - k_2 - p'')$

     - electron emergent from $x_1 \longrightarrow (2\pi)^{-3/2}\overline{v}_-^{(t)}(\vec{p})$

A similar procedure gives

$$S_{II}^{(2)} = \frac{1}{2} \frac{e^2}{(2\pi)^2} \frac{1}{\sqrt{4\omega_1\omega_2}} \int \overline{v}_-^{(t)}(\vec{p})\hat{e}_2 \frac{i\hat{p}'' - m}{p''^2 + m^2}\hat{e}_1 v_-^{(s)}(\vec{q})$$

$$\times \delta(p - k_2 - p'')\delta(p'' + q - k_1) \, dp''. \qquad (7.6.2)$$

Calculating the integrals in (7.6.1) and (7.6.2), and noting that

$$S^{(2)} = 2(S_I^{(2)} + S_{II}^{(2)}), \qquad (7.6.3)$$

one can write

$$S^{(2)} = M^{(2)} \, \delta(p + q - k_1 - k_2), \qquad (7.6.4)$$

221

where

$$M^{(2)} = \frac{ie^2}{(2\pi)^2} \frac{1}{\sqrt{4\omega_1\omega_2}} \overline{v}_-^{(t)}(\vec{p})$$

$$\times \left[\hat{e}_2 \frac{i(\hat{k}_1 - \hat{q}) - m}{(k_1 - q)^2 + m^2} \hat{e}_1 + \hat{e}_1 \frac{i(\hat{k}_2 - \hat{q}) - m}{(k_2 - q)^2 + m^2} \hat{e}_2 \right] v_-^{(s)}(\vec{q}), \qquad (7.6.5)$$

or, expressing the propagators in terms of p

$$M^{(2)} = \frac{ie^2}{(2\pi)^2} \frac{1}{\sqrt{4\omega_1\omega_2}} \overline{v}_-^{(t)}(\vec{p})$$

$$\times \left[\hat{e}_2 \frac{i(\hat{p} - \hat{k}_2) - m}{(p - k_2)^2 + m^2} \hat{e}_1 + \hat{e}_1 \frac{i(\hat{p} - \hat{k}_1) - m}{(p - k_1)^2 + m^2} \hat{e}_2 \right] v_-^{(s)}(\vec{q}), \qquad (7.6.6)$$

Using now the energy-momentum relation, the denominators in (7.6.5) and/or (7.6.6) can be written as

$$(k_1 - q)^2 + m^2 = (p - k_2)^2 + m^2 = -2k_1 q = -2p k_2 = -2\chi_1 \ ;$$
$$(k_2 - q)^2 + m^2 = (p - k_1)^2 + m^2 = -2k_2 q = -2p k_1 = -2\chi_2 \ ,$$
$$(7.6.7)$$

where χ_1 and χ_2 are shorthands. The explicit form of $M^{(2)}$ is then

$$M_{ts}^{(2)} = \frac{i e^2}{4(2\pi)^2 \sqrt{\omega_1\omega_2}} \overline{v}_{-i}^{(t)}(\vec{p})(\mathcal{M})_{ij} v_{-j}^{(s)}(\vec{q}), \qquad (7.6.8)$$

with

$$(\mathcal{M})_{ij} = \left\{ \frac{1}{\chi_1} \hat{e}_2 \left[i(\hat{p} - \hat{k}_2) - m \right] \hat{e}_1 + \frac{1}{\chi_2} \hat{e}_1 \left[i(\hat{p} - \hat{k}_1) - m \right] \hat{e}_2 \right\}_{ij}. \qquad (7.6.9)$$

Since the differential cross section is expressed in terms of the squared absolute value of (7.6.8), we need the Hermitian conjugate of $M_{ts}^{(2)}$, which is

$$M_{ts}^{(2)+} = -\frac{i e^2}{4(2\pi)^2 \sqrt{\omega_1\omega_2}} [v_{-k}^{(s)}(\vec{q})]^+ (\mathcal{M})_{kl}^+ [\overline{v}_{-l}^{(t)}(\vec{p})]^+. \qquad (7.6.10)$$

According to (5.3.33) and (5.3.41), we have

$$\begin{cases} [v_-^{(s)}]^+ = v_+^{(s)+} = v_+^{(s)+} \gamma_4^2 = \overline{v}_+^{(s)} \gamma_4 \\ [\overline{v}_-^{(t)}]^+ = [v_-^{(t)+} \gamma_4]^+ = \gamma_4 [v_-^{(t)+}]^+ = \gamma_4 v_+^{(t)} \end{cases}, \qquad (7.6.11)$$

Photon-photon scattering with electron-positron pair production

and (7.6.10) becomes

$$M_{ts}^{(2)+} = -\frac{i\,e^2}{4(2\pi)^2\sqrt{\omega_1\omega_2}}\overline{v}_{+k}^{(s)}(\vec{q})(\overline{\mathcal{M}})_{kl}\,v_{+l}^{(t)}(\vec{p}),\qquad (7.6.12)$$

with

$$(\overline{\mathcal{M}})_{kl} = \gamma_4(\mathcal{M})_{kl}^+\gamma_4.\qquad (7.6.13)$$

Using (7.6.8) and (7.6.12), we then have

$$|M_{ts}^{(t)}|^2 = \frac{e^4}{16(2\pi)^4\omega_1\omega_2}\left[\overline{v}_{+k}^{(s)}(\vec{q})(\overline{\mathcal{M}})_{kl}\,v_{+l}^{(t)}(\vec{p})\,\overline{v}_{-i}^{(t)}(\vec{p})(\mathcal{M})_{ij}\,v_{-j}^{(s)}(\vec{q})\right].$$
$$(7.6.14)$$

Due to the summation over spin states, the quantity between square brackets is a trace. Indeed, if we use (5.3.53) and (5.3.54), namely

$$\begin{cases} \sum_s v_{-j}^{(s)}(\vec{q})\,\overline{v}_{+k}^{(s)}(\vec{q}) = -\frac{i}{2E_+}[(\hat{q}-im)]_{jk}\ ; \\ \sum_t v_{+l}^{(t)}(\vec{p})\,\overline{v}_{-i}^{(t)}(\vec{p}) = -\frac{i}{2E_-}[(\hat{p}+im)]_{li}, \end{cases}\qquad (7.6.15)$$

then sum over t, permute $\overline{v}_{+k}^{(s)}(\vec{q})$ and $v_{-j}^{(s)}(\vec{q})$, and finally sum over s, we are left with

$$|M^{(2)}|^2 = \sum_{t,s}|M_{ts}^{(2)}|^2 = \frac{e^4}{64(2\pi)^4 E_+ E_-\omega_1\omega_2}\,X,\qquad (7.6.16)$$

where

$$X = -[(\overline{\mathcal{M}})_{kl}(\hat{p}+im)_{li}(\mathcal{M})_{ij}(\hat{q}-im)_{jk}] = -Tr[\overline{\mathcal{M}}(\hat{p}+im)\mathcal{M}(\hat{q}-im)].$$
$$(7.6.17)$$

Let us now write explicitly the quantity $\overline{\mathcal{M}}$. Making use of (7.1.31), (7.6.9), and (7.6.13), we can write

$$\overline{\mathcal{M}} = \gamma_4\left\{\frac{1}{\chi_1}\hat{e}_1^+[-i(\hat{p}^+ - \hat{k}_2^+) - m]\hat{e}_2^+ + \frac{1}{\chi_2}\hat{e}_2^+[-i(\hat{p}^+ - \hat{k}_1^+) - m]\hat{e}_1^+\right\}\gamma_4$$

$$= \left\{-\frac{1}{\chi_1}\hat{e}_1\gamma_4[-i(\hat{p}^+ - \hat{k}_2^+) - m]\hat{e}_2^+ - \frac{1}{\chi_2}\hat{e}_2\gamma_4[-i(\hat{p}^+ - \hat{k}_1^+) - m]\hat{e}_1^+\right\}\gamma_4$$

$$= \left\{-\frac{1}{\chi_1}\hat{e}_1[i(\hat{p} - \hat{k}_2) - m]\gamma_4\hat{e}_2^+ - \frac{1}{\chi_2}\hat{e}_2[i(\hat{p} - \hat{k}_1) - m]\gamma_4\hat{e}_1^+\right\}\gamma_4.$$

Since $\gamma_4^2 = 1$, we are left with

$$\overline{\mathcal{M}} = \frac{1}{\chi_1}\hat{e}_1[i(\hat{p} - \hat{k}_2) - m]\hat{e}_2 + \frac{1}{\chi_2}\hat{e}_2[i(\hat{p} - \hat{k}_1) - m]\hat{e}_1.\qquad (7.6.18)$$

Non-divergent second-order processes

By means of (7.6.9) and (7.6.18), formula (7.6.16) writes:

$$X = \frac{A}{\chi_1^2} + \frac{B}{\chi_2^2} + \frac{1}{\chi_1\chi_2}(C + D), \qquad (7.6.19)$$

with

$$
\begin{aligned}
A &= Tr\{\hat{e}_1[(\hat{p} - \hat{k}_2) + im]\hat{e}_2(\hat{p} + im)\hat{e}_2[(\hat{p} - \hat{k}_2) + im]\hat{e}_1(\hat{q} - im)\} \; ; \\
B &= Tr\{\hat{e}_2[(\hat{p} - \hat{k}_1) + im]\hat{e}_1(\hat{p} + im)\hat{e}_1[(\hat{p} - \hat{k}_1) + im]\hat{e}_2(\hat{q} - im)\} \; ; \\
C &= Tr\{\hat{e}_1[(\hat{p} - \hat{k}_2) + im]\hat{e}_2(\hat{p} + im)\hat{e}_1[(\hat{p} - \hat{k}_1) + im]\hat{e}_2(\hat{q} - im)\} \; ; \\
D &= Tr\{\hat{e}_2[(\hat{p} - \hat{k}_1) + im]\hat{e}_1(\hat{p} + im)\hat{e}_2[(\hat{p} - \hat{k}_2) + im]\hat{e}_1(\hat{q} - im)\} \; .
\end{aligned}
$$
$$(7.6.20)$$

Taking into account that photons and electrons have the same number of independent polarizations (namely two), we can average over one species in in $|M^{(2)}|^2$, and sum over the other, or vice-versa. We shall perform the average over the electron and positron polarizations (which introduces a factor $1/4$, i.e. $1/2$ for each species), and sum over the photon polarizations, because according to our assumption the particles are not polarized.

Consider, for example, a photon in the initial state, and let e_j^l be its polarization four-vector, which is summed over the index j of γ_j, corresponding to the vertex where the photon was absorbed. Using (4.2.18), the term appearing in the transition probability writes

$$\sum_{pol}(...\hat{e}...\hat{e}...) = \sum_l(...e_j^{(l)}\gamma_j...e_k^{(l)}\gamma_k...)$$

$$= \delta_{jk}(...\gamma_j...\gamma_k...) = (...\gamma_j...\gamma_j...),$$

which means basically that $\hat{e}$ has to be replaced by γ_j. Therefore, we have

$$\overline{X} = \frac{1}{2}\sum_{pol}X = \frac{1}{4}\left[\frac{A'}{\chi_1^2} + \frac{B'}{\chi_2^2} + \frac{1}{\chi_1\chi_2}(C' + D')\right], \qquad (7.6.21)$$

where

$$
\begin{aligned}
A' &= Tr\{\gamma_j[(\hat{p} - \hat{k}_2) + im]\gamma_k(\hat{p} + im)\gamma_k[(\hat{p} - \hat{k}_2) + im]\gamma_j(\hat{q} - im)\} \; ; \\
B' &= Tr\{\gamma_k[(\hat{p} - \hat{k}_1) + im]\gamma_j(\hat{p} + im)\gamma_j[(\hat{p} - \hat{k}_1) + im]\gamma_k(\hat{q} - im)\} \; ; \\
C' &= Tr\{\gamma_j[(\hat{p} - \hat{k}_2) + im]\gamma_k(\hat{p} + im)\gamma_j[(\hat{p} - \hat{k}_1) + im]\gamma_k(\hat{q} - im)\} \; ; \\
B' &= Tr\{\gamma_k[(\hat{p} - \hat{k}_1) + im]\gamma_j(\hat{p} + im)\gamma_k[(\hat{p} - \hat{k}_2) + im]\gamma_j(\hat{q} - im)\} \; ;
\end{aligned}
$$
$$(7.6.22)$$

Photon-photon scattering with electron-positron pair production

If we now interchange indices $j \leftrightarrow k$ in B' and D', and also perform the substitution $k_1 \leftrightarrow k_2$, we find that $A' \leftrightarrow B'$, and $C' \leftrightarrow D'$, as well as $C' = D'$. This means that it will suffice to calculate the traces A' and C'. To this end, we use relations (7.4.24). Introducing the notations

$$p' \equiv f_1 = p - k_1 = k_2 - q \; ; \quad p'' \equiv f_2 = p - k_2 = k_1 - q, \quad (7.6.23)$$

one obtains:

$$A' = Tr[(\gamma_j \hat{f}_2 \gamma_k \hat{p} + im\gamma_j \hat{f}_2 \gamma_k + im\gamma_j \gamma_k \hat{p} - m^2 \gamma_j \gamma_k)$$

$$\times (\gamma_k \hat{f}_2 \gamma_j \hat{q} - im\gamma_k \hat{f}_2 \gamma_j + im\gamma_k \gamma_j \hat{q} + m^2 \gamma_k \gamma_j)] \; ;$$

$$C' = Tr[(\gamma_j \hat{f}_2 \gamma_k \hat{p} + im\gamma_j \hat{f}_2 \gamma_k \hat{p} + im\gamma_j \gamma_k \hat{p} - m^2 \gamma_j \gamma_k)$$

$$\times (\gamma_j \hat{f}_1 \gamma_k \hat{q} - im\gamma_j \hat{f}_1 \gamma_k + im\gamma_j \gamma_k \hat{q} + m^2 \gamma_j \gamma_k)].$$

Due to $(7.3.24)_3$, the terms containing m and m^3 vanish. Writing explicitly $\hat{p} = p_a \gamma_a$, the trace A' can be expressed as

$$A' = \sum_{i=1}^{8} A_i', \quad (7.6.24)$$

where

$$A_1' = f_{2a}p_b f_{2l}q_r \, Tr(\gamma_j \gamma_a \gamma_k \gamma_b \gamma_k \gamma_l \gamma_j \gamma_r) \; ;$$
$$A_2' = m^2 \, f_{2a}p_b \, Tr(\gamma_j \gamma_a \gamma_k \gamma_b \gamma_k \gamma_j) \; ;$$
$$A_3' = m^2 f_{2a} f_{2b} \, Tr(\gamma_j \gamma_a \gamma_k \gamma_k \gamma_b \gamma_j) \; ;$$
$$A_4' = -m^2 f_{2a} \, q_b \, Tr(\gamma_j \gamma_a \gamma_k \gamma_k \gamma_j \gamma_b) \; ;$$
$$A_5' = m^2 p_a f_{2b} \, Tr(\gamma_j \gamma_k \gamma_a \gamma_k \gamma_b \gamma_j) \; ;$$
$$A_6' = -m^2 p_a q_b \, Tr(\gamma_j \gamma_k \gamma_a \gamma_k \gamma_j \gamma_b) \; ;$$
$$A_7' = -m^2 f_{2a} q_b \, Tr(\gamma_j \gamma_k \gamma_k \gamma_a \gamma_j \gamma_b) \; ;$$
$$A_8' = -m^4 \, Tr(\gamma_j \gamma_k \gamma_k \gamma_j) \; .$$

Evaluating these traces is an easy task, by means of (7.4.24). Recalling that the Dirac matrices anti-commute (but they can be cyclically permuted under a trace), and using the relation $\gamma_j \gamma_j = 4$, we have:

$$Tr(\gamma_j \gamma_a \gamma_k \gamma_b \gamma_k \gamma_l \gamma_j \gamma_r) = Tr \, (\gamma_j \gamma_r \gamma_j \gamma_a \gamma_k \gamma_b \gamma_k \gamma_l)$$

$$= Tr[\gamma_j (2\delta_{jr} - \gamma_j \gamma_r)\gamma_a \gamma_k \gamma_b \gamma_k \gamma_l]$$

Non-divergent second-order processes

$$= 2\,Tr(\gamma_r\gamma_a\gamma_k\gamma_b\gamma_k\gamma_l) - Tr(\gamma_j\gamma_j\gamma_r\gamma_a\gamma_k\gamma_b\gamma_k\gamma_l)$$

$$= -2Tr(\gamma_r\gamma_a\gamma_k\gamma_b\gamma_k\gamma_l) = -2\,Tr[\gamma_r\gamma_a\gamma_k(2\delta_{bk} - \gamma_k\gamma_b)\gamma_l]$$

$$= -4\,Tr(\gamma_r\gamma_a\gamma_b\gamma_l) + 2\,Tr(\gamma_r\gamma_a\gamma_k\gamma_b\gamma_k\gamma_l)$$

$$= 4\,Tr(\gamma_r\gamma_a\gamma_b\gamma_l) = 16\,(\delta_{ar}\delta_{bl} + \delta_{ab}\delta_{lr} - \delta_{al}\delta_{br})\ ;$$

$$Tr(\gamma_j\gamma_a\gamma_k\gamma_b\gamma_k\gamma_j) = Tr(\gamma_j\gamma_j\gamma_a\gamma_k\gamma_b\gamma_k) = 4\,Tr(\gamma_a\gamma_k\gamma_b\gamma_k)$$

$$= 16(\delta_{ak}\delta_{bk} + \delta_{ak}\delta_{bk} - \delta_{ab}\delta_{kk} = -32\delta_{ab}\ ;$$

$$Tr(\gamma_j\gamma_a\gamma_k\gamma_k\gamma_b\gamma_j) = 4\,Tr(\gamma_j\gamma_j\gamma_a\gamma_b) = 16\,Tr(\gamma_a\gamma_b) = 64\delta_{ab}\ ;$$

$$Tr(\gamma_j\gamma_a\gamma_k\gamma_k\gamma_j\gamma_b) = 4\,Tr(\gamma_j\gamma_a\gamma_j\gamma_b) = 4\,Tr(\gamma_a\gamma_j\gamma_b\gamma_j)$$

$$= 4\,Tr(\gamma_a\gamma_k\gamma_b\gamma_k) = -32\delta_{ab}\ ;$$

$$Tr(\gamma_j\gamma_k\gamma_a\gamma_k\gamma_b\gamma_j) = Tr(\gamma_j\gamma_j\gamma_k\gamma_a\gamma_k\gamma_b) = 4\,Tr(\gamma_k\gamma_a\gamma_k\gamma_b)$$

$$= 4\,Tr(\gamma_a\gamma_k\gamma_b\gamma_k) = -32\delta_{ab}\ ;$$

$$Tr(\gamma_j\gamma_k\gamma_a\gamma_k\gamma_j\gamma_b) = Tr[\gamma_j\gamma_k(2\delta_{ka} - \gamma_k\gamma_a)\gamma_j\gamma_b]$$

$$= 2\,Tr(\gamma_j\gamma_a\gamma_j\gamma_b) - Tr(\gamma_j\gamma_k\gamma_k\gamma_a\gamma_j\gamma_b)$$

$$= -2\,Tr(\gamma_j\gamma_a\gamma_j\gamma_b) = 16\delta_{ab}\ ;$$

$$Tr(\gamma_j\gamma_k\gamma_k\gamma_a\gamma_j\gamma_b) = 4\,Tr(\gamma_j\gamma_a\gamma_j\gamma_b) = -32\,\delta_{ab}\ ;$$

$$Tr(\gamma_j\gamma_k\gamma_k\gamma_j) = 4\,Tr(\gamma_j\gamma)j) = 16\,Tr(I) = 64\ .$$

Relation (7.6.24) then writes

$$A' = 32(f_2p)(f_2q) - 16\,f_2^2(pq) - 64\,m^2(f_2p)$$

$$+64\,m^4(f_2q) - 16\,m^2(pq) - 64\,m^4. \tag{7.6.25}$$

Making the change $f_2 \to f_1$, one obtains the trace B'

$$B' = 32(f_1p)(f_1q) - 16\,f_1^2(pq) - 64\,m^2(f_1p)$$

$$+64\,m^4(f_1q) - 16\,m^2(pq) - 64\,m^4. \tag{7.6.26}$$

A similar procedure leads to

$$C' = \sum_{i=1}^{8} C_i'\,, \tag{7.6.27}$$

226

where

$$C'_1 = f_{2a}p_b f_{1l}q_r \, Tr(\gamma_j\gamma_a\gamma_k\gamma_b\gamma_j\gamma_l\gamma_k\gamma_r) \; ;$$
$$C'_2 = m^2 f_{2a}p_b \, Tr(\gamma_j\gamma_a\gamma_k\gamma_b\gamma_j\gamma_k) \; ;$$
$$C'_3 = m^2 f_{2a}f_{1b} \, Tr(\gamma_j\gamma_a\gamma_k\gamma_j\gamma_b\gamma_k) \; ;$$
$$C'_4 = -m^2 f_{2a}q_b \, Tr(\gamma_j\gamma_a\gamma_k\gamma_j\gamma_k\gamma_b) \; ;$$
$$C'_5 = m^2 p_a f_{1b} \, Tr(\gamma_j\gamma_k\gamma_a\gamma_j\gamma_b\gamma_k) \; ;$$
$$C'_6 = -m^2 p_a q_b \, Tr(\gamma_j\gamma_k\gamma_a\gamma_j\gamma_k\gamma_b) \; ;$$
$$C'_7 = -m^2 f_{1a}q_b \, Tr(\gamma_j\gamma_k\gamma_j\gamma_a\gamma_k\gamma_b) \; ;$$
$$C'_8 = -m^4 \, Tr(\gamma_j\gamma_k\gamma_j\gamma_k) \; .$$

These traces have been calculated for the Möller scattering, and were found to be

$$\begin{cases} Tr(\gamma_j\gamma_a\gamma_k\gamma_b\gamma_j\gamma_l\gamma_k\gamma_r) = -32 \, \delta_{al}\delta_{br} \; ; \\ Tr(\gamma_j\gamma_a\gamma_k\gamma_b\gamma_j\gamma_k) = \ldots = Tr(\gamma_j\gamma_k\gamma_j\gamma_a\gamma_k\gamma_b) = 16 \, \delta_{ab} \; ; \\ Tr(\gamma_j\gamma_k\gamma_j\gamma_k) = -32 \; , \end{cases}$$

so that (7.6.27) yields

$$C' = -32(f_1 f_2)(pq) + 16 \, m^2 (f_1 p) - 16 \, m^2 (f_1 q) + 16 \, m^2 (f_2 p)$$

$$-16 \, m^2 (f_2 q) + 16 \, m^2 (f_1 f_2) - 16 \, m^2 (pq) + 32 \, m^4 = D', \qquad (7.6.28)$$

as easily proven.

On the other hand, (7.6.7) and (7.6.23) lead to:

$$\begin{cases} (f_1 p) = p^2 - pk_1 = -(\chi_2 + m^2) \; ; \quad (f_1 q) = qk_2 - q^2 = \chi_2 + m^2 \; ; \\ f_1^2 = p^2 - 2pk_1 = -(2\chi_2 + m^2) \; ; \\ (pq) = p(k_1 + k_2 - p) = \chi_1 + \chi_2 + m^2; \\ (f_1 f_2) = (p - k_1)(k_1 - q) = pk_1 + qk_1 - pq = -m^2 \; ; \\ (f_2 p) = p^2 - pk_2 = -(\chi_1 + m^2) \; ; \quad (f_2 q) = qk_1 - q^2 = \chi_1 + m^2 \; ; \\ f_2^2 = p^2 - 2pk_2 = -(2\chi_1 + m^2) \; . \end{cases}$$

$$(7.6.29)$$

Making allowance for (7.6.29), the traces (7.6.25), (7.6.26), and (7.6.28) become

$$\begin{cases} A' = 32(\chi_1\chi_2 - m^2\chi_1 - m^4) \; ; \\ B' = 32(\chi_1\chi_2 - m^2\chi_2 - m^4) \; ; \\ C' = D' = -16 \, m^2 (\chi_1 + \chi_2 + 2m^2) \; . \end{cases} \qquad (7.6.30)$$

Substituting (7.6.30) into (7.6.21), and the result into (7.6.16), we get

$$\left| M^{(2)} \right|^2 = \frac{e^4}{8(2\pi)^4 E_+ E_- \omega_1 \omega_2} \, Y, \qquad (7.6.31)$$

227

with

$$Y = \frac{\chi_1}{\chi_2} + \frac{\chi_2}{\chi_1} - 2m^2\left(\frac{1}{\chi_1} + \frac{1}{\chi_2}\right) - m^4\left(\frac{1}{\chi_1} + \frac{1}{\chi_2}\right)^2. \qquad (7.6.32)$$

To calculate the differential cross section, we shall use (6.7.22) and (6.7.24). In our case, p'_1 and E'_1 characterize the emergent electron, while v_1 is replaced by v_r, the relative velocity of the two photons. Since we are in the photon CMF, where velocities are opposite in direction but equal in magnitude, we have

$$|\vec{p}'_1| = |\vec{p}| = \beta_- E_- ; \quad E'_1 = E_- ; \quad v_r = 2c = 2 ;$$

$$f(E_-) = E_+ + E_- - \omega_1 - \omega_2 ; \quad f'(E'_1) = f'(E_-) ; \qquad (7.6.33)$$

$$|\vec{k}_1| = \omega_1 ; \quad |\vec{k}_2| = \omega_2.$$

To find the derivative of $f(E_-)$ with respect to E_-, we have to take into account that E_+ depends on E_-. In this respect, it is useful to define the four-vector $P = (\vec{P}, iE)$, associated to the centre-of-mass of the system, and assume that the polar axis of the coordinate system is oriented along $\vec{P}$. Then

$$\vec{P} \equiv \vec{k}_1 + \vec{k}_2 = \vec{p} + \vec{q} ;$$
$$E \equiv \omega_1 + \omega_2 = E_+ + E_- . \qquad (7.6.34)$$

By means of $(7.6.34)_1$, the momentum conservation law then writes

$$\vec{q} = \vec{P} - \vec{p}.$$

Using the energy-momentum relation and $(7.6.33)_1$, we also have

$$E_+^2 = \vec{q}^2 + m^2 = (\vec{P} - \vec{p})^2 + m^2 = \vec{P}^2 - 2\vec{p}\cdot\vec{P} + \vec{p}^2 + m^2$$

$$= \vec{P}^2 - 2|\vec{p}||\vec{P}|\cos(\vec{p}, \vec{P}) + E_-^2 = E_-^2 - 2\beta_- E_-|\vec{P}|\cos(\vec{p}, \vec{P}) + \vec{P}^2,$$

or

$$E_+ = \left[E_-^2 - 2\beta_- E_-|\vec{P}|\cos(\vec{p}, \vec{P}) + \vec{P}^2\right]^{1/2}. \qquad (7.6.35)$$

In view of $(7.6.33)_5$, $(7.6.34)_2$, and $(7.6.35)$, one can write

$$f'(E_-) = 1 + \frac{dE_+}{dE_-} = 1 + \frac{E_- - \beta_-|\vec{P}|\cos(\vec{p}, \vec{P})}{E_+} = \frac{E - \beta_-|\vec{P}|\cos\theta_-}{E_+}.$$

$$\qquad (7.6.36)$$

Photon-photon scattering with electron-positron pair production

The differential cross section (6.7.24), in view of (7.6.31), (7.6.33), and (7.6.36) then becomes

$$d\sigma = \frac{1}{4} \frac{m^2 r_0^2}{\omega_1 \omega_2} \frac{\beta_- E_- Y}{E - \beta_- |\vec{P}| \cos \theta_-} d\Omega_-, \qquad (7.6.37)$$

where $r_0 = \frac{e^2}{4\pi m}$ is the classical electron radius.

Our next objective is to particularize formula (7.6.37) for the centre-of-mass frame (CMF). In this frame, the following relations are valid:

$$\begin{cases} \vec{P} = \vec{k}_1 + \vec{k}_2 = \vec{p} + \vec{q} = 0 \ ; \\ \omega_1 = \omega_2 = E_+ = E_- = \omega \ ; \\ \beta = \frac{|\vec{p}|}{\omega} = \frac{\sqrt{\omega^2 - m^2}}{\omega} \ , \end{cases} \qquad (7.6.38)$$

yielding

$$d\sigma = \frac{1}{8} r_0^2 \left(\frac{m}{\omega}\right)^2 \beta \, Y \, d\Omega, \qquad (7.6.39)$$

where the elementary solid angle $d\Omega$ stands for any of the two emergent fermions

$$d\Omega = 2\pi \, \sin \theta \, d\theta. \qquad (7.6.40)$$

Here θ is the scattering angle, as shown in Fig.7.8.

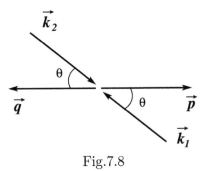

Fig.7.8

On the other hand, in view of (7.6.7), (7.6.33)$_6$, (7.6.33)$_7$, (7.6.38)$_2$, and (7.6.38)$_3$, one can write

$$\begin{cases} \chi_1 = |\vec{p}||\vec{k}_2| \cos \theta - E_- \omega_2 = -\omega^2(1 - \beta \cos \theta) \ ; \\ \chi_2 = |\vec{p}||\vec{k}_1| \cos(\pi - \theta) - E_- \omega_1 = -\omega^2(1 + \beta \cos \theta), \end{cases} \qquad (7.6.41)$$

and (7.6.32) becomes

$$Y = \frac{1 + \beta \cos \theta}{1 - \beta \cos \theta} + \frac{1 - \beta \cos \theta}{1 + \beta \cos \theta} + 2 \left(\frac{m}{\omega}\right)^2 \left(\frac{1}{1 - \beta \cos \theta} + \frac{1}{1 + \beta \cos \theta}\right)$$

229

Non-divergent second-order processes

$$-\left(\frac{m}{\omega}\right)^4 \left(\frac{1}{1-\beta\cos\theta} + \frac{1}{1+\beta\cos\theta}\right)^2,\qquad(7.6.42)$$

or, after some elementary calculations

$$Y = 2\,\frac{1-\beta^4\cos^4\theta + 2\left(\frac{m}{\omega}\right)^2\beta^2\sin^2\theta}{(1-\beta^2\cos^2\theta)^2}.\qquad(7.6.43)$$

In CMF, the differential cross section (7.6.39) associated with the production of a pair-electron-positron out of two non-polarized photons, is therefore

$$d\sigma = \frac{1}{2}\,\pi r_0^2\beta\left(\frac{m}{\omega}\right)^2\frac{1-\beta^2\cos^4\theta + 2(m/\omega)^2\beta^2\sin^2\theta}{(1-\beta^2\cos^2\theta)^2}\,\sin\theta\,d\theta.$$
$$(7.6.44)$$

As one can see, $d\sigma$ is symmetric about $\theta = \pi/2$. In the non-relativistic limit, $\beta \ll 1$ and $d\sigma$ is defined only above the threshold value $\omega = m$. In this case, if we neglect the terms proportional to β^2 and β^4 and use (7.6.38)$_3$, we are left with

$$d\sigma = \frac{1}{2}\,\pi r_0^2\sqrt{1-\left(\frac{m}{\omega}\right)^2}\,\sin\theta\,d\theta.\qquad(7.6.45)$$

At high energies (the ultra-relativistic limit) the electrons and positrons are produced under very small angles with respect to the direction of incidence. In this case $\beta \to 1$, and neglecting $(m/\omega)^2$ in the numerator of (7.6.44), we have

$$d\sigma = \frac{1}{2}\,\pi r_0^2\left(\frac{m}{\omega}\right)^2\frac{1+\cos^2\theta}{\sin^2\theta}\,\sin\theta\,d\theta.\qquad(7.6.46)$$

When $\theta \sim (0,\pi)$, we have $\sin^2\theta \sim \theta^2 \sim (m/\omega)^2$, $\cos^2\theta \sim 1$, and (7.6.46) becomes

$$d\sigma = \pi r_0^2\,\sin\theta\,d\theta.\qquad(7.6.47)$$

The total scattering cross section for non-polarized photons is obtained by integrating (7.6.44). To this end we make the substitution $x = 1-\beta\cos\theta$. Since $2-x = 1+\beta\cos\theta$, we get $dx = \beta\sin\theta\,d\theta$, and, as θ varies from 0 to π, x takes values from $1-\beta$ to $1+\beta$. Taking into account (7.6.38)$_3$ and (7.6.42), we finally have

$$\sigma = \frac{1}{2}\,\pi r_0^2(1-\beta^2)\,Z,\qquad(7.6.48)$$

230

Photon-photon scattering with electron-positron pair production

where
$$Z = \frac{1}{2} \int_{1-\beta}^{1+\beta} Y \, dx, \qquad (7.6.49)$$

with

$$\frac{1}{2} Y = \frac{1}{2} \left(\frac{2-x}{x} + \frac{x}{2-x} \right) + \left(\frac{m}{\omega} \right)^2 \left(\frac{1}{x} + \frac{1}{2-x} \right) - \frac{1}{2} \left(\frac{m}{\omega} \right)^4 \left(\frac{1}{x} + \frac{1}{2-x} \right)^2$$

$$= \frac{1}{x} + \frac{1}{2-x} - 1 + \left(\frac{m}{\omega} \right)^2 \left(\frac{1}{x} + \frac{1}{2-x} \right)$$

$$- \frac{1}{2} \left(\frac{m}{\omega} \right)^4 \left[\frac{1}{x^2} + \frac{1}{(2-x)^2} + \frac{1}{x} + \frac{1}{2-x} \right]$$

$$= -1 + \left[1 + \left(\frac{m}{\omega} \right)^2 - \frac{1}{2} \left(\frac{m}{\omega} \right)^4 \right] \left(\frac{1}{x} + \frac{1}{2-x} \right) - \frac{1}{2} \left(\frac{m}{\omega} \right)^4 \left[\frac{1}{x^2} + \frac{1}{(2-x)^2} \right].$$

Using $(7.6.38)_3$ as well as the last relation, then integrating (7.6.49) (the calculations are elementary), the total cross section (7.6.48) associated with the formation of an electron-positron pair out of two non-polarized photons writes

$$\sigma = \frac{1}{2} \pi r_0^2 (1 - \beta^2) \left[(3 - \beta^4) \ln \frac{1+\beta}{1-\beta} - 2\beta(2 - \beta^2) \right]. \qquad (7.6.50)$$

This result was first obtained by *Breit and Wheeler*[1].

Let us again consider the usual extreme cases.

i) *Non-relativistic approximation.* In this situation $\beta << 1$, and $\gamma \simeq 1$. Then

$$\ln \frac{1+\beta}{1-\beta} = \ln \left(\gamma + \sqrt{\gamma^2 - 1} \right)^2 = 2 \ln[\gamma(1+\beta)] = 2[\ln \gamma + \ln(1+\beta)] \simeq 2\beta.$$

Neglecting the terms proportional to β^2 and β^4 in (7.6.50), we obtain

$$\sigma = \pi r_0^2 \beta. \qquad (7.6.51)$$

ii) *Ultra-relativistic approximation.* This time we have $\beta \to 1$, and γ becomes very large. Recalling that $\gamma = 1/\sqrt{1 - \beta^2}$ and using $(7.6.38)_3$, we can write

$$\ln \frac{1+\beta}{1-\beta} = 2[\ln \gamma + \ln(1 + \beta)] \simeq 2 \ln(2\gamma) = 2 \ln 2 \left(\frac{\omega}{m} \right),$$

---

[1] G.Breit and J.A.Wheeler, Phys.Rev.**46**, 1934, pag.1087

and (7.6.50) becomes

$$\sigma = \pi\, r_0^2 \left(\frac{m}{\omega}\right)^2 \left(2\,\ln \frac{2\omega}{m} - 1\right). \qquad (7.6.52)$$

Note that, in this case, the total cross section increases slowly from $\omega = m$, attains a maximum, and then decreases again. Even if σ is of the same order of magnitude as some other cross sections, the photon energies leading to electron-positron pair production are much too small to make this phenomenon observable.

7.7. Electron-positron scattering

In quantum electrodynamics, the electron-positron scattering process is called *Bhabba scattering*, after the Indian physicist *Homi J. Bhabba*. By its nature, this process is very similar to Möller scattering. Let p, q and p', q' be the four-momenta of the electron and positron, in the initial and final states, respectively. There are two leading-order Feynman diagrams contributing to this interaction: an annihilation process and a scattering process, as shown in Fig.7.9.

A comparison between these diagrams and those given in Fig.7.3 and Fig.7.4b show the essential difference between the electron-positron scattering and the electron-electron scattering. While the first diagrams are equivalent, the diagram in Fig.7.4b appears due to the Pauli exclusion principle (which does not affect Bhabba scattering). On the other hand, Bhabba scattering implies a virtual annihilation followed by the creation of an electron-positron pair, and this effect has no analogue in the electron-electron interaction.

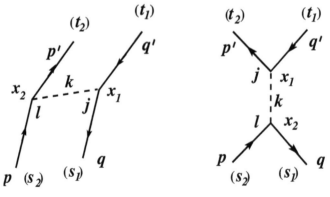

Fig.7.9

The kinematic equivalence between the two processes leads to the same general expression for the scattering cross section. However, there is a deeper topological connection of the associated diagrams, and the particle interchange and annihilation processes appear as two different consequences of the same mathematical structure. As we shall see, the matrix element associated with Bhabba scattering follows from the one associated with Möller scattering by applying the substitution law, the substitutions being valid for both diagrams. But, while in the first case the substitutions are trivial, in the second case they could be surprising, because they establish a connection between the symmetry properties demanded by the exclusion principle, on one hand, and the electron-positron creation-annihilation process, on the other.

To write the S-matrix element corresponding to the first graph, we use again the GRAPH TABLE and follow the already known procedure. So, we have:

- electron incident at $x_2 \longrightarrow (2\pi)^{-3/2} v_+^{(s_2)}(\vec{p})$
- vertex $x_2 \longrightarrow (2\pi)^4 e \gamma_l \delta(p' - p + k)$
- electron emergent from $x_2 \longrightarrow (2\pi)^{-3/2} \overline{v}_-^{(t_2)}(\vec{p}')$
- photonic propagator $(x_2 \to x_1) \longrightarrow -(2\pi)^{-4} \int \frac{\delta_{jl}}{k^2 - i\alpha} \, dk$
- positron incident at $x_1 \longrightarrow (2\pi)^{-3/2} \overline{v}_+^{(s_1)}(\vec{q})$
- vertex $x_1 \longrightarrow (2\pi)^4 e \gamma_j \delta(q' - q - k)$
- positron emergent from $x_1 \longrightarrow (2\pi)^{-3/2} v_-^{(t_1)}(\vec{q}')$

Let us take the product of these factors, placing the creation operators on the l.h.s. and the annihilation operators on the r.h.s. Taking into account the multiplication factor $1/2$ appearing in the $S^{(n)}$ expansion, we can write

$$S_I^{(2)} = -\frac{e^2}{2(2\pi)^2} \int \overline{v}_+^{(s_1)}(\vec{q}) \, \gamma_j \, v_-^{(t_1)}(\vec{q}') \, \delta(q' - q - k) \, \delta_{jl}$$

$$\times \frac{1}{k^2} \overline{v}_-^{(t_2)}(\vec{p}') \, \gamma_l \, v_+^{(s_2)}(\vec{p}) \, \delta(p' - p + k) \, dk. \qquad (7.7.1)$$

The elements of the second graph are
- electron incident at $x_2 \longrightarrow (2\pi)^{-3/2} v_+^{(s_2)}(\vec{p})$
- vertex $x_2 \longrightarrow (2\pi)^4 e \gamma_l \delta(k - p - q)$
- positron incident at $x_2 \longrightarrow (2\pi)^{-3/2} \overline{v}_+^{(s_1)}(\vec{q})$
- photonic propagator $(x_2 \to x_1) \longrightarrow -(2\pi)^{-4} \int \frac{\delta_{jl}}{k^2 - i\alpha} \, dk$
- electron emergent from $x_1 \longrightarrow (2\pi)^{-3/2} \overline{v}_-^{(t_2)}(\vec{p}')$
- vertex $x_1 \longrightarrow (2\pi)^4 e \gamma_j \delta(p' + q' - k)$
- positron emergent from $x_1 \longrightarrow (2\pi)^{-3/2} v_-^{(t_1)}(\vec{q}')$

which gives

$$S_{II}^{(2)} = -\frac{e^2}{2(2\pi)^2} \int \overline{v}_-^{(t_2)}(\vec{p}') \gamma_j v_-^{(t_1)}(\vec{q}') \delta(p' + q' - k) \delta_{jl}$$

$$\times \frac{1}{k^2} \overline{v}_+^{(s_1)}(\vec{q}) \gamma_l v_+^{(s_2)}(\vec{p}) \delta(k - p - q) \, dk. \tag{7.7.2}$$

Since in the case of fermions the state vectors are antisymmetric, the total matrix element is equal to the difference between (7.7.1) and (7.7.2). Performing the calculations, one obtains

$$S^{(2)} = 2(S_I^{(2)} - S_{II}^{(2)}), \tag{7.7.3}$$

or

$$S^{(2)} = M^{(2)} \delta(p' + q' - p - q), \tag{7.7.4}$$

where

$$M^{(2)} = -\frac{i e^2}{(2\pi)^2} \left\{ \frac{1}{(q' - q)^2} \left[\overline{v}_+^{(s_1)}(\vec{q}) \gamma_j v_-^{(t_1)}(\vec{q}')\right] \left[\overline{v}_-^{(t_2)}(\vec{p}') \gamma_j v_+^{(s_2)}(\vec{p})\right] \right.$$

$$\left. -\frac{1}{(p' + q')^2} \left[\overline{v}_-^{(t_2)}(\vec{p}') \gamma_j v_-^{(t_1)}(\vec{q}')\right] \left[\overline{v}_+^{(s_1)}(\vec{q}) \gamma_j v_+^{(s_2)}(\vec{p})\right] \right\}. \tag{7.7.5}$$

If the propagators are put in a different form, we also can write:

$$M^{(2)} = -\frac{i e^2}{(2\pi)^2} \left\{ \frac{1}{(p - p')^2} \left[\overline{v}_+^{(s_1)}(\vec{q}) \gamma_j v_-^{(t_1)}(\vec{q}')\right] \left[\overline{v}_-^{(t_2)}(\vec{p}') \gamma_j v_+^{(s_2)}(\vec{p})\right] \right.$$

$$\left. -\frac{1}{(p + q)^2} \left[\overline{v}_-^{(t_2)}(\vec{p}') \gamma_j v_-^{(t_1)}(\vec{q}')\right] \left[\overline{v}_+^{(s_1)}(\vec{q}) \gamma_j v_+^{(s_2)}(\vec{p})\right] \right\}. \tag{7.7.6}$$

Comparing (7.7.5) and (7.7.6) with (7.4.6) and (7.4.7), one observes that the Möller scattering matrix element becomes the Bhabba matrix element following the substitution

$$p_1 \to -q' \; ; \quad p_1' \to -q \; ; \quad p_2 \to p \; ; \quad p_2' \to p' \; ;$$
$$\overline{v}_-^{(t_1)}(\vec{p}_1') \to \overline{v}_+^{(s_1)}(\vec{q}) \; ; \quad v_+^{(s_1)}(\vec{p}_1) \to v_-^{(t_1)}(\vec{q}'). \tag{7.7.7}$$

Therefore, we do not have to calculate any new traces, only substitute (7.7.7) into the results for Möller scattering. We also note that (6.4.34), (7.4.35), and (7.4.36) become

$$\begin{aligned} -m^2\kappa &= -p q' = -p' q \; ; \\ -m^2\mu &= -p'q' = -pq \; ; \\ -m^2\lambda &= q q' = p p' \; , \end{aligned} \tag{7.7.8}$$

while relations (7.4.37) remain unchanged. With all these observations, (7.4.39) takes the form

$$X = \frac{1}{64}\left[\frac{A}{(q'-q)^4} + \frac{B}{(p'+q')^4} - \frac{C+D}{(q'-q)^2(p'+q')^2}\right]$$

$$= \frac{m^4}{2}\left[\frac{\kappa^2 + \mu^2 - 2\lambda + 2}{(q'-q)^4} + \frac{\kappa^2 + \lambda^2 - 2\mu + 2}{(p'+q')^4} + 2\frac{\kappa^2 - \kappa - \mu - \lambda + 1}{(q'-q)^2(p'+q')^2}\right],$$

$$(7.7.9)$$

and (7.4.38) becomes

$$\left|M^{(2)}\right|^2 = \frac{e^4}{(2\pi)^4\, E_+ E_- E'_+ E'_-}\, X, \qquad .(7.7.10)$$

where $E_\pm$ and $E'_\pm$ are the electron (-) and positron (+) energies, corresponding to the initial and final states, respectively.

To calculate the differential cross section, we use the expression (6.7.24). In our case, $\vec{p}'_1$, E'_1, and $d\Omega'$ characterize the emergent electron, v_1 is replaced by v_r, the relative velocity of the positron with respect to the electron, while $f(E'_1)$ is given by (6.7.22), that is:

$$\begin{cases} |\vec{p}'_1| \equiv |\vec{p}'| = \beta'_- E'_- \;; \quad E'_1 \equiv E'_- = m\gamma'_- \;; \quad \gamma'_- = \frac{1}{\sqrt{1-\beta_-^2}} \;; \\ f(E'_-) = E'_+ + E'_- - E_+ - E_- \;; \quad f'(E'_1) = f'(E'_-);\, . \end{cases}$$

$$(7.7.11)$$

Define now the four-vector $P \equiv (\vec{P},\ iE)$ which describes the motion of the centre-of-mass of the system

$$\vec{P} = \vec{p} + \vec{q}\;; \qquad E = E_+ + E_-, \qquad (7.7.12)$$

and assume that the polar axis of the reference frame is oriented along $\vec{P}$. Following the same procedure as in Section 7.5, and using (7.5.6), we have:

$$f'(E'_-) = 1 + \frac{dE'_+}{dE'_-} = 1 + \frac{E'_- - \beta'_-|\vec{P}|\cos(\vec{p}',\vec{P})}{E'_+} = \frac{E - \beta'_-|\vec{P}|\cos\theta'_-}{E'_+}.$$

$$(7.7.13)$$

Using (7.7.10), (7.7.11), and (7.7.13), the differential cross section (6.7.24) writes

$$d\sigma = \frac{4\,r_0^2 m^2 \beta'_- E'_-\, X\, d\Omega'_-}{v_r E_+ E_- (E - \beta'_-|\vec{P}|\cos\theta'_-)}. \qquad (7.7.14)$$

Let us now move to the centre-of-mass frame (CMF), characterized by

$$\vec{p} = -\vec{q} ; \quad \vec{p}\,' = -\vec{q}\,' ; \quad |\vec{p}| = |\vec{p}\,'| ;$$

$$E_+ = E_- = E'_+ = E'_- = \epsilon ; \tag{7.7.15}$$

$$|\vec{p}| = \beta\epsilon ; \quad \epsilon = \gamma\, m ; \quad \gamma = \frac{1}{\sqrt{1 - \beta^2}}.$$

Making use of (7.7.15) and Fig.7.10, we also have

$$\vec{p} \cdot \vec{q}\,' = -\vec{p} \cdot \vec{p}\,' = -|\vec{p}||\vec{p}\,'| \cos\theta = -\beta^2\epsilon^2 \cos\theta$$

$$= -\beta^2\gamma^2\, m^2 \cos\theta = -(\gamma^2 - 1)m^2 \cos\theta. \tag{7.7.16}$$

Since we are working in CMF, the electron and positron velocities are equal and opposite, therefore $v_r = 2\beta$. In view of (7.7.12) and (7.7.15), the differential cross section (7.7.14) becomes

$$d\sigma = \frac{r_0^2\, X}{\gamma^2}\, d\Omega, \tag{7.7.17}$$

with X given by (7.7.9).

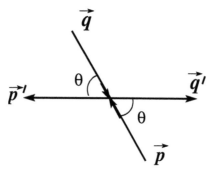

Fig.7.10

Let us find the expression for κ, μ, and λ in the CMF. Using (7.7.8), (7.7.15), and (7.7.16), one obtains:

$$\kappa = \frac{1}{m^2}(\vec{p} \cdot \vec{q}\,' - E_- E'_+) = -\frac{1}{m^2}[(\gamma^2 - 1)m^2 \cos\theta + \epsilon^2]$$

$$= -\gamma^2 - (\gamma^2 - 1)\cos\theta = -\gamma^2(1 + \beta^2 \cos\theta) ; \tag{7.7.18}$$

$$\mu = \frac{1}{m^2}(\vec{p} \cdot \vec{q} - E_- E_+) = -\frac{1}{m^2}(\vec{p}^{\,2} + \epsilon^2)$$

$$= -\gamma^2(1+\beta^2) = -(2\gamma^2 - 1) \; ; \tag{7.7.19}$$

$$\lambda = -\frac{1}{m^2}(\vec{p}\cdot\vec{p}' - E_-E'_-) = \frac{1}{m^2}(\vec{p}\cdot\vec{q}' + \epsilon^2)$$

$$= \frac{1}{m^2}[-(\gamma^2-1)m^2\cos\theta + \epsilon^2] = \gamma^2 - (\gamma^2-1)\cos\theta = \gamma^2(1-\beta^2\cos\theta). \tag{7.7.20}$$

Since

$$(q'-q)^2 = q'^2 + q^2 - 2qq' = -2m^2 - 2(\vec{q}\cdot\vec{q}' - E_+E'_+) = 2m^2 + 2(\vec{p}\cdot\vec{q}' + \epsilon^2)$$

$$= -2m^2 - 2[(\gamma^2-1)\,m^2\cos^2\theta - m^2\gamma^2]$$

$$= 2m^2(\gamma^2-1)(1-\cos\theta) = m^2(2\gamma\beta)^2\sin^2\frac{\theta}{2} \; ; \tag{7.7.21}$$

$$(p'+q')^2 = p'^2 + q'^2 + 2p'q' = -2m^2 + 2(\vec{p}'\cdot\vec{q}' - E'_-E'_+) =$$

$$-2m^2 - 2(p'^2 + \epsilon^2) = -2m^2 - 2\epsilon^2(1+\beta^2)$$

$$= -2m^2[1 + \gamma^2(1+\beta^2)] = -m^2(2\gamma)^2, \tag{7.7.22}$$

substituting the last two relations into (7.7.9), and the result into (7.7.17), we are left with the following result for the differential cross section in the CMF:

$$d\sigma = \frac{r_0^2\,d\Omega}{64\,m^4\gamma^2(2\gamma)^4}\left\{\frac{A}{[\beta\sin(\theta/2)]^4} + B + \frac{2C}{[\beta\sin(\theta/2)]^2}\right\}. \tag{7.7.23}$$

Comparing this result with (7.5.15) and (7.5.20) we come to the conclusion that Möller scattering and Bhabba scattering are similar processes. It can be easily shown that the trace A is the same in both cases, which means that the first two terms of the differential cross sections, respectively, are identical. The term B represents the exchange term in Möller scattering, and the annihilation term in Bhabba scattering. The terms containing C are interference terms, since both the exchange and annihilation phenomena contribute to the scattering amplitude more than do in the scattering probability.

One can obtain an alternative expression for the differential cross section (7.7.17) by substituting (7.7.18)-(7.7.22) into (7.7.9). Since $\gamma^2 = 1 + \gamma^2\beta^2$, we can write

$$d\sigma = \frac{1}{2}\frac{r_0^2\,d\Omega}{\gamma^2}\left\{\frac{1}{(2\beta\gamma)^4\sin^4\frac{\theta}{2}}\left[\gamma^4(1+\beta^2\cos\theta)^2\right.\right.$$

$$+(2\gamma^2-1)^2-2\gamma^2(1-\beta^2\cos\theta)+2\Big]$$

$$+\frac{1}{(2\gamma)^4}\Big[\gamma^4(1+\beta^2\cos\theta)^2+\gamma^4(1-\beta^2\cos\theta)^2+2(2\gamma^2-1)+2\Big]$$

$$-\frac{2}{(2\gamma)^2(2\beta\gamma)^2\sin^2\frac{\theta}{2}}\Big[\gamma^4(1+\beta^2\cos\theta)^2+\gamma^2(1+\beta^2\cos\theta)+(2\gamma^2-1)$$

$$-\gamma^2(1-\beta^2\cos\theta)+1\Big]\Big\}=\frac{r_0^2}{16\gamma^2}\Big[\frac{N}{(\beta\gamma\sin\frac{\theta}{2})^4}+\frac{P}{\gamma^4}-\frac{R}{\gamma^2(\beta\gamma\sin\frac{\theta}{2})^2}\Big].$$

Using the formula $1+\cos\theta=2\cos^2\frac{\theta}{2}$, we have:

$$N=\frac{1}{2}\Big\{(1+\beta^2\gamma^2)^2(1+\beta^2\cos\theta)^2+[2(1+\beta^2\gamma^2)-1]^2$$

$$-2(1+\beta^2\gamma^2)^2(1-\beta^2\cos\theta)+2\Big\}$$

$$=\frac{1}{2}[2+4\beta^2\gamma^2(1+\cos\theta)+4\beta^4\gamma^4+\beta^4\gamma^4(1+\cos\theta)^2]$$

$$=1+(2\beta\gamma\cos\frac{\theta}{2})^2+2(\beta\gamma)^4(1+\cos^4\frac{\theta}{2})\ ;$$

$$P=\frac{1}{2}\Big\{(1+\beta^2\gamma^2)^2(1+\beta^2\cos\theta)^2+(1+\beta^2\gamma^2)$$

$$\times(1-\beta^2\cos\theta)^2+2[2(1+\beta^2\gamma^2)-1]+2\Big\}$$

$$=3+4(\beta\gamma)^2+(\beta\gamma)^4(1+\cos^2\theta)\ ;$$

$$R=(1+\beta^2\gamma^2)^2(1+\beta^2\cos\theta)^2+(1+\beta^2\gamma^2)(1+\beta^2\cos\theta)+[2(1+\beta^2\gamma^2)-1]$$

$$-(1+\beta^2\gamma^2)(1-\beta^2\cos\theta)+1=3+4(\beta\gamma)^2(1+\cos\theta)+\beta^4\gamma^4(1+\cos\theta)^2$$

$$=3+2(2\beta\gamma\cos\frac{\theta}{2})^2+\frac{1}{4}(2\beta\gamma\cos\frac{\theta}{2})^4.$$

The differential cross section in CMF then writes:

$$d\sigma=\frac{r_0^2\,d\Omega}{16\gamma^2}\Big\{\frac{1}{[\beta\gamma\sin\frac{\theta}{2}]^4}\Big[1+(2\beta\gamma\cos\frac{\theta}{2})^2+2(\beta\gamma)^4(1+\cos^4\frac{\theta}{2})\Big]$$

$$+\frac{1}{\gamma^4}\Big[3+4(\beta\gamma)^2+(\beta\gamma)^4(1+\cos^2\theta)\Big]-\frac{1}{\gamma^2[\beta\gamma\sin\frac{\theta}{2}]^2}$$

$$\times \left[3 + 2\left(2\beta\gamma \cos\frac{\theta}{2}\right)^2 + \frac{1}{4}\left(2\beta\gamma \cos\frac{\theta}{2}\right)^4 \right] \right\}. \tag{7.7.24}$$

To write the differential cross section in LS, we use the already known transformation relations (see(7.5.28)-(7.5.30))

$$\gamma^2 = \frac{1}{2}(\gamma^* + 1) \ ;$$

$$\cos\theta = x = \frac{2 - (\gamma^* + 3)\sin^2\theta^*}{2 + (\gamma^* - 1)\sin^2\theta^*} \ ; \tag{7.7.25}$$

$$d\Omega = \frac{8\,\cos\theta^*(\gamma^* + 1)}{[2 + (\gamma^* - 1)sin^2\theta^*]^2}\,d\Omega^* \ ,$$

where variables marked with "*" are measured in LF.
These formulas have to be supplemented with

$$(\beta\gamma)^2 = \gamma^2 - 1 = \frac{1}{2}(\gamma^* + 1) \ ;$$

$$\cos^2\frac{\theta}{2} = \frac{1}{2}(1 + x) \ ; \tag{7.7.26}$$

$$\sin^2\frac{\theta}{2} = \frac{1}{2}(1 - x) \ ,$$

and

$$\frac{\gamma^*}{\gamma^* - 1} = \frac{\gamma^* + 1}{\beta^{*2}\gamma^*}. \tag{7.7.27}$$

Using (7.7.24), the differential cross section in LF then writes (the calculations are elementary, but tedious):

$$d\sigma^* = r_0^2 \left(2\frac{\gamma^* + 1}{\beta^{*2}\gamma^*}\right)^2 \frac{\cos\theta^*\,d\Omega^*}{[2 + (\gamma^* - 1)\sin^2\theta^*]^2}$$

$$\times \left\{ \frac{4}{(1-x)^2}\left[1 - \frac{\gamma^{*2} - 1}{4\gamma^{*2}}(1-x) + \frac{1}{2}\left(\frac{\gamma^* - 1}{2\gamma^*}\right)^2(1-x^2)\right] \tag{7.7.28}$$

$$+ \left(\frac{\gamma^* - 1}{\gamma^* + 1}\right)^2 \left[\frac{1}{2} + \frac{1}{\gamma^*} + \frac{3}{2\gamma^{*2}} - \left(\frac{\gamma^* - 1}{2\gamma^*}\right)^2(1-x^2)\right]$$

$$- \frac{2}{1-x}\left(\frac{\gamma^* - 1}{\gamma^* + 1}\right)\left[\frac{2\gamma^* + 1}{\gamma^{*2}} + \frac{\gamma^{*2} - 1}{\gamma^{*2}}x + \left(\frac{\gamma^{*2} - 1}{2\gamma^*}\right)^2(1-x)^2\right] \right\}.$$

The total scattering cross section is found by integrating (7.7.24), or (7.7.28), over all possible angles. But, unlike Möller scattering, the factor $1/2$ does not appear anymore, because the particles are discernible by their charge. As in the case of Möller scattering, the differential cross section becomes divergent for very small scattering angles, so that it cannot be fully integrated. However, since the particles are discernible, there is no divergence for $\theta = \pi/2$, unlike for Möller scattering.

Let us finally consider the non-relativistic limit of the angular distribution. As in the case of Möller scattering, the scattering angle in CMF is twice the angle in LS. In the non-relativistic approximation ($\beta \ll 1$, $\gamma \to 1$) the terms containing β^2 and β^4 in (7.7.24) can be neglected, and we are left with

$$d\sigma = \frac{r_0^2}{16\beta^4} \left(\frac{1}{\sin^4 \frac{\theta}{2}} - \frac{\beta^2}{sin^2 \frac{\theta}{2}} + 3\beta^4 \right) d\Omega.$$

If the last two terms are neglected, we arrive at

$$d\sigma = \frac{r_0^2}{16\,\beta^4\,\sin^4 \frac{\theta}{2}}, \tag{7.7.29}$$

which is precisely the first term in the differential cross section for Möller scattering.

It is worthwhile mentioning that all these theoretical considerations are in excellent agreement with the experimental data.

CHAPTER VIII
DIVERGENT SECOND-ORDER PROCESSES

The more success the quantum theory has, the sillier it looks.

Albert Einstein

The study of some second-order processes is very difficult, since they lead to divergent integrals. The divergences are of several kinds. In this chapter we shall focus on the so-called "ultraviolet divergences", which arrise at very high energies, due to the fact that the product of two integrable singular functions does not lead - in general - to an integrable singular function. In principle, these kinds of divergences can be eliminated by subtracting from the matrix elements some finite quantities, well-justified from the physical point of view. As we shall see, the physical interpretation of eliminating divergences leads to *mass and charge renormalization.*

8.1. Electron self-energy diagram

As a first example of a second-order divergent renormalizable process we consider the electron self-energy graph (see Section 6.5):

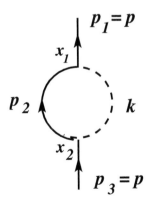

Divergent second-order processes

As we shall see, there exists a modification of the electron mass due to its "self-energy", which can be pictured as a continuous exchange of photons with itself (the Coulomb interaction between two electrons can be imagined as a continuous exchange of virtual photons). For this reason, the corresponding graph is called the *self-energy diagram*. This can also appear as an intermediate process in a higher order effect, such as:

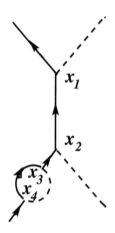

The momentum conservation law demands

$$p_1 = p ; \quad p_2 = p - k \text{ (the intermediate state) ;} \quad p_3 = p. \quad (8.1.1)$$

The segments of the graph, followed in chronological order, contribute to the S-matrix element as follows (see the **GRAPH TABLE**):

- electron incident at $x_2 \longrightarrow (2\pi)^{-3/2} v_+^{(s)}(\vec{p}_1)$
- vertex $x_2 \longrightarrow (2\pi)^4 e \gamma_j \delta(p_2 - p_1 + k)$
- fermionic propagator $\longrightarrow (2\pi)^{-4} \int \frac{i\hat{p}_2 - m}{p_2^2 + m^2 - i\alpha} dp_2$
- photonic propagator $\longrightarrow -(2\pi)^{-4} \int \frac{1}{k^2 - i\alpha} \delta_{jl} dk$
- vertex $x_1 \longrightarrow (2\pi)^4 e \gamma_l \delta(p_3 - p_2 - k)$
- electron emergent from $x_1 \longrightarrow (2\pi)^{-3/2} \overline{v}_-^{(t)}(\vec{p}_3)$

with $dp_2 = i \, d\vec{p}_2 \, dE_{p_2}; \quad dk = i \, d\vec{k} \, d\omega.$

Recalling that the creation operators always appear to the left of the annihilation operators, the corresponding matrix element is

$$S^{(2)} = -\frac{e^2}{(2\pi)^3} \overline{v}_-^{(t)}(\vec{p}_3) \left\{ \int dp_2 \, dk \, \gamma_l \frac{\delta_{jl}}{k^2 - i\alpha} \frac{i\hat{p}_2 - m}{p_2^2 + m^2 - i\alpha} \right.$$

$$\left. \times \gamma_j \, \delta(p_3 - p_2 - k) \, \delta(p_2 - p_1 + k) \right\} v_+^{(s)}(\vec{p}_1),$$

or, if we perform the integration with respect to p_2

$$S^{(2)} = -\frac{e^2}{(2\pi)^3 i}\, \bar{v}_-^{(t)}(\vec{p}_3)\Big\{\int d\vec{k}\, d\omega\, \frac{1}{k^2 - i\alpha}\, \gamma_l\, \frac{(\hat{p}_1 - \hat{k}) + im}{(p_1 - k)^2 + m^2 - i\alpha}$$

$$\times\, \gamma_l\, \delta(p_3 - p_1)\Big\} v_+^{(s)}(\vec{p}_1). \tag{8.1.2}$$

Denoting

$$\Sigma(p_1) = \frac{e^2}{(2\pi)^4 i}\int d\vec{k}\, d\omega\, \frac{1}{k^2 - i\alpha}\, \gamma_l\, \frac{(\hat{p}_1 - \hat{k}) + im}{(p_1 - k)^2 + m^2 - i\alpha}\, \gamma_l, \tag{8.1.3}$$

we can write

$$S^{(2)} = -\,(2\pi)\, \bar{v}_-^{(t)}(\vec{p}_3)\Sigma(p_1)\, v_+^{(s)}(\vec{p}_1)\, \delta(p_3 - p_1). \tag{8.1.4}$$

As can be seen, in the configuration representation (8.1.3) writes

$$\Sigma(x_1 - x_2) = e^2\, \gamma_l\, S^{(c)}(x_1 - x_2)\, \gamma_l\, D_0^{(c)}(x_1 - x_2), \tag{8.1.5}$$

where $S^{(c)}(x)$ and $D_0^{(c)}(x)$ are the electronic and photonic causal functions, respectively (see (A.129) and (A.131)). These functions are responsible for the divergence mentioned at the beginning of this chapter, due to the fact that the product of two singular integrable functions is sometimes meaningless. Indeed, for $|k| \to \infty$, the integrand in (8.1.3) behaves like $|k|^{-3}$, therefore the integral behaves like

$$\int d\vec{k}\, d\omega\, |k|^{-3} \sim \int_0^{|k|} |k|^3\, d|k|\, |k|^{-3} = \int_0^{|k|} d|k| = |k| \to \infty.$$

Consequently, the integral (8.1.3) is linearly divergent for large values of $|k|$, and the formula makes no sense.

A method of eliminating divergences of this nature was given by W.Pauli and F.Willars.[1] In order to isolate the divergent terms from the finite parts, these authors define the regularized causal functions

$$(2\pi)^2 i\, S_R^{(c)}(p) = (i\hat{p} - m)\Big\{\frac{1}{p^2 + m^2 - i\alpha} + \sum_j C_j \frac{1}{p^2 + M_j^2 - i\alpha}\Big\}; \tag{8.1.6}$$

---

[1] W.Pauli, F.Willars, Rev.Mod.Phys.,**21**, 1929, p.434.

$$(2\pi)^2 i\, D_{0R}^{(c)}(k) = \frac{1}{k^2 - i\alpha} + \sum_j C_j \frac{1}{k^2 + M_j^2 - i\alpha}, \qquad (8.1.7)$$

where the quantities M_j are called *auxiliary masses*, while the coefficients C_j are chosen in such a way so as to compensate the regularities of the functions $S^{(c)}(p)$ and $D_0^{(c)}(k)$.

As one observes, in the limit $M_j \to \infty$ the regularized causal functions (8.1.6) and (8.1.7) converge to the corresponding causal functions, that is

$$\lim_{M_j \to \infty} S_R^{(c)}(p) = S^{(c)}(p) ; \quad \lim_{M_j \to \infty} D_{0R}^{(c)}(k) = D_0^{(c)}(k). \qquad (8.1.8)$$

Relation (8.1.3) can also be written as

$$\Sigma(p) = e^2 \int d\vec{k}\, d\omega\, D_0^{(c)}(k)\, \gamma_l\, S^{(c)}(p - k)\, \gamma_l. \qquad (8.1.9)$$

In view of (8.1.6) and (8.1.7) let us define

$$\Sigma_R(p) = e^2 \int d\vec{k}\, d\omega\, D_{0R}^{(c)}(k)\, \gamma_l\, S_R^{(c)}(p - k)\, \gamma_l. \qquad (8.1.10)$$

Obviously, we also have

$$\lim_{M_j \to \infty} \Sigma_R(p) = \Sigma(p). \qquad (8.1.11)$$

In the following, we shall consider a single auxiliary mass M, and choose $C = -1$. In this case, (8.1.6) and (8.1.7) yield

$$(2\pi)^2 i\, S_R^{(c)}(p) = (i\hat{p} - m)\left\{ \frac{1}{p^2 + m^2 - i\alpha} - \frac{1}{p^2 + M^2 - i\alpha} \right\} ; \quad (8.1.12)$$

$$(2\pi)^2 i\, D_{0R}^{(c)}(k) = \frac{1}{k^2 - i\alpha} - \frac{1}{k^2 + M^2 - i\alpha}. \qquad (8.1.13)$$

To calculate $\Sigma_R(p)$ given by (8.1.10), with $D_{0R}^{(c)}(k)$ and $S_R^{(c)}(p)$ defined by (8.1.12) and (8.1.13), we first observe that

$$i \int_0^\infty e^{-i(x - i\alpha)\lambda}\, d\lambda = \frac{1}{x - i\alpha} ; \quad \alpha > 0,$$

which allows us to write

$$(2\pi)^2 S_R^{(c)}(p) = (i\hat{p} - m) \int_0^\infty \left[e^{-i(p^2 + m^2 - i\alpha)\lambda} - e^{-i(p^2 + M^2 - i\alpha)\lambda} \right] d\lambda$$

$$= (i\hat{p} - m) \int_0^\infty e^{-i(p^2 - i\alpha)\lambda} \left(e^{-im^2\lambda} - e^{-iM^2\lambda} \right) d\lambda \; ; \qquad (8.1.14)$$

and

$$(2\pi)^2 D_{0R}^{(c)}(k) = \int_0^\infty \left[e^{-i(k^2 - i\alpha)\mu} - e^{-i(k^2 + M^2 - i\alpha)\mu} \right] d\mu$$

$$= \int_0^\infty e^{-i(k^2 - i\alpha)\mu} \left(1 - e^{-iM^2\mu} \right) d\mu. \qquad (8.1.15)$$

Substituting (8.1.14) and (8.1.15) into (8.1.10), one obtains

$$\Sigma_R(p) = \frac{e^2}{(2\pi)^4} \int d\vec{k} \; d\omega \; d\lambda \; d\mu \; \gamma_l [i(\hat{p} - \hat{k}) - m] \gamma_l \, e^{-ik^2\mu}$$

$$\times e^{-(\lambda + \mu)\alpha} \left(1 - e^{-iM^2\mu} \right) e^{-i(p-k)^2\lambda} \left(e^{-im^2\lambda} - e^{-iM^2\lambda} \right). \qquad (8.1.16)$$

Denoting

$$I = \int dk \; \gamma_l(\hat{p} - \hat{k} + im) \gamma_l \, e^{-i[k^2(\lambda + \mu) - 2\lambda \, p \, k]}, \qquad (8.1.17)$$

we can write

$$\Sigma_R(p) = \frac{e^2}{(2\pi)^4} \int_0^\infty d\mu \int_0^\infty d\lambda \; I \; e^{-ip^2\lambda} \, e^{-(\lambda + \mu)\alpha}$$

$$\times \left(1 - e^{-iM^2\mu} \right) \left(e^{-im^2\lambda} - e^{-iM^2\lambda} \right). \qquad (8.1.18)$$

We shall adopt the following succession of steps: first determine I given by (8.1.17), then substitute the result into (8.1.18) and find $\Sigma_R(p)$.

Calculation of the integral I

Consider the Gauss-type integral

$$I_1 = \int_{-\infty}^{+\infty} e^{i(at^2 + bt) - \eta t^2} \, dt, \qquad (8.1.19)$$

where the factor $e^{-\eta t^2}$ insures the convergence of the integral. If t is a complex variable, then for any closed contour we have

$$\oint e^{i(at^2 + bt) - \eta t^2} \, dt = 0,$$

because the integrand is a holomorphic function. Therefore, we can write

$$\int_{-\infty}^{+\infty} e^{i(at^2+bt)-\eta t^2} \, dt = \int_{(d)} e^{i(at^2+bt)-\eta t^2} \, dt,$$

where (d) is a suitably chosen straight line in the complex plane t.

The change of variable

$$t = \frac{1-i}{\sqrt{2}} x - \frac{b}{2a} \; ; \quad dt = \frac{1-i}{\sqrt{2}} \, dx \qquad (8.1.20)$$

gives

$$i(at^2 + bt) = ax^2 - i\frac{b^2}{4a}. \qquad (8.1.21)$$

Using (8.1.21), one observes that the integral (8.1.19) converges even without the contribution of the factor $e^{-\eta t^2}$, meaning that we may choose $\eta = 0$. Then

$$I_1 = e^{-\frac{ib^2}{4a}} \frac{1-i}{\sqrt{2}} \int_{-\infty}^{+\infty} e^{ax^2} \, dx = \frac{1-i}{\sqrt{2}} \sqrt{\frac{\pi}{|a|}} e^{-\frac{ib^2}{4a}}.$$

Let us set now $t \equiv k_i$. In this case

$$I_i = \int_{-\infty}^{+\infty} e^{i(ak_i^2 + b_i k_i)} \, dk_i = \frac{1-i}{\sqrt{2}} \sqrt{\frac{\pi}{|a|}} e^{-\frac{ib_i^2}{4a}},$$

and we can to write

$$I' = \prod_{i=1}^{4} I_i = \int_{-\infty}^{+\infty} e^{i(ak^2 + bk)} \, dk$$

$$= \left(\frac{1-i}{\sqrt{2}}\right)^4 \left(\sqrt{\frac{\pi}{|a|}}\right)^4 e^{-\frac{ib^2}{4a}} = -\frac{\pi^2}{a^2} e^{-\frac{ib^2}{4a}}, \qquad (8.1.22)$$

where b is a constant four-vector, bk is a scalar product of two four-vectors, and $k^2 = \sum_{i=1}^{4} k_i^2$.

Taking derivative of (8.1.22) with respect to b_n,

$$\int_{-\infty}^{+\infty} k_n e^{i(ak^2 + bk)} \, dk = \frac{\pi^2 b_n}{2a^3} e^{-\frac{ib^2}{4a}}$$

and multiplying by γ_n, we get

$$\int_{-\infty}^{+\infty} \hat{k} e^{i(ak^2 + bk)} \, dk = \frac{\pi^2 \hat{b}}{2a^3} e^{-\frac{ib^2}{4a}}. \qquad (8.1.23)$$

Electron self-energy diagram

Noting that

$$\gamma_l \hat{p} \gamma_l = p_n \gamma_l \gamma_n \gamma_l = p_n (2\delta_{ln} - \gamma_n \gamma_l) \gamma_l = -2\hat{p} ;$$
$$\gamma_l (\hat{p} - \hat{k} + im) \gamma_l = -2\hat{p} + 2\hat{k} + 4im, \tag{8.1.24}$$

let us consider the integral (see (8.1.17))

$$I = \int dk \; \gamma_l \; (\hat{p} - \hat{k} + im) \gamma_l \; e^{i(ak^2 + bk)}, \tag{8.1.25}$$

where

$$-a = \lambda + \mu ; \qquad b = 2\lambda p. \tag{8.1.26}$$

In view of (8.1.23), (8.1.24), and (8.1.26), the integral (8.1.25) yields

$$I = \frac{\pi^2}{a^2} \left(2\hat{p} - 4im + \frac{\hat{b}}{a} \right) e^{-\frac{ib^2}{4a}}$$

$$= \frac{\pi^2}{(\lambda + \mu)^2} \left[2\hat{p} - 4im - \frac{2\lambda \hat{p}}{(\lambda + \mu)} \right] e^{\frac{i\lambda^2 p^2}{(\lambda + \mu)}}, \tag{8.1.27}$$

and (8.1.18) finally reads

$$\Sigma_R(p) = \frac{e^2}{8\pi^2} \int_0^\infty d\mu \int_0^\infty d\lambda \; \frac{1}{(\lambda + \mu)^2} \left(\frac{\mu}{\lambda + \mu} \hat{p} - 2im \right) e^{-\frac{ip^2 \lambda \mu}{(\lambda + \mu)}}$$

$$\times e^{-(\lambda + \mu)\alpha} \left(1 - e^{-iM^2\mu} \right) \left(e^{-im^2\lambda} - e^{-iM^2\lambda} \right). \tag{8.1.28}$$

Calculation of $\Sigma_R(p)$

Let us make the following change of variables:

$$\mu = \xi \eta ; \qquad \lambda = (1 - \xi)\eta. \tag{8.1.29}$$

This leads to

$$\eta = \lambda + \mu ; \qquad \xi = \frac{\mu}{\lambda + \mu}, \tag{8.1.30}$$

with the new limits $0 \leq \eta \leq \infty$; $0 \leq \xi \leq 1$. Noticing that the functional determinant of the transformation is

$$\frac{D(\lambda, \mu)}{D(\xi, \eta)} = \begin{vmatrix} \eta & \xi \\ -\eta & 1 - \xi \end{vmatrix} = \eta,$$

relation (8.1.28) writes

$$\Sigma_R(p) = -\frac{e^2}{8\pi^2} \int_0^1 d\xi \int_0^\infty d\eta \, (2im - \xi \hat{p}) \frac{1}{\eta} e^{-\alpha \eta} e^{-i\xi(1-\xi)\eta p^2}$$

247

$$\times \left(1 - e^{-iM^2 \xi \eta}\right)\left[e^{-im^2(1-\xi)\eta} - e^{-iM^2(1-\xi)\eta}\right]. \tag{8.1.31}$$

Denoting

$$J_\alpha = \int_0^\infty \frac{1}{\eta} \, d\eta \, e^{-\alpha\eta} \left(e^{-ia_1\eta} + e^{-ia_2\eta} - e^{-ib_1\eta} - e^{-ib_2\eta}\right), \tag{8.1.32}$$

where

$$a_1 = \xi(1-\eta)p^2 + m^2(1-\xi) = (1-\xi)(\xi p^2 + m^2) \; ;$$

$$a_2 = \xi(1-\eta)p^2 + M^2(1-\xi) + M^2\xi = \xi(1-\xi)p^2 + M^2 \; ; \tag{8.1.33}$$

$$b_1 = \xi(1-\eta)p^2 + M^2(1-\xi) = (1-\xi)(\xi p^2 + M^2) \; ;$$

$$b_2 = \xi(1-\eta)p^2 + m^2(1-\xi) + M^2\xi = (1-\xi)(\xi p^2 + m^2) + M^2\xi \; ,$$

formula (8.1.31) becomes

$$\Sigma_R(p) = -\frac{e^2}{8\pi^2} \int_0^1 d\xi \, (2im - \xi\hat{p}) \, J_\alpha. \tag{8.1.34}$$

Let us first calculate J_α. To eliminate the divergence due to the presence of the variable of integration η in the denominator, we resort to a trick. Considering $a_1, \; a_2, \; b_1, and \; b_2$ as independent variables, one can write

$$dJ_\alpha = \frac{\partial J_\alpha}{\partial a_1} \, da_1 + \frac{\partial J_\alpha}{\partial a_2} \, da_2 + \frac{\partial J_\alpha}{\partial b_1} \, db_1 + \frac{\partial J_\alpha}{\partial b_2} \, db_2.$$

We now proceed to integrate with respect to the new variables $a_1, \; a_2, \; b_1, and \; b_2$, choosing a constant value A in the lower limit, the same for all integrals. In view of (8.1.32), we have:

$$J_\alpha = \sum_{s=1}^2 \int_A^{a_s} \frac{\partial J_\alpha}{\partial a_s} \, da_s + \sum_{s=1}^2 \int_A^{b_s} \frac{\partial J_\alpha}{\partial b_s} \, db_s$$

$$= (-i) \int_0^\infty \sum_{s=1}^2 \left\{\int_A^{a_s} e^{-(\alpha+ia_s)\eta} \, da_s - \int_A^{b_s} e^{-(\alpha+ib_s)\eta} \, db_s\right\} d\eta$$

$$\tag{8.1.35}$$

$$= (-i) \sum_{s=1}^2 \left\{\int_A^{a_s} \frac{e^{-(\alpha+ia_s)\eta}}{-(\alpha+ia_s)}\Big|_0^\infty \, da_s - \int_A^{b_s} \frac{e^{-(\alpha+ib_s)\eta}}{-(\alpha+ib_s)}\Big|_0^\infty \, db_s\right\}$$

$$= (-i) \sum_{s=1}^{2} \left\{ \int_{A}^{a_s} \frac{da_s}{\alpha + ia_s} - \int_{A}^{b_s} \frac{db_s}{\alpha + ib_s} \right\}.$$

Recalling that the causal functions $D_0^{(c)}(k)$ and $S^{(c)}(p)$ are defined as limits for $\alpha \to 0$ (see (A.129) and (A.131)), we can take

$$J_0 = \lim_{\alpha \to 0} J_\alpha = -(\ln|a_1| + \ln|a_2| - \ln|b_1| - \ln|b_2|) = \ln \left| \frac{b_1 b_2}{a_1 a_2} \right|, \quad (8.1.36)$$

where the vanishing terms have been omitted. Replacing (8.1.36) into (8.1.34), and taking into account (8.1.33), we are left with:

$$\Sigma_R(p) = -\frac{e^2}{8\pi^2} \int_0^1 d\xi \, (2im - \xi\hat{p})$$

$$\times \ln \frac{(\xi p^2 + M^2)[\xi(1-\xi)p^2 + m^2(1-\xi) + M^2 \xi]}{(\xi p^2 + m^2)[\xi(1-\xi)p^2 + M^2]}. \quad (8.1.37)$$

The inspection of (8.1.37) shows that the integral is divergent for $M \to \infty$. To eliminate this divergence, one splits the integral in two parts, one convergent and the other divergent. This procedure will be justified later on.

Separation of the convergent part of $\Sigma_R(p)$

Let us expand $\Sigma_R(p)$ in Maclaurin series, in terms of p, about the origin. We then have

$$\Sigma_R(p) = \Sigma_R(0) + \left. \frac{\partial \Sigma_R(p)}{\partial p_l} \right|_{p=0} p_l + \tilde{\Sigma}_R(p) ; \quad (l = 1, 2, 3, 4) \quad (8.1.38)$$

where $\tilde{\Sigma}_R(p)$ stands for the remaining terms in the expansion (8.1.38) (beginning with the third one).

According to (8.1.37), we can write

$$\Sigma_R(0) = -\frac{e^2}{8\pi^2} \int_0^1 2\,i\,m \, \ln \frac{m^2(1-\xi) + M^2\xi}{m^2} \, d\xi \, ;$$

$$\left. \frac{\partial \Sigma_R(p)}{\partial p_l} \right|_{p=0} p_l = \frac{e^2}{8\pi^2} \int_0^1 \xi \, \hat{p} \, \ln \frac{m^2(1-\xi) + M^2\xi}{m^2} \, d\xi \, ,$$

and (8.1.38) becomes

$$\Sigma_R(p) = -\frac{e^2}{8\pi^2} \int_0^1 d\xi \, (2im - \hat{p}\xi)$$

$$\times \ln \frac{m^2(1-\xi) + M^2\xi}{m^2} + \tilde{\Sigma}_R(p) = \Sigma'_R(p) + \tilde{\Sigma}_R(p), \qquad (8.1.39)$$

where the notation $\Sigma'_R(p)$ is transparent. Relation (8.1.39) allows one to write

$$\tilde{\Sigma}_R(p) = \Sigma_R(p) - \Sigma'_R(p) = -\frac{e^2}{8\pi^2} \int_0^1 d\xi \, (2im - \hat{p}\xi)$$

$$\times \ln \frac{(\xi p^2 + M^2)[\xi(1-\xi)p^2 + m^2(1-\xi) + M^2\xi]m^2}{(\xi p^2 + m^2)[\xi(1-\xi)p^2 + M^2][(1-\xi)m^2 + M^2\xi]}. \qquad (8.1.40)$$

One observes that $\tilde{\Sigma}_R(p)$ is convergent for $M \to \infty$, so that

$$\tilde{\Sigma}(p) = \lim_{M \to \infty} \tilde{\Sigma}_R(p) = -\frac{e^2}{8\pi^2} \int_0^1 d\xi \, (2im - \hat{p}\xi) \ln \frac{m^2}{\xi p^2 + m^2}. \qquad (8.1.41)$$

Combining (8.1.41) with (8.1.39) and (8.1.11), we can write

$$\Sigma(p) = \lim_{M \to \infty} \Sigma_R(p) = \lim_{M \to \infty} \Sigma'_R(p) + \tilde{\Sigma}(p), \qquad (8.1.42)$$

where the first term on the r.h.s. is divergent, and the second one convergent, according to (8.1.41).

We mention that the separation of the convergent part in (8.1.42) is not unique. But, observing that only the part linear in p is divergent, we conclude that we can add to $\tilde{\Sigma}(p)$ any function linear in p, e.g. a first degree polynomial in p. Due to the requirement of relativistic invariance, this polynomial should have the form

$$c_1(i\hat{p} + m) + c_2 m. \qquad (8.1.43)$$

Consequently, the most general form of $\tilde{\Sigma}(p)$ is

$$\tilde{\Sigma}(p) = -\frac{e^2}{8\pi^2} \left[\int_0^1 d\xi \, (2im - \hat{p}\xi) \ln \frac{m^2}{\xi p^2 + m^2} + c_1(i\hat{p} + m) + c_2 m \right]. \qquad (8.1.44)$$

Justification of removing the divergent part

As shown in (8.1.39), the divergent part of $\Sigma_R(p)$ is

$$\Sigma'_R(p) = -\frac{e^2}{8\pi^2} \int_0^1 d\xi \, (2im - \hat{p}\xi) \ln \frac{m^2(1-\xi) + M^2\xi}{m^2}.$$

Electron self-energy diagram

For large values of M, we have

$$m^2(1 - \xi) << M^2\xi,$$

which means that

$$\lim_{M\to\infty} \Sigma'_R(p) = \lim_{M\to\infty} \left\{ -\frac{e^2}{8\pi^2 i} \int_0^1 d\xi \, (2im - \hat{p}\xi) \, \ln \xi \frac{M^2}{m^2} \right\}$$

$$= \lim_{M\to\infty} \left\{ -\frac{e^2}{8\pi^2} \left[(2im\xi - \frac{\xi^2}{2}\hat{p}) \ln \xi \frac{M^2}{m^2} \Big|_0^1 - \int_0^1 d\xi \, (2im\xi - \frac{\xi^2}{2}\hat{p})\frac{1}{\xi} \right] \right\}$$

$$= \lim_{M\to\infty} \left\{ -\frac{e^2}{8\pi^2} \left[(2im - \frac{1}{2}\hat{p}) \ln \frac{M^2}{m^2} - 2im + \frac{1}{4}\hat{p} \right] \right\} \qquad (8.1.45)$$

$$= \lim_{M\to\infty} \left\{ -\frac{e^2}{(4\pi)^2 \, i} \left[(4m + i\hat{p}) \, \ln \frac{M^2}{m^2} - (4m + \frac{1}{2}i\hat{p}) \right] \right\}.$$

Using the Fourier transform, let us go from momentum to configuration representation. This is done by

$$\lim_{M\to\infty} \Sigma'_R(x) = \lim_{M\to\infty} \left\{ -\frac{e^2}{(4\pi)^2 i} \left[(4m + \hat{\partial}) \ln \frac{M^2}{m^2} - (4m + \frac{1}{2}\hat{\partial}) \right] \delta(x) \right\},$$

$$(8.1.46)$$

where the delta function has a singularity (divergence) at the origin ($x = 0$). This divergence is explained by the fact that our model is a point particle. This kind of divergence could be eliminated by using the finite-size model, but this theory is very complicated and we shall resort to a different method. Using the GRAPH TABLE, let us write the matrix element associated with the electron self-energy, in configuration representation. This is

$$S^{(2)} = \int \overline{\psi}_-(x_1)\Sigma(x_1 - x_2)\psi_+(x_2) \, dx_1 \, dx_2, \qquad (8.1.47)$$

where

$$\Sigma(x_1 - x_2) = e^2 \, D_0^{(c)}(x_1 - x_2)\gamma_l \, S^{(c)}(x_1 - x_2) \, \gamma_l$$

$$= \tilde{\Sigma}(x_1 - x_2) + \lim_{M\to\infty} \left[\frac{e^2}{(4\pi)^2 i} \left[(4m + \hat{\partial}) \ln \frac{M^2}{m^2} - (4m + \frac{1}{2}\hat{\partial}) \right] \delta(x_1 - x_2) \right.$$

$$= \tilde{\Sigma}(x_1 - x_2) + \lim_{M\to\infty} \left[\frac{e^2 m}{(4\pi)^2 i} (4 \ln \frac{M^2}{m^2} - 4) \, \delta(x_1 - x_2) \right] \qquad (8.1.48)$$

$$+ \lim_{M\to\infty} \left[\frac{e^2 m}{(4\pi)^2 i} (\ln \frac{M^2}{m^2} - \frac{1}{2})\hat{\partial} \, \delta(x_1 - x_2) \right].$$

Divergent second-order processes

Let us add and subtract the quantity

$$\lim_{M\to\infty}\left[\frac{e^2 m}{(4\pi)^2 i}\left(\ln\frac{M^2}{m^2}-\frac{1}{2}\right)\right].$$

Then (8.1.11) becomes

$$\Sigma(x_1-x_2)=\tilde{\Sigma}(x_1-x_2)+A\,\delta(x_1-x_2)+B\,(\hat{\partial}+m)\delta(x_1-x_2),\quad (8.1.49)$$

with

$$A=\lim_{M\to\infty}\left[\frac{e^2 m}{(4\pi)^2 i}\left(3\ln\frac{M^2}{m^2}-\frac{7}{2}\right)\right]\ ; \qquad (8.1.50)$$

$$B=\lim_{M\to\infty}\left[\frac{e^2 m}{(4\pi)^2 i}\left(\ln\frac{M^2}{m^2}-\frac{1}{2}\right)\right]. \qquad (8.1.51)$$

Using (8.1.49), the matrix element (8.1.47) writes

$$S^{(2)}=e^2\int\overline{\psi}_-(x_1)\tilde{\Sigma}(x_1-x_2)\psi_+(x_2)\,dx_1\,dx_2$$

$$+e^2 A\int\overline{\psi}_-(x_1)\psi_+(x_2)\delta(x_1-x_2)\,dx_1\,dx_2 \qquad (8.1.52)$$

$$+e^2 B\int\overline{\psi}_-(x_1)(\hat{\partial}+m)\psi_+(x_2)\delta(x_1-x_2)\,dx_1\,dx_2.$$

Since $\tilde{\Sigma}(x_1-x_2)$ is convergent, the first term in (8.1.52) is also convergent. Integrating with respect to x_2 ($dx_2 = i\,d\vec{x}_2\,dx_{20}$), the second term in (8.1.52) writes

$$Ae^2\,i\int\overline{\psi}_-(x_1)\psi_+(x_1)\,dx_1. \qquad (8.1.53)$$

To interpret this term, we use the Dirac equation, written for an electron of mass m and charge e, in the presence of an electromagnetic field characterized by the four-potential A_l:

$$(\hat{p}+e\hat{A}-im)\psi=0. \qquad (8.1.54)$$

Suppose that the observable electron mass is composed of two parts: the rest mass, appearing in special relativity and denoted by the letter m, and the mass δm associated with the interaction between the electron and the field. (Some authors use $m=m_0+\delta m$, where m denotes the observable mass of the particle, m_0 stands for the so-called

252

bare mass, as the mass in the limit of zero size, and δm represents the increase in mass owing to the interaction between the particle and the field or its environment). Equation (8.1.54) then yields

$$[\hat{p} + e\hat{A} - i(m + \delta m)]\psi = [\hat{p} + e(\hat{A} - \frac{i}{c}\delta m) - im]\psi = 0, \quad (8.1.55)$$

where we brought out the fact that a mass variation δm is equivalent to a variation $\frac{i}{c}\delta m$ in $\hat{A}$. Using (6.3.6), the corresponding S-matrix is therefore

$$S = \sum_{n=0}^{\infty} S^{(n)} = \sum_{n=0}^{\infty} \frac{(-ie)^n}{n!} \int \dots \int T\{N[\overline{\psi}(x_1)\hat{A}(x_1)\psi(x_1)$$

$$\times \dots N[\overline{\psi}(x_n)\hat{A}(x_n)\psi(x_n)]\} \, dx_1 \dots dx_n \qquad (8.1.56)$$

$$\longrightarrow 1 - ie\, N \int \overline{\psi}(x_1)\hat{A}(x_1)\psi(x_1)\,dx_1 - \delta m \int \underline{\overline{\psi}(x_1)\psi(x_1)\,dx_1} + \dots$$

It can be seen that the underlined term in (8.1.56), corresponding to the first order matrix element, or, in other words, to the interaction Hamiltonian density

$$H'(x_1) = \delta m\,\overline{\psi}(x_1)\psi(x_1)$$

represented by the Feynman graph

δm

has the same form as the divergent term (8.1.53), obtained in the second approximation. The comparison between these two terms yields

$$\delta m = -i\,e^2 A = -\lim_{M\to\infty}\left[\frac{e^4 m}{(4\pi)^2}\left(3\ln\frac{M^2}{m^2} - \frac{7}{2}\right)\right]. \qquad (8.1.57)$$

Therefore, δm diverges logarithmically for $M \to \infty$. This result leads to the following interpretation: the total mass of the electron in the Dirac equation is infinite. It is composed of the rest mass, which is finite and can be determined, and the infinite mass δm, due to the interaction between the electron and the field. In other words, the electron rest mass could be obtained by subtracting δm from the total mass. According to the S-matrix theory, this amounts to subtracting the divergent term (8.1.53) obtained in second-order approximation. This procedure is called *mass renormalization.*

Assuming that the mass renormalization was performed, let us now focus our attention on eliminating the divergence introduced by the third term in (8.1.52). Using the momentum representation, we can write

$$\Sigma'(p) = \tilde{\Sigma}(p) + B(i\hat{p} + m). \tag{8.1.58}$$

At this stage, we have to mention that the electron self-energy graph appears in high-order processes, being accompanied by other graphic elements. We can have the following possibilities:

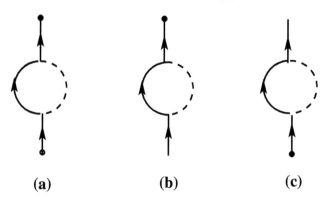

(a)   **(b)**   **(c)**

where vertexes are marked by circles. The contributions of these diagrams to the total graph can be obtained from the second-order diagrams, with the following substitutions:

$$a) \; S^{(c)}(p) \rightarrow S^{(c)}(p) + S^{(c)}(p)\Sigma'(p)S^{(c)}(p) \; ;$$

$$b) \; \psi(p) \rightarrow \psi(p) + S^{(c)}(p)\Sigma'(p)\psi(p) \; ; \tag{8.1.59}$$

$$c) \; \overline{\psi}(p) \rightarrow \overline{\psi}(p) + \overline{\psi}(p)\Sigma'(p)S^{(c)}(p).$$

But, in view of (8.1.58),

$$S^{(c)}(p)\Sigma'(p) = S^{(c)}(p)\tilde{\Sigma}(p) + \frac{1}{(2\pi)^2 i}\frac{i\hat{p} - m}{p^2 + m^2 + i\alpha}B(i\hat{p} + m)$$

$$= S^{(c)}(p)\tilde{\Sigma}(p) - B' \quad (\alpha \rightarrow 0), \tag{8.1.60}$$

Electron self-energy diagram

with $B' = B/(2\pi)^2 i$, and (8.1.59) yields

$$a)\ S^{(c)}(p) \rightarrow (1 - B')S^{(c)}(p) + S^{(c)}(p)\tilde{\Sigma}(p)S^{(c)}(p)\ ;$$

$$b)\ \psi(p) \rightarrow (1 - B')\psi(p) + S^{(c)}(p)\tilde{\Sigma}(p)\psi(p)\ ;\qquad (8.1.61)$$

$$c)\ \overline{\psi}(p) \rightarrow (1 - B')\overline{\psi}(p) + \overline{\psi}(p)\tilde{\Sigma}(p)S^{(c)}(p).$$

Since $\tilde{\Sigma}(p)\psi(p)$ is convergent, we conclude that divergences are eliminated in the higher-order terms.

To get rid of the divergence introduced by B', we renormalize the functions $\psi(p)$, $\overline{\psi}(p)$, and $S^{(c)}(p)$ as follows:

$$\psi(p) \longrightarrow \frac{\psi(p)}{\sqrt{1 - B'}}\ ;\quad \overline{\psi}(p) \longrightarrow \frac{\overline{\psi}(p)}{\sqrt{1 - B'}}\ ;\quad S^{(c)}(p) \longrightarrow \frac{S^{(c)}(p)}{1 - B'}.$$
$$(8.1.62)$$

Transformation (8.1.62) eliminates the divergence in some element of the S-matrix. Indeed, let us consider a diagram with n nodes and f external electronic lines. In this case, each vertex has two incoming or two outgoing electronic lines, which means a total of $2n$ electronic lines for the entire graph. If one subtracts f external electronic lines from the total of $2n$ electronic lines, and take into account that to each pair corresponds one propagator, one obtains $n - \frac{f}{2}$ internal electronic lines, i.e. $n - \frac{f}{2}$ functions $S^{(c)}(p)$. We also have $\frac{f}{2}$ external lines associated with $\psi(p)$ and $\frac{f}{2}$ external lines associated with $\overline{\psi}(p)$. Consequently, if we perform the substitutions (8.1.62), $(1 - B')$ enters to the power $-\frac{f}{4} - \frac{f}{4} - \left(n - \frac{f}{2}\right) = -n$, meaning that the matrix element $S^{(n)}$ is multiplied by a factor of $(1 - B')^{-n}$. For $n = 2$, this factor can be compensated if we take

$$e \longrightarrow e_0 = e(1 - B') = e + \delta e, \qquad (8.1.63)$$

with

$$\delta e = -e\,B'. \qquad (8.1.64)$$

This procedure is similar to the one used for mass renormalization, and is called *charge renormalization*.

Observation. In our case, the "renormalization" method consists in omitting some divergent terms from the electron self-energy. In fact, the mass and charge of the electron maintain their experimental meaning. Generally speaking, *renormalization* is a technique used to treat infinite quantities in various fields of science. At present, there are some attempts to use non-linear methods in quantum electrodynamics, but these methods demonstrate large computational difficulties, and the results are still unsatisfactory.

255

8.2. Photon self-energy diagram.
Vacuum polarization

Another second-order diagram leading to divergent integrals in quantum electrodynamics, is represented by (see (6.5.21))

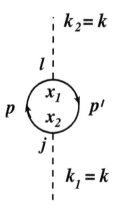

Intuitively, this diagram represents the successive creation and annihilation of a virtual pair in vacuum, leading to the photon self-energy, sometimes called *vacuum polarization*. This name comes from *Dirac's hole theory*. In Dirac's hole theory the vacuum state is the state where each negative energy state is occupied by a single electron, and each positive energy state is unoccupied. The energy of the vacuum state is given by summing over the energies of all negative energy states. Assuming that all negative energy levels are occupied, then under the action of an electromagnetic field the vacuum can be "polarized", meaning that the non-observable negative charges move with respect to the "holes" where they were located at the beginning.

The graph displayed above can arise as an intermediate process in a higher-order effect, such as:

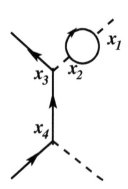

Photon self-energy diagram. Vacuum polarization

According to (6.5.21), this graph contains the operator products

$$\Omega = N\{ < T\big[\overline{\psi}_a(x_1)\psi_b(x_2)\big] >_0 A_l(x_1)(\gamma_l)_{aa}$$

$$\times \ < T\big[\psi_a(x_1)\overline{\psi}_b(x_2)\big] >_0 A_j(x_2)(\gamma_j)_{bb}\}.$$

Using (6.3.7) and the spinor anti-commutation rule, after some term rearrangements we obtain

$$\Omega = (-1)\Big\{(\gamma_l)_{aa} S_{ab}^{(c)}(x_1 - x_2)(\gamma_j)_{bb} S_{ba}^{(c)}(x_2 - x_1) A_{l-}(x_1) A_{j+}(x_2)$$

$$= (-1) A_{l-}(x_1)\, Tr\big[\gamma_l S^{(c)}(x_1 - x_2)\gamma_j S^{(c)}(x_2 - x_1)\big] A_{j+}(x_2). \quad (8.2.1)$$

This relation shows that whenever a graph contains a closed fermionic loop, one takes the trace over the loop spinors, and the result gets multiplied by (-1). Using this observation, as well as the GRAH TABLE, and following the photon self-energy graph in chronological order, we have the following segments:

- photon incident at $x_2 \longrightarrow (2\pi)^{-3/2}\frac{1}{\sqrt{2\omega_1}}(e_2)_j$
- vertex $x_2 \longrightarrow (2\pi)^4\, e\, \gamma_j\delta(p - k_1 - p')$
- positron propagator $(x_2 \to x_1) \longrightarrow (2\pi)^{-4}\int \frac{i\hat{p}'-m}{p'^2+m^2-i\alpha}\, dp'$
- electron propagator $(x_2 \to x_1) \longrightarrow (2\pi)^{-4}\int \frac{i\hat{p}-m}{p^2+m^2-i\alpha}\, dp$
- vertex $x_1 \longrightarrow (2\pi)^4\, e\, \gamma_l\delta(k_2 + p' - p)$
- photon emergent from $x_1 \longrightarrow (2\pi)^{-3/2}\frac{1}{\sqrt{2\omega_2}}(e_1)_l$

The S-matrix element is obtained by using an already known procedure:

$$S^{(2)} = -\frac{e^2}{4(2\pi)^3\sqrt{\omega_1\omega_2}}(e_1)_l\Big\{ \int dp\, dp'$$

$$Tr\Big[\gamma_l\frac{i\hat{p}-m}{p^2+m^2-i\alpha}\gamma_j\frac{i\hat{p}'-m}{p'^2+m^2-i\alpha}\Big]\delta(k_2+p'-p)\,\delta(p-k_1-p')\Big\}(e_2)_j,$$

or, if we perform the integration with respect to p'

$$S^{(2)} = -\frac{e^2}{4(2\pi)^3\sqrt{\omega_1\omega_2}}(e_1)_l\Big\{ \int d\vec{p}\, dE_p\, \delta(k_2 - k_1)$$

$$\times Tr\Big[\gamma_l\frac{\hat{p}+im}{p^2+m^2-i\alpha}\gamma_j\frac{\hat{p}-\hat{k}_1+im}{(p-k_1)^2+m^2-i\alpha}\Big]\Big\}(e_2)_j. \quad (8.2.2)$$

According to the momentum conservation law, the final four-momentum k_2 equals the initial four-momentum k_1. Denoting $k_1 =$

Divergent second-order processes

$k_2 = k$, the matrix element between the photonic states of momentum $\vec{k}$ and energy ω is therefore

$$S^{(2)} = 2\pi i \frac{(e_1)_l}{\sqrt{2\omega}} \Pi_{lj}(k) \frac{(e_2)_j}{\sqrt{2\omega}} \delta(k_2 - k_1), \qquad (8.2.3)$$

with

$$\Pi_{lj} = \frac{ie^2}{(2\pi)^4} \int d\vec{p}\, dE_p\, Tr\left[\gamma_l \frac{\hat{p} + im}{p^2 + m^2 - i\alpha} \gamma_j \frac{\hat{p} - \hat{k} + im}{(p-k)^2 + m^2 - i\alpha}\right]. \qquad (8.2.4)$$

As one can see, (8.2.4) is the Fourier transform of the function

$$\Pi_{lj}(x_1 - x_2) = ie^2 \, Tr\left[\gamma_l S^{(c)}(x_1 - x_2)\gamma_j \, S^{(c)}(x_2 - x_1)\right]. \qquad (8.2.5)$$

For $|p| \to \infty$, the integrand in (8.2.4) behaves like $|p|^{-2}$, and we have

$$\int d\vec{p}\, dE_p\, |p|^{-2} \sim \int_0^{|p|} |p|^3 |p|^{-2}\, d|p| = \int_0^{|p|} |p|\, d|p| = \frac{1}{2}|p|^2 \to \infty,$$

that is, for large $|p|$ the integral (8.2.4) diverges as $|p|^2$.

To separate the divergent part in (8.2.4), we shall use the procedure developed in the previous section. This relation can be written as

$$\Pi_{lj}(k) = -ie^2 \int d\vec{p}\, dE_p\, Tr[\gamma_l \, S^{(c)}(p)\, \gamma_j \, S^{(c)}(p - k)]. \qquad (8.2.6)$$

In view of (8.1.6), we can write

$$\Pi_{Rlj} = -ie^2 \int d\vec{p}\, dE_p\, Tr[\gamma_l \, S_R^{(c)}(p)\, \gamma_j \, S_R^{(c)}(p - k)]. \qquad (8.2.7)$$

As one can see

$$\lim_{M_j \to \infty} \Pi_{Rlj}(k) = \Pi_{lj}(k). \qquad (8.2.8)$$

Let us consider a single auxiliary mass M, and choose $C = -1$ in (8.1.6). Then (8.1.6) becomes (see (8.1.12))

$$(2\pi)^2 i\, S_R^{(c)}(p) = (i\hat{p} - m)\left\{\frac{1}{p^2 + m^2 - i\alpha} - \frac{1}{p^2 + M^2 - i\alpha}\right\} ; \qquad (8.2.9)$$

Photon self-energy diagram. Vacuum polarization

Let us calculate $\Pi_{Rlj}(k)$ given by (8.2.7), with $S_R^{(c)}(p)$ given by (8.2.9). To this end, we use again the formula (see previous section)

$$i \int_0^\infty e^{-i(x-i\alpha)\lambda} \, d\lambda = \frac{1}{x - i\alpha} \; ; \quad \alpha > 0, \tag{8.2.10}$$

and transform a quadruple integral into a sextuple one. In view of (8.2.10), $S_R^{(c)}(p)$ and $S_R^{(c)}(p-k)$ become

$$-2\pi^2 i \, S_R^{(c)}(p)$$

$$= (\hat{p} + im) \int_0^\infty \left\{ e^{-i(p^2+m^2-i\alpha)\lambda_1} - e^{-i(p^2+M^2-i\alpha)\lambda_1} \right\} d\lambda_1 \; ;$$

$$-2\pi^2 i \, S_R^{(c)}(p-k) = (\hat{p} - \hat{k} + im) \int_0^\infty \left\{ e^{-i[(p-k)^2+m^2-i\alpha)]\lambda_2} \right.$$

$$\left. - e^{-i[(p-k)^2+M^2-i\alpha)]\lambda_2} \right\} d\lambda_2 \; ,$$

or

$$-2\pi^2 i \, S_R^{(c)}(p) = (\hat{p} + im) \int_0^\infty e^{-i(p^2-i\alpha)\lambda_1} \left(e^{-im^2\lambda_1} - e^{-iM^2\lambda_1} \right) d\lambda_1 \; ;$$

$$-2\pi^2 i \, S_R^{(c)}(p-k) = (\hat{p} - \hat{k} + im) \int_0^\infty e^{-i[(p-k)^2-i\alpha)]\lambda_2} \tag{8.2.11}$$

$$\times \left(e^{-im^2\lambda_2} - e^{-iM^2\lambda_2} \right) d\lambda_2 \; ,$$

and (8.2.7) becomes

$$\Pi_{Rlj} = -\frac{ie^2}{(2\pi)^4} \int d\vec{p} \, dE_p \int_0^\infty d\lambda_1 \int_0^\infty d\lambda_2$$

$$\times Tr\left[\gamma_l(\hat{p} + im)\gamma_j(\hat{p} - \hat{k} + im) \right]$$

$$\times e^{-i\{(p^2-i\alpha)\lambda_1+[(p-k)^2-i\alpha]\lambda_2\}} \tag{8.2.12}$$

$$\times \left(e^{-im^2\lambda_1} - e^{-iM^2\lambda_1} \right)\left(e^{-im^2\lambda_2} - e^{-iM^2\lambda_2} \right).$$

Since

$$-i\{(p^2 - i\alpha)\lambda_1 + [(p-k)^2 - i\alpha]\lambda_2\}$$

$$= -\alpha(\lambda_1 + \lambda_2) - i\lambda_2 k^2 - i[(\lambda_1 + \lambda_2)p^2 - 2\lambda_2 kp], \tag{8.2.13}$$

259

we have

$$
\Pi_{Rlj} = -\frac{ie^2}{(2\pi)^4} \int_0^\infty d\lambda_1 \int_0^\infty d\lambda_2 \left(e^{-im^2\lambda_1} - e^{-iM^2\lambda_1} \right)
$$

$$
\times \left(e^{-im^2\lambda_2} - e^{-iM^2\lambda_2} \right) e^{-\alpha(\lambda_1+\lambda_2)-i\lambda_2 k^2} \tag{8.2.14}
$$

$$
\times \int d\vec{p}\, dE_p\, Tr[\gamma_l(\hat{p}+im)\gamma_j(\hat{p}-\hat{k}+im)]\, e^{-i[(\lambda_1+\lambda_2)p^2 - 2\lambda_2\, p\, k]}.
$$

Obviously, here we do not take $k^2 = 0$, because here the photonic lines do not correspond to free photons.

To calculate the trace in (8.2.14) we use (5.1.30), (5.1.33) and (5.1.34). Thus,

$$
Tr\left[\gamma_l(\hat{p}+im)\gamma_j(\hat{p}-\hat{k}+im)\right] = Tr[\gamma_l\hat{p}\gamma_j(\hat{p}-\hat{k})] - m^2\, Tr(\gamma_l\gamma_j)
$$

$$
= p_m(p_n - k_n)\, Tr(\gamma_l\gamma_m\gamma_j\gamma_n) - 4m^2\delta_{lj} \tag{8.2.15}
$$

$$
= 4[(\delta_{lm}\delta_{jn} + \delta_{ln}\delta_{mj} - \delta_{lj}\delta_{mn})p_m(p_n - k_n) - m^2\delta_{lj}]
$$

$$
= 4[2p_l p_j - (p_l k_j + p_j k_l) - \delta_{lj}(p^2 - pk + m^2)],
$$

and (8.2.14) becomes

$$
\Pi_{Rlj} = -\frac{e^2}{(2\pi)^4} \int_0^\infty d\lambda_1 \int_0^\infty d\lambda_2 \left(e^{-im^2\lambda_1} - e^{-iM^2\lambda_1} \right)
$$

$$
\times \left(e^{-im^2\lambda_2} - e^{-iM^2\lambda_2} \right) e^{-\alpha(\lambda_1+\lambda_2)-i\lambda_2 k^2}\, I_{lj}, \tag{8.2.16}
$$

where

$$
I_{lj} = 4 \int dp\, e^{-i[(\lambda_1+\lambda_2)p^2 - 2\lambda_2 pk]}
$$

$$
\times [2p_l p_j - (p_l k_j + p_j k_l) - \delta_{lj}(p^2 - pk + m^2)], \tag{8.2.17}
$$

with $dp = i\, d\vec{p}\, dE_p$.

The integral (8.2.17) is worked out using the Gauss-type integral (see the previous section)

$$
I_1 = \int_{-\infty}^{+\infty} e^{i(ap^2+bp)}\, dp = -\left(\frac{\pi}{a}\right)^2 e^{-i\frac{b^2}{4a}}. \tag{8.2.18}
$$

Keeping in mind that $bp = b_l p_l = b_j p_j$, $b^2 = b_l b_l = b_j b_j$, we first take the derivative of (8.2.18) with respect to b_l:

$$
I_l' = \int_{-\infty}^{+\infty} p_l e^{i(ap^2+bp)}\, dp = \frac{b_l}{2a}\left(\frac{\pi}{a}\right)^2 e^{-i\frac{b^2}{4a}}, \tag{8.2.19}
$$

Photon self-energy diagram. Vacuum polarization

and then the derivative of (8.2.19) with respect to b_j:

$$I'_{lj} = \int p_l p_j \, e^{i(ap^2 + bp)} \, dp = -\frac{2ia\delta_{lj} + b_l b_j}{4a^2} \left(\frac{\pi}{a}\right)^2 e^{-i\frac{b^2}{4a}}. \qquad (8.2.20)$$

One also observes that

$$I' = I'_{ll} = \int p^2 \, e^{i(ap^2 + bp)} \, dp = -\frac{8ia + b^2}{4a^2} \left(\frac{\pi}{a}\right)^2 e^{-i\frac{b^2}{4a}}. \qquad (8.2.21)$$

The last four relations can now be used to put (8.2.17) in a condensed form. Indeed, denoting

$$\lambda_1 + \lambda_2 = -a \; ; \qquad 2\lambda_2 k = b, \qquad (8.2.22)$$

we have

$$I_{lj} = 4[2I'_{lj} - (I'_l k_j + I'_j k_l) - \delta_{lj}(I' - I'_s k_s + m^2 I_1)],$$

and, after some algebra,

$$I_{lj} = \left(\frac{2\pi}{\lambda_1 + \lambda_2}\right)^2 e^{i\frac{\lambda_2^2 k^2}{\lambda_1 + \lambda_2}}$$

$$\times \left[\frac{\lambda_1 \lambda_2}{(\lambda_1 + \lambda_2)^2}(2k_l k_j - k^2 \delta_{lj}) - \left(\frac{i}{\lambda_1 + \lambda_2} - m^2\right)\delta_{lj}\right]. \qquad (8.2.23)$$

Substituting (8.2.23) into (8.2.16), we have:

$$\Pi_{Rlj}(k) = -\frac{e^2}{(2\pi)^2} \int_0^\infty d\lambda_1 \int_0^\infty d\lambda_2 \, e^{-\alpha(\lambda_1 + \lambda_2) - i\frac{\lambda_1 \lambda_2}{\lambda_1 + \lambda_2}k^2}$$

$$\times \frac{\left(e^{-im^2\lambda_1} - e^{-iM^2\lambda_1}\right)\left(e^{-im^2\lambda_2} - e^{-iM^2\lambda_2}\right)}{(\lambda_1 + \lambda_2)^2} \qquad (8.2.24)$$

$$\times \left[\frac{\lambda_1 \lambda_2}{(\lambda_1 + \lambda_2)^2}(2k_l k_j - k^2 \delta_{lj}) - \left(\frac{i}{\lambda_1 + \lambda_2} - m^2\right)\delta_{lj}\right].$$

Let us now make the change of variables

$$\lambda_1 = \xi\eta \; ; \qquad \lambda_2 = (1 - \xi)\eta \; ; \quad 0 \le \xi \le 1 \; ; \; 0 \le \eta \le \infty \qquad (8.2.25)$$

and observe that the Jacobian of the transformation is

$$\frac{D(\lambda_1, \lambda_2)}{D(\xi, \eta)} = \begin{vmatrix} \partial\lambda_1/\partial\xi & \partial\lambda_1/\partial\eta \\ \partial\lambda_2/\partial\xi & \partial\lambda_2/\partial\eta \end{vmatrix} = \begin{vmatrix} \eta & \xi \\ -\eta & 1-\xi \end{vmatrix} = \eta,$$

261

while $d\lambda_1\,d\lambda_2 = \eta\,d\xi\,d\eta$. Then

$$\Pi_{Rlj}(k) = -\frac{e^2}{(2\pi)^2}\int_0^\infty d\xi \int_0^\infty \frac{d\eta}{\eta} e^{-\alpha\eta}$$

$$\times \left(e^{-i\xi\eta m^2} - e^{-i\xi\eta M^2}\right)\left(e^{-i(1-\xi)\eta m^2} - e^{-i(1-\xi)\eta M^2}\right)e^{-i\xi(1-\xi)\eta k^2}$$

$$(8.2.26)$$

$$\times \left[\xi(1-\xi)(2k_lk_j - k^2\delta_{lj}) - (\frac{i}{\eta} - m^2)\delta_{lj}\right].$$

To integrate with respect to η, it is convenient to write

$$\Pi_{Rlj}(k) = -\frac{e^2}{(2\pi)^2}\int_0^1 d\xi\, F_\alpha, \qquad (8.2.27)$$

where

$$F_\alpha = \int_0^\infty \frac{d\eta}{\eta} e^{-\alpha\eta} f(\eta)\left(N_{lj} - \frac{i}{\eta}\delta_{lj}\right), \qquad (8.2.28)$$

with

$$f(\eta) = e^{-iA\eta} - e^{-iB\eta} - e^{-iC\eta} + e^{-iD\eta}, \qquad (8.2.29)$$

and

$$A = \xi(1-\xi)k^2 + m^2\;;\quad B = \xi(1-\xi)k^2 + \xi m^2 + (1-\xi)M^2\;;$$

$$C = \xi(1-\xi)k^2 + (1-\xi)m^2 + \xi M^2\;;\quad D = \xi(1-\xi)k^2 + M^2\;; \quad (8.2.30)$$

$$N_{lj} = \xi(1-\xi)(2k_lk_j - k^2\delta_{lj}) + m^2\delta_{lj}.$$

On the other hand, if we take $u = e^{-\alpha\eta}f(\eta)$, $dv = \frac{1}{\eta^2}d\eta$, and integrate by parts, we obtain:

$$\int_0^\infty \frac{d\eta}{\eta^2}e^{-\alpha\eta}f(\eta) = -\frac{1}{\eta}e^{-\alpha\eta}f(\eta)\Big|_0^\infty$$

$$+ \int_0^\infty \frac{d\eta}{\eta}e^{-\alpha\eta}\frac{\partial f(\eta)}{\partial\eta} - \alpha\int_0^\infty \frac{d\eta}{\eta}e^{-\alpha\eta}f(\eta)$$

$$\xrightarrow[\alpha\to 0]{} \int_0^\infty \frac{d\eta}{\eta}e^{-\alpha\eta}\frac{\partial f(\eta)}{\partial\eta}. \qquad (8.2.31)$$

Indeed, the first term vanishes at the upper limit due to the finite character of $f(\eta)$ for $\eta\to\infty$, and also at the lower limit by l'Hospital's rule for indeterminate limits: $\lim_{\eta\to 0}\frac{f(\eta)}{\eta} = \lim_{\eta\to 0}\frac{\partial f(\eta)}{\partial\eta} = 0$. The last term is not taken into account, since it vanishes in the limit $\alpha\to 0$.

Photon self-energy diagram. Vacuum polarization

In this case (8.2.28) writes

$$F_\alpha = \int_0^\infty \frac{d\eta}{\eta} e^{\alpha \eta} \left[N_{lj} \, F(\eta) - i\, \delta_{lj} \frac{\partial f(\eta)}{\partial \eta} \right], \qquad (8.2.32)$$

with

$$\frac{\partial f(\eta)}{\partial \eta} = -i \left(A e^{-iA\eta} - B e^{-iB\eta} - C e^{-iC\eta} + D e^{-iD\eta} \right). \qquad (8.2.33)$$

To calculate the integral in (8.2.32), in the limit $\alpha \to 0$, one observes that each term has the form

$$I(a) = \lim_{\alpha \to 0} \int_0^\infty \frac{d\eta}{\eta} e^{-ia\eta} e^{-\alpha\eta}.$$

Using (8.2.10), let us take the derivative with respect to a

$$\frac{dI(a)}{da} = -i \lim_{\alpha \to 0} \int_0^\infty d\eta \, e^{-i\eta(a-i\alpha)} = -\frac{1}{a},$$

that is

$$I(a) = - \int \frac{da}{a} = - \ln |a|. \qquad (8.3.34)$$

By means of (8.2.29), (8.2.33), and (8.2.34), in the limit $\alpha \to 0$ the integral (8.2.32) writes

$$\lim_{\alpha \to 0} F_\alpha = F_0 = N_{lj} \, \ln \left| \frac{BC}{AD} \right|$$

$$+ \delta_{lj} (A \ln |A| - B \ln |B| - C \ln |C| + D \ln |D|), \qquad (8.2.35)$$

where A, B, C, D, and N_{lj} are given by (8.2.30).

To calculate the expression in parentheses, we add and subtract the quantities $m^2 \xi \ln |A|$ and $M^2 \xi \ln |D|$. The result is

$$A \ln |A| - B \ln |B| - C \ln |C| + D \ln |D| = \xi (1 - \xi) k^2 \ln \left| \frac{AD}{BC} \right|$$

$$+ \xi \left(a^2 \ln \left| \frac{A}{B} \right| + M^2 \ln \left| \frac{D}{C} \right| \right) + (1 - \xi) \left(m^2 \ln \left| \frac{A}{C} \right| + M^2 \ln \left| \frac{D}{E} \right| \right), \qquad (8.2.36)$$

and (8.2.35) becomes

$$F_0 = P + Q, \qquad (8.2.37)$$

263

Divergent second-order processes

with

$$P = [2\xi(1-\xi)(k_l k_j - k^2 \delta_{lj}) + m^2 \delta_{lj}]$$

$$\times \left\{ \ln \left| \frac{(1-\xi)M^2 + \xi m^2 + \xi(1-\xi)k^2}{m^2 + \xi(1-\xi)k^2} \right| \right.$$

$$\left. - \ln \left| \frac{M^2 + \xi(1-\xi)k^2}{\xi M^2 + (1-\xi)m^2 + \xi(1-\xi)k^2} \right| \right\}$$

$$-\delta_{lj}\,\xi\,m^2 \ln \left| \frac{(1-\xi)M^2 + \xi m^2 + \xi(1-\xi)k^2}{m^2 + \xi(1-\xi)k^2} \right| ;$$

$$Q = \delta_{lj}\,\xi\,M^2 \ln \left| \frac{M^2 + \xi(1-\xi)k^2}{\xi M^2 + (1-\xi)m^2 + \xi(1-\xi)k^2} \right|$$

$$-\delta_{lj}(1-\xi)\left\{ m^2 \ln \left| \frac{\xi M^2 + (1-\xi)m^2 + \xi(1-\xi)k^2}{m^2 + \xi(1-\xi)k^2} \right| \right.$$

$$\left. -M^2 \ln \left| \frac{M^2 + \xi(1-\xi)k^2}{(1-\xi)M^2 + \xi m^2 + \xi(1-\xi)k^2} \right| \right\}.$$

Separation of the divergent part
Let us calculate

$$\lim_{M\to\infty} F_0 = [2\xi(1-\xi)(k_l k_j - k^2 \delta_{lj}) + m^2 \delta_{lj}](E+G)$$

$$-\delta_{lj}[\xi(H+J) + (1-\xi)(K+L)], \qquad (8.2.38)$$

where the notations E, G, H, and K are obvious:

$$E = \lim_{M\to\infty} \ln \left| \frac{M^2}{m^2} \frac{(1-\xi)m^2 + (\xi m^4/M^2) + [(1-\xi)m^2 k^2/M^2]}{m^2 + \xi(1-\xi)k^2} \right|$$

$$= \ln \left| \frac{M^2}{m^2} \frac{(1-\xi)m^2}{m^2 + \xi(1-\xi)k^2} \right| ; \qquad (8.2.39)$$

$$G = \lim_{M\to\infty} \ln \left| \frac{\xi + [(1-\xi)m^2/M^2 + \xi(1-\xi)k^2/M^2]}{1 + \xi(1-\xi)k^2/M^2} \right| = \ln \xi ; \quad (8.2.40)$$

$$H = m^2 \lim_{M\to\infty} \ln \left| \frac{M^2}{m^2} \frac{(1-\xi)m^2 + \xi m^4/M^2 + \xi(1-\xi)m^2 k^2/M^2}{m^2 + \xi(1-\xi)k^2} \right|$$

$$= m^2 \ln \left| \frac{M^2}{m^2} \frac{(1-\xi)m^2}{m^2 + \xi(1-\xi)k^2} \right| ; \qquad (8.2.41)$$

Photon self-energy diagram. Vacuum polarization

$$K = m^2 \lim_{M \to \infty} \ln \left| \frac{M^2}{m^2} \frac{\xi\, m^2 + (1-\xi)m^4/M^2 + \xi(1-\xi)k^2 m^2/M^2}{m^2 + \xi(1-\xi)k^2} \right|$$

$$= m^2 \ln \left| \frac{M^2}{m^2} \frac{\xi m^2}{m^2 + \xi(1-\xi)k^2} \right|. \tag{8.2.42}$$

Using the series expansion $\ln(1+x) = x + ...$, for very small x, we also have:

$$J = M^2 \lim_{M \to \infty} \ln \left| \xi \frac{1 + \{[(1-\xi)m^2/\xi] + (1-\xi)k^2\}/M^2}{1 + \xi(1-\xi)k^2/M^2} \right|$$

$$= M^2 \left[\ln \xi + \frac{(1-\xi)m^2/\xi + (1-\xi)k^2}{M^2} - \frac{\xi(1-\xi)k^2}{M^2} \right]$$

$$= M^2 \ln \xi + \frac{1-\xi}{\xi}[m^2 + \xi(1-\xi)k^2] \ ; \tag{8.2.43}$$

$$L = M^2 \lim_{M \to \infty} \ln \left| (1-\xi) \frac{1 + [\frac{\xi}{1-\xi}m^2 + \xi k^2]/M^2}{1 + \xi(1-\xi)k^2/M^2} \right|$$

$$= M^2 \left[\ln(1-\xi) + \frac{\frac{\xi}{1-\xi}m^2 + \xi k^2}{M^2} - \frac{\xi(1-\xi)k^2}{M^2} \right]$$

$$= M^2 \ln(1-\xi) + \frac{\xi}{1-\xi}[m^2 + \xi(1-\xi)k^2]. \tag{8.2.44}$$

Using the expressions (8.2.39)-(8.2.44), relation (8.2.38) writes

$$\lim_{M \to \infty} F_0 = 2\xi(1-\xi)(k_l k_j - k^2 \delta_{lj}) \ln \frac{M^2}{m^2}$$

$$-\delta_{lj}(M^2 - m^2)[\xi \ln \xi + (1-\xi)\ln(1-\xi)] + 2\xi(1-\xi)(k_l k_j - k^2 \delta_{+} lj) \tag{8.2.45}$$

$$\times \ln \left| \frac{\xi(1-\xi)m^2}{m^2 + (1-\xi)k^2} \right| - \delta_{lj}[m^2 + \xi(1-\xi)k^2].$$

As one can see, in view of (8.2.45), (8.2.27) reduces to elementary integrals of the type $\int_0^1 \xi(1-\xi)\, d\xi$, $\int_0^1 \xi \ln \xi \, d\xi$, and $\int_0^1 (1-\xi)\ln(1-\xi)\, s\xi$. The last two are integrated by parts, setting $u = \ln \xi$, $dv = \xi \, d\xi$, and $u = \ln(1-\xi)$, $dv = (1-\xi)\, d\xi$, respectively. The result is:

$$\Pi_{Rlj}(k) = -\frac{e^2}{12\pi^2}(k_l k_j - k^2 \delta_{lj}) \ln \frac{M^2}{m^2} - \delta_{lj}\frac{e^2}{8\pi^2}(M^2 - m^2)$$

$$-\frac{e^2}{2\pi^2}(k_l k_j - k^2\delta_{lj})\int_0^1 d\xi\, \xi(1-\xi)\ln\left|\frac{\xi(1-\xi)m^2}{m^2+\xi(1-\xi)k^2}\right|$$

$$+\delta_{lj}\frac{e^2}{4\pi^2}\left(m^2+\frac{1}{6}k^2\right).$$

The first two terms are logarithmically and quadratically divergent, respectively, while the last two are convergent. The term finite in m^2 was included in the second term due to the fact that it is not gauge invariant (as we shall see further on). For the same reason, we also omit the last term (it can also be included in one of the first two divergent terms). Thus, for large values of M, (8.2.27) becomes

$$\Pi_{Rlj}(k) = \delta_{lj}\frac{e^2}{8\pi^2}(m^2 - M^2)$$

$$-\frac{e^2}{12\pi^2}(k_l k_j - \delta_{lj}k^2)\ln\frac{M^2}{m^2}+\tilde{\Pi}_{lj}(k),\qquad(8.2.46)$$

where

$$\tilde{\Pi}_{lj}(k) = \frac{e^2}{2\pi^2}(k_l k_j - \delta_{lj}k^2)\int_0^1 d\xi\, \xi(1-\xi)\ln\left|\frac{m^2+\xi(1-\xi)k^2}{\xi(1-\xi)m^2}\right|$$

$$(8.2.47)$$

is a regulated function.

Going now to the configuration representation ($k_l \to \frac{1}{i}\frac{\partial}{\partial x_l}$), for M large enough we find

$$\Pi_{Rlj}(x) = \delta_{lj}\frac{e^2}{8\pi^2}(m^2 - M^2)\,\delta(x)$$

$$+\frac{e^2}{12\pi^2}\left(\frac{\partial}{\partial x_l}\frac{\partial}{\partial x_j}-\delta_{lj}\Box\right)\ln\frac{M^2}{m^2}\delta(x)+\tilde{\Pi}_{Mlj}(x),\qquad(8.2.48)$$

where the Dirac delta function displays a singularity (divergence) at the origin. In the limit $M\to\infty$, the term $\tilde{\Pi}_{Mlj}(x)$ improperly converges to the integrable function $\tilde{\Pi}_{lj}(x)$:

$$\lim_{M\to\infty}\tilde{\Pi}_{Mlj}(x)=\tilde{\Pi}_{lj}(x)$$

whose Fourier transform is given by (8.2.47). We mention that decomposition of $\Pi_{Rlj}(x)$ into two parts, one singular and the other finite, cannot be done without ambiguity. Consequently, the finite part $\tilde{\Pi}_{lj}$ is not uniquely determined: one can add to $\tilde{\Pi}_{lj}$ any second-order polynomial in k_l (because the singular part is also a quadratic polynomial in k_l).

Photon self-energy diagram. Vacuum polarization

Let us now express the gauge invariance condition that has to be satisfied by Π_{lj}. In configuration representation, the term in the S-matrix expansion corresponding to the photon is proportional to

$$N[A_l(x_1)\Pi_{lj}(x_1 - x_2)A_j(x_2)],$$

where $\Pi_{lj}(x_1 - x_2)$ is given by (8.2.5). In momentum representation, this term is proportional to

$$N[A_l(k)\Pi_{lj}(k)A_j(k)].$$

As well-known, the electromagnetic four-potential demands that all physical quantities remain invariant, when the four-potential A_m undergoes the gauge transformation

$$A_m(x) \longrightarrow A'_m = A_m + \frac{\partial g(x)}{\partial x_m},$$

or, in momentum representation

$$A_m(k) \longrightarrow A'_m(k) + ik_m \, g(k). \tag{8.2.49}$$

Since $\Pi_{lj}(k)$ does not contain electromagnetic potentials, the invariance of the term in the S-matrix expansion

$$N[A'_l(k)\Pi_{lj}(k)A'_j(k)] = N[A_l(k)\Pi_{lj}(k)A_j(k)]$$

leads, in view of (8.2.49), to the transversality condition for the operator $\Pi_{lj}(\mathrm{k})$:

$$k_l\Pi_{lj}(k) = 0, \tag{8.2.50}$$

This shows that Π_{lj} has to have the form

$$\Pi_{lj} = (k_l k_j - \delta_{lj} \, k^2)\Pi(k^2), \tag{8.2.51}$$

which can be easily verified.

Returning to (8.2.46) and (8.2.47), one observes that the gauge invariance condition (8.2.50) can only be used for regulated part $\bar{\Pi}_{lj}(k)$ of $\Pi_{Rlj}(k)$, because the function $\Pi_{lj}(k)$ is divergent. We can then add to $\tilde{\Pi}_{lj}(k)$ a second order polynomial in k

$$C_3\frac{e^2}{2\pi^2}(k_l k_j - \delta_{lj}k^2),$$

where C_3 is an arbitrary constant. This term is taken out of the logarithmically divergent term in (8.2.46). Thus, (8.2.48) becomes

$$\Pi_{Rlj}(x) = \delta_{lj}\frac{e^2}{8\pi^2}(m^2 - M^2)\,\delta(x)$$

$$+\frac{e^2}{12\pi^2}\left(\ln\frac{M^2}{m^2} - 6C_3\right)\left(\frac{\partial}{\partial x_l}\frac{\partial}{\partial x_j} - \delta_{lj}\Box\right)\delta(x) + \tilde{\Pi}_{lj}^i(x), \quad (8.2.52)$$

where $\tilde{\Pi}_{lj}^i(x)$ is the convergent (finite) part, and is invariant with respect to gauge transformations. In momentum representation, (8.2.46) writes

$$\Pi_{Rlj}(k) = \delta_{lj}\frac{e^2}{8\pi^2}(m^2 - M^2)$$

$$+\frac{e^2}{12\pi^2}\left(\ln\frac{M^2}{m^2} - 6C_3\right)\left(k_l k_j - \delta_{lj}k^2\right) + \tilde{\Pi}_{lj}^i(k), \quad (8.2.53)$$

where the convergent part $\tilde{\Pi}_{lj}^i(k)$ is given by

$$\tilde{\Pi}_{lj}^i(k) = \frac{e^2}{2\pi^2}(k_l k_j - \delta_{lj}k^2)\left(C_3 + \int_0^1 d\xi\,\xi(1-\xi)\ln\left|\frac{m^2 + \xi(1-\xi)k^2}{\xi(1-\xi)m^2}\right|\right).$$
$$(8.2.54)$$

The constant C_3 can be determined by imposing the following conditions:

$$\tilde{\Pi}_{lj}^i(0) = 0\ ; \quad \frac{\partial}{\partial x_i}\left[\tilde{\Pi}_{lj}^i(0)\right] = 0\ ; \quad \frac{\partial^2}{\partial x_i \partial x_k}\left[\tilde{\Pi}_{lj}^i(0)\right] = 0. \quad (8.2.55)$$

Consequently, the function corresponding to the photon self-energy, in configuration representation, is:

$$\Pi_{lj}(x_1 - x_2) = \tilde{\Pi}_{lj}^i(x_1 - x_2) + C_1\delta_{lj}\delta(x_1 - x_2)$$

$$-C_2\left(\frac{\partial}{\partial x_l}\frac{\partial}{\partial x_j} - \delta_{lj}\Box\right)\delta(x_1 - x_2), \quad (8.2.56)$$

where the constants

$$C_1 = \frac{e^2}{8\pi^2}\lim_{M\to\infty}(m^2 - M^2)\ ; \quad C_2 = \frac{e^2}{12\pi^2}\lim_{M\to\infty}\left(6C_3 - \ln\frac{M^2}{m^2}\right)$$
$$(8.2.57)$$

are divergent, while $\tilde{\Pi}_{lj}^i(x_1 - x_2)$ is convergent. In momentum representation (8.2.56) writes

$$\Pi_{lj}(k) = \tilde{\Pi}_{lj}^i(k) + C_1\delta_{lj} + C_2(k_l k_j - k^2\delta_{lj}). \quad (8.2.58)$$

8.3. Mass and charge renormalization

The S-matrix element associated with the photon self-energy, in configuration representation, is

$$S^{(2)} = -i \int A_{l-}(x_1)\Pi_{lj}(x_1-x_2)A_{j+}(x_2)\, dx_1\, dx_2 = S_1^{(2)} + S_2^{(2)} + S_3^{(2)},$$
(8.3.1)

where $\Pi_{lj}(x_1 - x_2)$ is given by (8.2.56). We are mainly interested in determining $S_2^{(2)}$, which is

$$S_2^{(2)} = C_1 \int A_{l-}(x_1)A_{l+}(x_1)\, dx_1,$$
(8.3.2)

where the integration with respect to x_2 ($dx_2 = i\, d\vec{x}_2\, dx_{20}$) has been performed.

Let us now assume that the photon is described by a vector field of mass μ. In zeroth order, the photon mass is defined by

$$\mu_0 = <\vec{k} = 0|H_0|\vec{k} = 0>,$$

where H_0 is the corresponding Hamiltonian, and $|\vec{k} = 0>$ is the single-photon rest state. In higher orders, we have

$$\mu = <\vec{k} = 0|H_0 + H'|\vec{k} = 0> = \mu_0 - \delta\mu'$$

or

$$\mu^2 = \mu_0^2 - \delta\mu^2,$$
(8.3.3)

where μ is the experimentally measured mass, μ_0 is the infinite non-renormalized (bare) mass, and $\delta\mu^2$ is the mass correction term, which is also infinite. In this case, the mass μ_0 in the total Lagrangian density

$$\mathcal{L} = -\frac{1}{2}A_{l,k}A_{l,k} - \frac{1}{2}\mu_0^2 A_l A_l - \bar{\psi}(\hat{p}+m)\psi - im\bar{\psi}\,\hat{A}\psi$$

has to be replaced by the experimental (observed) value μ. This action has to be compensated by a mass term counterpart. If the mass variation obeys (8.3.3), then the additional interaction is given by the Lagrangian density

$$\mathcal{L}' = -\frac{1}{2}\delta\mu^2\, A_l A_l.$$
(8.3.4)

The corresponding Hamiltonian density is

$$H'(x_1) = \frac{1}{2}\delta\mu^2\, A_l(x_1)A_l(x_1),$$
(8.3.5)

269

which leads to the following first-order term in the S-matrix

$$S^{(1)} = -\int N[H'(x_1)]\, dx_1 = -\frac{1}{2}\delta\mu^2 \int A_{l-}(x_1)A_{l+}(x_1)\, dx_1. \quad (8.3.6)$$

with the associated Feynman graph

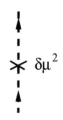

In order to make the Lagrangian invariant under the considered gauge transformation, one must set $C_1 = 0$, and (8.2.58) reduces to

$$\Pi_{lj}(k) = \tilde{\Pi}^i_{lj}(k) + C_2(k_l k_j - k^2 \delta_{lj}). \quad (8.3.7)$$

Here $k^2 \neq 0$, since there are no free photons, but the gauge condition $k_i A_i(k) = 0$ is still valid.

The self-energy of the photon can be inserted into either an internal or an external photonic line, such as

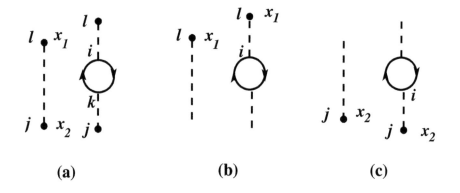

(a)   (b)   (c)

which can be written as:

$$a)\quad \delta_{lj}D_0^{(c)}(k) \longrightarrow \delta_{lj}D_0^{(c)}(k) + \delta_{li}D_0^{(c)}(k)\,\Pi_{ik}(k)\delta_{kj}D_0^{(c)}(k),$$

with $D_0^{(c)}(k)$ given by (A.129), or

$$\delta_{lj}D_0^{(c)}(k) \longrightarrow \delta_{lj}D_0^{(c)}(k) + D_0^{(c)}(k)\,\Pi_{lj}(k)D_0^{(c)}(k)\ ; \quad (8.3.8)$$

b) $\quad A_l(k) \longrightarrow A_l(k) + \delta_{li} D_0^{(c)}(k)\, \Pi_{ij}(k) A_j(k),$

or

$$A_l(k) \longrightarrow A_l(k) + D_0^{(c)}(k)\, \Pi_{lj}(k) A_j(k) \; ; \qquad (8.3.9)$$

c) $\quad A_j(k) \longrightarrow A_j(k) + A_l(k)\, \Pi_{li}(k)\delta_{ij} D_0^{(c)}(k),$

or

$$A_j(k) \longrightarrow A_j(k) + D_0^{(c)}(k)\, \Pi_{lj}(k) A_l(k). \qquad (8.3.10)$$

On the other hand, (8.3.7) allows one to write

$$D_0^{(c)}(k)\, \Pi_{lj}(k) = D_0^{(c)}(k)\, \tilde{\Pi}^i_{lj}(k) - \delta_{lj}\, C_2\, D_0^{(c)}(k)\, k^2,$$

because the term containing the product $k_l k_j$ vanishes according to the gauge condition. Substituting the last relation into (8.3.8) and (8.3.9), and taking into account that

$$D_0^{(c)}(k) \to \frac{1}{(2\pi)^2 i}\, \frac{1}{k^2} \qquad \text{for } \alpha \to 0,$$

one obtains

$$\delta_{lj} D_0^{(c)}(k) \longrightarrow (1 - C_2')\delta_{lj} D_0^{(c)}(k) + D_0^{(c)}(k)\, \tilde{\Pi}^i_{lj}(k) D_0^{(c)}(k) \; ; \quad (8.3.11)$$

$$A_l(k) \longrightarrow (1 - C_2') A_l(k) + D_0^{(c)}(k)\, \tilde{\Pi}^i_{lj}(k) A_j(k), \qquad (8.3.12)$$

where $C_2' = C_2/(2\pi)^2 i$.

Let us consider a diagram with n vertices and f external photonic lines. Since one photonic line enters or leaves each vertex, we have n photonic lines in total. Subtracting f external photonic lines from the total of n photonic lines, and taking into account that one propagator is associated with each pair, we are left with $(n-f)/2$ internal photonic lines, that is $(n-f)/2$ functions $D_0^{(c)}(k)$, and f external lines associated with $A_l(k)$. It is convenient to define the renormalized functions

$$D_0^{(c)}(k) \longrightarrow \frac{D_0^{(c)}(k)}{1 - C_2'} \; ; \quad A_l(k) \longrightarrow \frac{A_l(k)}{\sqrt{1 - C_2'}}. \qquad (8.3.13)$$

Since the factor $(1 - C_2')^{-f/2-(n-f)/2} = (1 - C_2')^{-n/2}$ in front of the matrix element $S^{(n)}$ can be compensated by taking (n=2)

$$e \longrightarrow e_0 = \sqrt{1 - C_2'}\, e \simeq \left(1 - \frac{1}{2}C_2'\right) e = e + \delta e, \qquad (8.3.14)$$

271

where

$$\delta e = -\frac{1}{2} C_2' \, e, \qquad (8.3.15)$$

then the two divergent quantities e_0 and δe balance each other out, and we are left with the finite electron charge e, observed experimentally. This way we have eliminated C_2', and we are only left with $\tilde{\Pi}_{lj}^i(k)$, which is calculable and finite. Relation (8.3.7) then becomes

$$\Pi_{lj}(k) = \tilde{\Pi}_{lj}^i(k), \qquad (8.3.16)$$

and we finally obtain

$$S_1^{(2)} = -i \int A_{l-}(x_1) \, \tilde{\Pi}_{lj}^i(x_1 - x_2) A_{j+}(x_2) \, dx_1 \, dx_2. \qquad (8.3.17)$$

To conclude, our analysis has shown that the photon self-energy graph (or the vacuum polarization diagram), represents the interaction between the vacuum and the fermionic field, mediated by a virtual electron-positron pair, in other words a self-interaction of the electromagnetic field. To eliminate the divergence related to the polarization operator one proceeds in the same manner as in the case of the mass operator, except that this time one does not need to perform the photon mass renormalization, since its rest mass is zero, and instead one uses the invariance of the electromagnetic field under gauge transformations.

APPENDIX. DISTRIBUTIONS

In physics and engineering many problems naturally lead to differential equations whose solutions or initial conditions are *distributions* (*generalized functions*).

This Appendix is concerned with the so-called Dirac delta function and several other functions related to it, used in Quantum Field Theory as distribution models. Our purpose is to give only a general survey on some singular functions employed in our investigations, without any claim of deep rigorousness or exhaustivity.

A.1. Unidimensional delta function

Let $\delta(x, \alpha)$ be a function that depends on a variable x, and a parameter $\alpha > 0$, satisfying the following conditions:

$$
\begin{cases}
\lim_{\alpha \to 0} \delta(x, \alpha) = \begin{cases} 0 & \text{for } x \neq 0 \ ; \\ +\infty & \text{for } x = 0 \ , \end{cases} \\
\lim_{\alpha \to 0} \int_{-\infty}^{+\infty} \delta(x, \alpha) \ dx = 1.
\end{cases} \tag{A.1}
$$

Let us denote

$$
\begin{cases}
\delta(x) = \lim_{\alpha \to 0} \delta(x, \alpha) \ ; \\
\int_{-\infty}^{+\infty} \delta(x) \ dx = \lim_{\alpha \to 0} \int_{-\infty}^{+\infty} \delta(x, \alpha) \ dx = 1.
\end{cases} \tag{A.2}
$$

We mention that in general $\delta(x, \alpha)$ is not uniformly convergent for $\alpha \to 0$ and $x = 0$, therefore the limit and integral cannot be reversed, hence (A.2) is only a symbolic notation.

Whenever $\delta(x)$ appears under the integral sign, we have to imply the following operation

$$
\int_{-\infty}^{+\infty} f(x) \ \delta(x) \ dx = \lim_{\alpha \to 0} \int_{-\infty}^{+\infty} f(x) \ \delta(x, \alpha) \ dx. \tag{A.3}
$$

Let $f(x)$ be a continuous function at some point $x = a$. Then

$$
\int_{a-u}^{a+v} f(x) \ \delta(x - a) \ dx = f(a), \tag{A.4}
$$

273

Appendix. Distributions

where u and v are two positive arbitrary constants. In particular, taking $a = 0$ and extending the integral over the whole x-axis, we have

$$\int_{-\infty}^{+\infty} f(x)\, \delta(x)\, dx = f(0). \qquad (A.5)$$

The last two relations yield several properties of $\delta(x)$. For example,

$$\int_{-\infty}^{+\infty} f(x)\, \delta(-x)\, dx = -\int_{+\infty}^{-\infty} f(-x)\, \delta(x)\, dx$$

$$= \int_{-\infty}^{+\infty} f(-x)\, \delta(x)\, dx = f(0),$$

which gives

$$\delta(-x) = \delta(x), \qquad (A.6)$$

thus $\delta(x)$ is an *even* function. Using the substitution $y = cx$ and (A.6), we have

$$\int_{-\infty}^{+\infty} f(x)\, \delta(cx)\, dx = \frac{1}{|c|} \int_{-\infty}^{+\infty} f(\frac{y}{|c|})\delta(y)\, dy = \frac{1}{|c|} f(0),$$

therefore

$$\delta(cx) = \frac{1}{|c|}\delta(x). \qquad (A.7)$$

If $\varphi(x)$ has only simple roots $x_1, x_2, ..., x_n$, i.e. $\varphi(x_i) = 0$, and $\varphi'(x_i) \neq 0$ $(i = 1, 2, ..., n)$, then

$$\int_{-\infty}^{+\infty} f(x)\, \delta[\varphi(x)]\, dx = \sum_{i=1}^{n} \int_{n_{i-1}}^{n_i} f(x)\, \delta[\varphi(x)]\, dx,$$

where

$$a_o < x_1 < a_1 < x_2 < a_2 < ... < a_{n-1} < x_n < a_n.$$

But, according to our previous results,

$$\int_{n_{i-1}}^{n_i} f(x)\, \delta[\varphi(x)]\, dx = \int_{x_i-\epsilon}^{x_i+\epsilon} f(x)\, \delta[\varphi'(\xi)(x - x_i)]\, dx = \frac{f(x_i)}{|\varphi'(x_i)|},$$

so that

$$\int_{-\infty}^{+\infty} f(x)\, \delta[\varphi(x)]\, dx = \sum_{i=1}^{n} \frac{f(x_i)}{|\varphi'(x_i)|} = \int_{-\infty}^{+\infty} f(x) \sum_{i=1}^{n} \frac{\delta(x - x_i)}{|\varphi'(x_i)|}\, dx$$

274

which gives

$$\delta[\varphi(x)] = \sum_{i=1}^{n} \frac{\delta(x - x_i)}{|\varphi'(x_i)|}. \qquad (A.8)$$

In particular, if $\varphi(x) = x^2 - a^2$, then

$$\delta(x^2 - a^2) = \frac{\delta(x - a) + \delta(x + a)}{2|a|}. \qquad (A.9)$$

A.2. Various representations of the delta function

a) The Dirac delta function can be completely described by a complete orthonormalized set of functions $\psi_l(x)$. Indeed, expanding $f(x)$ in terms of $\psi_l(x)$, we obtain

$$f(x_0) = \sum_l a_l \psi_l(x_0),$$

where

$$a_l = \int_{-\infty}^{+\infty} f(x)\, \psi_l^*(x)\, dx,$$

so that

$$f(x_0) = \int_{-\infty}^{+\infty} \left[\sum_l \psi_l^*(x)\, \psi_l(x_0)\right] f(x)\, dx. \qquad (A.10)$$

Comparing (A.10) with (A.4), one can write

$$\delta(x - x_0) = \sum_l \psi_l^*(x)\psi_l(x_0), \qquad (A.11)$$

which is the representation of $\delta(x - x_0)$ by means of the complete orthonormalized set $\psi(x)$.

b) As well-known, the Fourier integral expansion of $f(x_0)$ is given by

$$f(x_0) = \frac{1}{\sqrt{2\pi}} \int_{-\infty}^{+\infty} e^{ikx_0} f(k)\, dk, \qquad (A.12)$$

where the Fourier transform $f(k)$ of $f(x)$ is

$$f(k) = \frac{1}{\sqrt{2\pi}} \int_{-\infty}^{+\infty} e^{-ikx} f(x)\, dx. \qquad (A.13)$$

Appendix. Distributions

The last two relations yield

$$f(x_0) = \int_{-\infty}^{+\infty} \left\{ \frac{1}{2\pi} \int_{-\infty}^{+\infty} e^{ik(x_0-x)} \right\} f(x)\, dx. \qquad (A.14)$$

Setting $x_0 = 0$ in (A.14) and using (A.6), we have

$$\delta(x) = \frac{1}{2\pi} \int_{-\infty}^{+\infty} e^{ikx}\, dk = \lim_{K\to\infty} \int_{-K}^{+K} e^{ikx}\, dk, \qquad (A.15)$$

which is called the *Fourier integral representation* of the Dirac delta function.

Since $\sin kx$ is an odd function, and $\cos kx$ an even function, in view of (A.15) one can also write

$$\delta(x) = \lim_{K\to\infty} \frac{1}{\pi} \int_0^K \cos kx\, dk = \lim_{K\to\infty} \frac{1}{\pi} \frac{\sin Kx}{x}. \qquad (A.16)$$

Expressions (A.15) and (A.16) can also be used as definitions of the delta function. For example, using (A.16) one obtains

$$\int_{-\infty}^{+\infty} f(x)\,\delta(x)\, dx = \lim_{K\to\infty} \frac{1}{\pi} \int_{-\infty}^{+\infty} f(x) \frac{\sin Kx}{x}\, dx$$

$$= \lim_{K\to\infty} \frac{1}{\pi} \int_{-\infty}^{+\infty} f\left(\frac{x}{K}\right) \frac{\sin x}{x}\, dx = f(0),$$

and we recover formula (A.5).

c) Let us consider the complex plane (x, y), and let C be a closed contour around $x = 0$. According to Cauchy's formula

$$\frac{1}{2\pi i} \oint \frac{f(x)}{x}\, dx = f(0),$$

so that we can formally write

$$\delta(x) = \frac{1}{2\pi i} \left.\frac{1}{x}\right|_C, \qquad (A.17)$$

where the contour is traversed in the positive (counterclockwise) direction.

A.3. Some functions related to delta

a) The basic properties of $\delta(x)$ allow one to write

$$\epsilon(x) = \int_{-x}^{+x} \delta(\xi)\,d\xi = \begin{cases} +1 & \text{for} \quad x>0\ ; \\ -1 & \text{for} \quad x<0\ . \end{cases} \tag{A.18}$$

Due to (A.6) we then have

$$\frac{1}{2}\epsilon(x) = \int_0^x \delta(\xi)\,d\xi, \tag{A.19}$$

and we can formally write

$$\delta(x) = \frac{1}{2}\,\epsilon'(x). \tag{A.20}$$

According to (A.18) we have

$$\epsilon(x) = \frac{d}{dx}\,|x|, \tag{A.21}$$

and (A.20) becomes

$$\delta(x) = \frac{1}{2}\,\frac{d^2}{dx^2}\,|x|, \tag{A.22}$$

which is another representation of $\delta(x)$.

We can also define the derivative of $\delta(x)$. Performing a formal integration by parts, we can write

$$\int_{-\infty}^{+\infty} f(x)\,\delta'(x)\,dx = f(x)\,\delta(x)\Big|_{-\infty}^{+\infty} - \int_{-\infty}^{+\infty} f'(x)\,\delta(x)\,dx,$$

so that

$$\int_{-\infty}^{+\infty} f(x)\,\delta'(x)\,dx = -f'(0). \tag{A.23}$$

Similarly, we have

$$\int_{-\infty}^{+\infty} x\,f(x)\,\delta'(x)\,dx = x\,f(x)\,\delta(x)\Big|_{-\infty}^{+\infty} - \int_{-\infty}^{+\infty} [f(x)+x\,f'(x)]\,\delta(x)\,dx,$$

therefore

$$\int_{-\infty}^{+\infty} f(x)\,[-x\,\delta'(x)]\,dx = f(0). \tag{A.24}$$

Appendix. Distributions

By means of (A.24) and (A.5), one then obtains

$$\delta'(x) = -\frac{\delta(x)}{x}. \tag{A.25}$$

Using the same procedure, the nth derivative of $\delta(x)$ writes

$$\delta^{(n)}(x) = (-1)^n \frac{n!}{x^n}\, \delta(x). \tag{A.26}$$

It is easily noted that $\epsilon(x)$, $\delta^{(2n+1)}(x)$ are odd functions, while $\delta(x)$, $\delta^{(2n)}(x)$ are even functions.

b) Another important singular function is

$$\zeta(x) = -i \lim_{K\to\infty} \int_0^K e^{ikx}\, dx = \lim_{K\to\infty} \frac{1 - e^{-iKx}}{x}. \tag{A.27}$$

By means of (A.16), one can also write

$$\zeta(x) = i\pi\,\delta(x) = \lim_{K\to\infty} \frac{1 - \cos Kx}{x}. \tag{A.28}$$

The function

$$\frac{\mathcal{P}}{x} = \lim_{K\to\infty} \frac{1 - \cos Kx}{x} = \lim_{K\to\infty} \int_0^K \sin kx\, dk = \frac{1}{2\pi} \int_{-\infty}^{+\infty} e^{ikx}\, \epsilon(k)\, dk \tag{A.29}$$

is called the *principal part of* $1/x$. To determine the behavior of $\mathcal{P}/x$, consider the integral

$$\int_{-\infty}^{+\infty} f(x)\frac{\mathcal{P}}{x}\, dx = \lim_{K\to\infty} \int_{-\infty}^{+\infty} f(x)\frac{1 - \cos Kx}{x}\, dx. \tag{A.30}$$

The integral on the r.h.s. can be decomposed as

$$\int_{-\infty}^{+\infty} f(x)\frac{1 - \cos Kx}{x}\, dx = \int_{-\infty}^{-\epsilon} \frac{f(x)}{x}\, dx + \int_{\epsilon}^{+\infty} \frac{f(x)}{x}\, dx$$

$$- \int_{-\infty}^{-\epsilon} \frac{f(x)}{x}\cos Kx\, dx - \int_{\epsilon}^{+\infty} \frac{f(x)}{x}\cos Kx\, dx \tag{A.31}$$

$$+ \int_{-\epsilon}^{+\epsilon} f(x)\frac{1 - \cos Kx}{x}\, dx.$$

278

Some functions related to delta

But, on one hand

$$\lim_{K\to\infty}\int_{-\infty}^{-\epsilon}\frac{f(x)}{x}\cos Kx\,dx = \lim_{K\to\infty}\int_{\epsilon}^{+\infty}\frac{f(x)}{x}\cos Kx\,dx = 0,$$

(A.32)

because the period of $\cos Kx$ decreases as K increases, while $f(x)/x$ is non-singular over the intervals $(-\infty,-\epsilon)$ and $(\epsilon,+\infty)$. On the other hand, since ϵ is arbitrarily small and

$$\lim_{x\to 0}\frac{1-\cos Kx}{x}=0$$

for any K , we have

$$\lim_{K\to\infty}\lim_{\epsilon\to 0}\int_{-\epsilon}^{+\epsilon} f(x)\frac{1-\cos Kx}{x}\,dx = 0.$$

(A.33)

Using (A.30)-(A.33), we get

$$\int_{-\infty}^{+\infty} f(x)\frac{\mathcal{P}}{x}\,dx = \lim_{\epsilon\to 0}\left\{\int_{-\infty}^{-\epsilon}\frac{f(x)}{x}\,dx + \int_{\epsilon}^{+\infty}\frac{f(x)}{x}\,dx\right\},\quad (A.34)$$

which also justifies the designation "principal part".

Relations (A.28) and (A.29) allow us to write

$$\zeta(x) = \frac{\mathcal{P}}{x} - i\pi\,\delta(x),$$

(A.35)

as well as

$$\zeta^*(x) = \frac{\mathcal{P}}{x} + i\pi\,\delta(x) = -\zeta(-x).$$

(A.36)

Sometimes is more convenient to use the functions

$$\begin{cases}\delta_+(x) = \frac{1}{2\pi i}\zeta^*(x) = \frac{1}{2\pi i}\frac{\mathcal{P}}{x} + \frac{1}{2}\delta(x)\;;\\ \delta_-(x) = -\frac{1}{2\pi i}\zeta(x) = -\frac{1}{2\pi i}\frac{\mathcal{P}}{x} + \frac{1}{2}\delta(x).\end{cases}$$

(A.37)

Using expressions(A.35)-(A.37), we have

$$\begin{cases}\delta(x) = \frac{1}{2\pi i}\left[\zeta^*(x)-\zeta(x)\right] = \delta_+(x)+\delta_-(x)\;;\\ \frac{\mathcal{P}}{x} = \frac{1}{2}\left[\zeta^*(x)+\zeta(x)\right] = i\pi\left[\delta_+(x)-\delta_-(x)\right].\end{cases}$$

(A.38)

c) All generalized functions introduced so far can be written as functions of $1/x$ if the integration contour is suitably chosen. Indeed,

Appendix. Distributions

from (A.34) it follows that the integration contour for $\mathcal{P}/x$ is given in Fig.A.1(a), while according to (A.17), the integration contour for $2\pi i\delta(x)$ is the one shown in Fig.A.1(b). Following (A.35) and (A.36), the integration paths for $\zeta(x)$ and $\zeta^*(x)$ are given in Fig.A.1(c) and Fig.A.1(d), respectively.

It is apparent that instead of contour (c) we can take a horizontal line above the real axis, while contour (d) can be replaced by a straight line under the real axis. These replacements are equivalent to the transformations: $x \to x + i\alpha$ and $x \to x - i\alpha$, respectively, and we can formulate the following representations

$$\zeta(x) = \lim_{\alpha \to 0} \frac{1}{x + i\alpha} \; ; \quad \zeta^*(x) = \lim_{\alpha \to 0} \frac{1}{x - i\alpha}. \qquad (A.39)$$

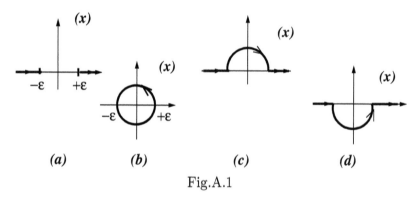

Fig.A.1

Obviously, when $\zeta(x)$ and $\zeta^*(x)$ appear under the integral sign, the limit is taken *after* the integration.

Using (A.38) and (A.39), we find

$$\delta(x) = \lim_{\alpha \to 0} \frac{1}{\pi} \frac{\alpha}{x^2 + \alpha^2} \; ; \quad \frac{\mathcal{P}}{x} = \lim_{\alpha \to 0} \frac{x}{x^2 + \alpha^2}. \qquad (A.40)$$

Comparing $(A.1)_1$ with $(A.40)_1$, we can then take

$$\delta(x, \alpha) = \frac{1}{\pi} \frac{\alpha}{x^2 + \alpha^2}. \qquad (A.41)$$

Using this choice,

$$\int_{-\infty}^{+\infty} \delta(x, \alpha) \, dx = \frac{1}{\pi} \int_{-\infty}^{+\infty} \frac{\alpha \, dx}{x^2 + \alpha^2} = \frac{1}{\pi} \arctan \frac{x}{\alpha} \Big|_{-\infty}^{+\infty} = 1,$$

for any given α. Condition $(A.1)_2$ is, therefore, fulfilled.

We conclude that (A.41) is a concrete example of a function $\delta(x, \alpha)$ which satisfies our initial definition (A.1).

A.4. Functions D_{m+} and D_{m-}

A multidimensional generalization of $\delta(x)$ can be obtained as a product of several functions $\delta(x_1)$, $\delta(x_2)$, ...$\delta(x_n)$. For example, the three-dimensional Dirac delta function, as a generalization of (A.15), writes

$$\delta(\vec{x}) = \frac{1}{(2\pi)^3} \int e^{i\vec{k}\cdot\vec{x}} \, d\vec{k}, \qquad (A.42)$$

where the integral extends over the whole space $\vec{k}$, $d\vec{k}$ is a volume element in this space, and $\vec{k} \cdot \vec{x} = k_1 x_1 + k_2 x_2 + k_3 x_3$.

In view of (A.6) and (A.7), one can write

$$\delta(t) = \frac{1}{2\pi} \int e^{-itk_0} \, dk_9 = \frac{1}{2\pi i} \int e^{ix_4 k_4} \, dk_4, \qquad (A.43)$$

where $x_4 = it$, $k_4 = ik_0$.

Using (A.42) and (A.43), one obtains the four-dimensional generalization of the delta function as

$$\delta(x) = \frac{1}{(2\pi)^4} \int e^{ikx} \, d\vec{k} \, dk_0 = \frac{1}{(2\pi)^4 i} \int e^{ikx} \, dk, \qquad (A.44)$$

where the four-vectors x and k have the components ($\vec{x}$, $x_4 = it$), and ($\vec{k}$, $k_4 = ik_0$), respectively, while $kx = \vec{k} \cdot \vec{x} + k_4 x_4 = \vec{k} \cdot \vec{x} - k_0 t$, and $dk = i\, d\vec{k}\, dk_0$.

We have to mention that the four-dimensional delta function (A.44) plays a secondary role compared to other singular four-dimensional functions, since we have to take into account that the momentum four-vector k obeys the energy-momentum relation

$$k^2 + m^2 = 0, \qquad (A.45)$$

where m is the rest mass of the particle. In addition, an important role is played by the separation of the k space into positive energy states ($k_0 > 0$), and negative energy states ($k_0 < 0$). This separation can be done with the help of the *Heaviside step function*

$$\theta(k_0) = \begin{cases} +1 & \text{for} \quad k_0 > 0 \, ; \\ 0 & \text{for} \quad k_0 < 0 \, . \end{cases} \qquad (A.46)$$

The Heaviside step function is related to the function $\epsilon(x)$ defined by (A.18). Indeed,

$$\begin{aligned} \theta(x) + \theta(-x) &= 1 \, ; \\ \theta(x) - \theta(-x) &= \epsilon(x), \end{aligned} \qquad (A.47)$$

Appendix. Distributions

so that (see (A.20)) we have

$$\theta(x) = \frac{1}{2}[\epsilon(x) + 1] \; ; \quad \theta'(x) = \frac{1}{2}\epsilon'(x) = \delta(x). \qquad (A.48)$$

Using the above considerations, let us now define the functions

$$D_{m\pm}(x) = \pm \frac{1}{(2\pi)^3 i} \int e^{ikx} \, \theta(\pm k_0)\delta(k^2 + m^2) \, d\vec{k} \, dk_0. \qquad (A.49)$$

By convention, either the upper or the lower sign is used.

Relation (A.12) can be generalized as

$$f(x) = \frac{1}{(2\pi)^2} \int e^{ikx} \, f(k) \, d\vec{k} \, dk_0, \qquad (A.50)$$

where x and k are four-vectors, and $f(k)$ is the Fourier transform of $f(x)$. Then (A.49) yields

$$D_{m\pm}(x) = \pm \frac{1}{2\pi \, i} \theta(\pm k_0) \, \delta(k^2 + m^2). \qquad (A.51)$$

Due to the presence of the delta function under the integral in (A.49), one can easily reduce the quadruple integral to a triple integral. Since

$$k^2 + m^2 = (\vec{k}^2 + m^2) - k_0^2 = -(k_0^2 - \omega^2),$$

where

$$\omega = +\sqrt{\vec{k}^2 + m^2}, \qquad (A.52)$$

then taking into account (A.6) and (A.9), we get

$$\delta(k^2 + m^2) = \delta(k_0^2 - \omega^2) = \frac{\delta(k_0 - \omega) + \delta(k_0 + \omega)}{2\omega}, \qquad (A.53)$$

so that

$$\int_{-\infty}^{+\infty} e^{-ik_0 t} \, \theta(k_0) \, \frac{\delta(k_0 - \omega) + \delta(k_0 + \omega)}{2\omega} \, dk_0$$

$$= \int_0^{+\infty} \frac{e^{-ik_0 t}}{2\omega} \delta(k_0 - \omega) \, dk_0 \; ;$$

$$\int_{-\infty}^{+\infty} e^{-ik_0 t} \, \theta(-k_0) \, \frac{\delta(k_0 - \omega) + \delta(k_0 + \omega)}{2\omega} \, dk_0$$

$$= \int_0^{+\infty} \frac{e^{-ik_0t}}{2\omega} \delta(k_0 + \omega)\, dk_0 \;,$$

and therefore

$$\int_{-\infty}^{+\infty} e^{-ik_0t}\, \theta(\pm k_0)\, \delta(k^2 + m^2)\, dk_0 = \frac{e^{\mp i\omega t}}{2\omega}. \qquad (A.54)$$

In view of (A.49) and (A.54), one can write:

$$D_{m+}(x) = \frac{1}{(2\pi)^3 i} \int e^{i\vec{k}\cdot\vec{x}}\, \frac{e^{-i\omega t}}{2\omega}\, d\vec{k}, \qquad (A.55)$$

and

$$D_{m-}(x) = \frac{-1}{(2\pi)^3 i} \int e^{i\vec{k}\cdot\vec{x}}\, \frac{e^{i\omega t}}{2\omega}\, d\vec{k}. \qquad (A.55)$$

If in the last integral we make the change of variable $\vec{k} \to -\vec{k}$, which results in $\omega \to \omega$, and $d\vec{k} \to d\vec{k}$ (see (A.52)), the absolute value of the Jacobian being $+1$, the last two formulas can be combined into

$$D_{m\pm}(x) = \pm\frac{1}{(2\pi)^3 i} \int \frac{e^{\pm ikx}}{2\omega}\, d\vec{k}. \qquad (A.57)$$

We shall come back to these definitions. For the moment, we just emphasize that in (A.57) $kx = \vec{k}\cdot\vec{x} - \omega t$, i.e. $k_0 = \omega$.

A.5. Functions D_m and $\tilde{D}_m$

These functions are defined as

$$\begin{cases} D_m(x) = D_{m+}(x) + D_-(x) \;; \\ \tilde{D}_m(x) = -i[D_{m+}(x) - D_-(x)]. \end{cases} \qquad (A.58)$$

In view of (A.47), (A.49), and (A.58), one can write

$$\begin{cases} D_m(x) = \frac{1}{(2\pi)^3 i} \int e^{ikx} \epsilon(k_0)\, \delta(k^2 + m^2)\, d\vec{k}\, dk_0 \;; \\ \tilde{D}_m(x) = \frac{-1}{(2\pi)^3} \int e^{ikx} \delta(k^2 + m^2)\, d\vec{k}\, dk_0 \;. \end{cases} \qquad (A.59)$$

The Fourier transforms of these relations are found by comparison with (A.50):

$$\begin{cases} D_m(k) = \frac{1}{2\pi i} \epsilon(k_0)\delta(k^2 + m^2) \;; \\ \tilde{D}_m(x) = \frac{-1}{(2\pi)} \delta(k^2 + m^2). \end{cases} \qquad (A.60)$$

Using (A.55) and (A.56), relations (A.58) can also be written as

$$\begin{cases} D_m(x) = -\frac{1}{(2\pi)^3} \int e^{i\vec{k}\cdot\vec{x}} \frac{\sin \omega t}{\omega} \, d\vec{k} \; ; \\ \tilde{D}_m(x) = -\frac{1}{(2\pi)^3} \int e^{i\vec{k}\cdot\vec{x}} \frac{\cos \omega t}{\omega} \, d\vec{k} \end{cases} \tag{A.61}$$

where ω is given by (A.52).

To perform the integration in (A.61), we choose the x_3-axis oriented along $\vec{x}$. Then we have

$$\begin{cases} \vec{k} \cdot \vec{x} = \kappa r \cos \theta \; ; \quad \kappa = |\vec{k}|, \quad r = |\vec{x}| \; ; \\ d\vec{k} = \kappa^2 \sin \theta \, d\kappa \, d\theta \, d\varphi, \end{cases} \tag{A.62}$$

and the relations (A.61) yield

$$D_m(x) = -\frac{1}{(2\pi)^3} \int_0^\infty \frac{\sin \omega t}{\omega} \left\{ \int_0^{2\pi} d\varphi \int_0^\pi e^{i\kappa r \cos \theta} \sin \theta \, d\theta \right\} \kappa^2 \, d\kappa \; ;$$

$$\tilde{D}_m(x) = -\frac{1}{(2\pi)^3} \int_0^\infty \frac{\cos \omega t}{\omega} \left\{ \int_0^{2\pi} d\varphi \int_0^\pi e^{i\kappa r \cos \theta} \sin \theta \, d\theta \right\} \kappa^2 \, d\kappa \; .$$

Since

$$\int_0^\pi e^{i\kappa r \cos \theta} \sin \theta \, d\theta = -\left.\frac{e^{i\kappa r \cos \theta}}{i\kappa r}\right|_0^\pi = \frac{2}{\kappa r} \sin \kappa r, \tag{A.63}$$

we also have

$$\begin{cases} D_m(x) = -\frac{2}{(2\pi)^2 r} \int_0^\infty \frac{\sin \omega t \, \sin \kappa r}{\omega} \, \kappa \, d\kappa \; ; \\ \tilde{D}_m(x) = -\frac{2}{(2\pi)^2 r} \int_0^\infty \frac{\cos \omega t \, \sin \kappa r}{\omega} \, \kappa \, d\kappa \; . \end{cases} \tag{A.64}$$

Denoting

$$\begin{cases} I = \int_{-\infty}^{+\infty} \frac{\sin \omega t \, \cos \kappa r}{\omega} \, d\kappa \; ; \\ \tilde{I} = \int_{-\infty}^{+\infty} \frac{\cos \omega t \, \cos \kappa r}{\omega} \, d\kappa \; , \end{cases} \tag{A.65}$$

we can write (A.64) as

$$D_m(x) = \frac{1}{(2\pi)^2 r} \frac{\partial I}{\partial r} \; ; \quad \tilde{D}_m(x) = \frac{1}{(2\pi)^2 r} \frac{\partial \tilde{I}}{\partial r}. \tag{A.66}$$

Next, we shall consider the integral $(A.65)_1$. Making the substitution

$$\kappa = m \sinh z \; ; \quad d\kappa = m \cosh z \, dz, \tag{A.67}$$

from (A.52) we get

$$\omega = m \cosh z, \tag{A.68}$$

so that

$$I = \int_{-\infty}^{+\infty} \sin(mt \cosh z) \, \cos(mr \sinh z) \, dz, \qquad (A.69)$$

or

$$I = \frac{1}{2} \int_{-\infty}^{+\infty} [\sin(mt \cosh z + mr \cosh z) + \sin(mt \cosh z - mr \cosh z).$$

If one takes

$$mt = A \cosh \alpha \; ; \quad mr = A \sinh \alpha \; ; \quad A = m\sqrt{t^2 - r^2}, \qquad (A.70)$$

then

$$I = \frac{1}{2} \int_{-\infty}^{+\infty} \sin[A \cosh(z + \alpha)] \, dz + \frac{1}{2} \int_{-\infty}^{+\infty} \sin[A \cosh(z - \alpha)] \, dz.$$

If in the first integral we make the substitution $z + \alpha \to z$, and in the second $z - \alpha \to z$, the two integrals become equal and we can write

$$I = \int_{-\infty}^{+\infty} \sin(A \cosh z) \, dz. \qquad (A.71)$$

According to $(A.70)_1$, sgn $A =$ sgn t (here *sgn* stands for "signum"), because m and $\cosh \alpha$ are non-negative. This can be shown by taking

$$I = \epsilon(t) \int_{-\infty}^{+\infty} \sin(|A| \cosh z) \, dz. \qquad (A.72)$$

Using the same procedure for $(A.65)_2$, we also obtain

$$\tilde{I} = \int_{-\infty}^{+\infty} \cos(A \cosh z) \, dz. \qquad (A.73)$$

The last two formulas can be expressed in terms of the zero-order Hankel function of the first kind $H_0^{(1)}$

$$H_0^{(1)}(|A|) = \frac{1}{\pi i} \int_{-\infty}^{+\infty} e^{i|A| \cosh z} \, dz \qquad (A.74)$$

as

$$I = \pi \epsilon \, Re[H_0^{(1)}(|A|)] \; ; \quad \tilde{I} = \pi \, Re[i H_0^{(1)}(|A|)]. \qquad (A.75)$$

Appendix. Distributions

To derive an explicit form for (A.66) we need the derivative of $H_0^{(1)}$, which is

$$\frac{d}{dx} H_0^{(1)}(x) = -H_1^{(1)}(x). \tag{A.76}$$

If we also take into account that, in our case, the Hankel function has a singularity for $A = 0$ (i.e. on the light cone), then (A.66), (A.70), (A.74), and (A.75) yield:

$$\begin{cases} D_m(x) = \frac{m}{4\pi} \epsilon(t) \, Re\left[\frac{H_1^{(1)}(m\sqrt{t^2-r^2})}{\sqrt{t^2-r^2}}\right] + D_0(x)\Big|_{x=0} \; ; \\ \tilde{D}_m(x) = \frac{m}{4\pi} \, Re\left[i \frac{H_1^{(1)}(m\sqrt{t^2-r^2})}{\sqrt{t^2-r^2}}\right] + \tilde{D}_0(x)\Big|_{x=0} \; , \end{cases} \tag{A.77}$$

where $D_0(x)\Big|_{x=0}$ and $\tilde{D}_0(x)\Big|_{x=0}$ show the behavior of $D_m(x)$ and $\tilde{D}_m(x)$ on the light cone. The notation used for these corrective functions is justified by the fact that A vanishes not only on the light cone, but also for $m = 0$. This means that the behavior of $D_m(x)$ and $\tilde{D}_m(x)$ on the light cone is similar to that of the corresponding singular functions, written for $m = 0$, which will be presented in the next section.

A.6. Functions D_0, $\tilde{D}_0$, D_{0+}, D_{0-}

If the field quanta have zero mass, then we need the functions $D_0(x)$ and $\tilde{D}_0(x)$ which, in analogy with (A.59), are defined as

$$\begin{cases} D_0(x) = \frac{1}{(2\pi)^3 i} \int e^{ikx} \epsilon(k_0) \, \delta(k^2) \, d\vec{k} \, dk_0 \; ; \\ \tilde{D}_0(x) = -\frac{1}{(2\pi)^3} \int e^{ikx} \, \delta(k^2) \, d\vec{k} \, dk_0 \; , \end{cases} \tag{A.78}$$

as well as their Fourier transforms

$$D_0(k) = \frac{1}{2\pi i} \epsilon(k_0) \delta(k^2) \; ; \quad \tilde{D}_0(k) = -\frac{1}{2\pi} \delta(k^2). \tag{A.79}$$

Since $m = 0$, (A.52) and (A.62) give

$$\kappa = |\vec{k}| = \omega. \tag{A.80}$$

Integrating with respect to k_0, θ, and φ in (A.78), then rather than (A.64) we have

$$\begin{cases} D_0(x) = -\frac{2}{(2\pi)^r} \int_0^\infty \sin \omega t \, \sin \omega r \, d\omega \; ; \\ \tilde{D}_0(x) = -\frac{2}{(2\pi)^r} \int_0^\infty \cos \omega t \, \sin \omega r \, d\omega \; . \end{cases} \tag{A.81}$$

It then follows that

$$
\begin{cases}
D_0(x) = \frac{1}{(2\pi)^2 r}\left\{ \int_0^\infty \cos[\omega(t+r)]\, d\omega - \int_0^\infty \cos[\omega(t-r)]\, d\omega \right\}; \\
\tilde{D}_0(x) = -\frac{1}{(2\pi)^2 r}\left\{ \int_0^\infty \sin[\omega(t+r)]\, d\omega - \int_0^\infty \sin[\omega(t-r)]\, d\omega \right\},
\end{cases}
$$
(A.82)

or, in view of (A.16) and (A.29),

$$
\begin{cases}
D_0 = \frac{1}{(2\pi)^2 r}[\delta(t+r) - \delta(t-r)]; \\
\tilde{D}_0(x) = -\frac{1}{(2\pi)^2 r}\left[\frac{\mathcal{P}}{r+t} + \frac{\mathcal{P}}{r-t}\right].
\end{cases}
$$
(A.83)

Since, due to its physical meaning, r is a positive quantity, using (A.18) we can write

$$
\delta(t+r) - \delta(t-r) = -\epsilon(t)[\delta(t-r) + \delta(t+r)].
$$
(A.84)

Indeed, if $t > 0$, we have $\delta(t+r) = 0$ and $\epsilon(t) = 1$, therefore (A.84) leads to the identity $-\delta(t-r) \equiv -\delta(t-r)$. If $t < 0$, we obtain $\delta(t-r) = 0$, and $\epsilon(t) = -1$, thus (A.84) yields $\delta(t+r) \equiv \delta(t+r)$. Using (A.6) and (A.9), we have

$$
\delta(t+r) - \delta(t-r) = -2\pi\epsilon(t)\,\delta(t^2 - r^2) = -2\pi\epsilon(t)\,\delta(x^2).
$$
(A.85)

Relation (A.34) shows that $\mathcal{P}/x$ is additive, and we can write

$$
\frac{\mathcal{P}}{r+t} + \frac{\mathcal{P}}{r-t} = 2r\frac{\mathcal{P}}{r^2 - t^2} = 2r\frac{\mathcal{P}}{x^2}.
$$
(A.86)

Substituting (A.85) and (A.86) into (A.83), we are left with

$$
D_0(x) = -\frac{1}{2\pi}\epsilon(t)\,\delta(x^2); \quad \tilde{D}_0(x) = -\frac{1}{2\pi^2}\frac{\mathcal{P}}{x^2}.
$$
(A.87)

We can now complete the analysis developed in the previous section. In view of (A.77) and (A.87), we have:

$$
\begin{cases}
D_m(x) = \frac{m}{4\pi}\epsilon(t)\, Re\left[\frac{H_1^{(1)}(m\sqrt{t^2-r^2})}{\sqrt{t^2-r^2}}\right] - \frac{1}{2\pi}\epsilon(t)\delta(x^2); \\
\tilde{D}_m(x) = \frac{m}{4\pi}\epsilon(t)\, Re\left[\frac{i\,H_1^{(1)}(m\sqrt{t^2-r^2})}{\sqrt{t^2-r^2}}\right],
\end{cases}
$$
(A.88)

because outside the light cone $D_0 = 0$, so that the definition $D_0(x)\big|_{x=0}$ becomes meaningless, while on the light cone $\tilde{D}_0(x) = 0$.

Appendix. Distributions

For $m = 0$, relations (A.58) become

$$\begin{cases} D_0(x) = D_{0+}(x) + D_{0-}(x) \ ; \\ \tilde{D}_0(x) = -i[D_{0+}(x) - D_{0-}(x)] \ . \end{cases} \qquad (A.89)$$

By means of (A.87) and (A.89) one obtains

$$D_{0\pm}(x) = \frac{1}{2}[D_0(x) \pm i\tilde{D}_0(x)] = -\frac{1}{4\pi}\left[\epsilon(t)\,\delta(x^2) \pm \frac{1}{\pi}\frac{\mathcal{P}}{x^2}\right]. \qquad (A.90)$$

Obviously, these relations can also be obtained from

$$D_{0\pm}(x) = \pm\frac{1}{(2\pi)^3 i}\int e^{ikx}\theta(\pm k_0)\,\delta(k^2)\,d\vec{k}\,dk_0, \qquad (A.91)$$

or

$$D_{0\pm}(x) = \pm\frac{1}{(2\pi)^3 i}\int \frac{e^{\pm ikx}}{2\omega}\,d\vec{k}, \qquad (A.92)$$

which have been written as particular cases of (A.49) and (A.57).

A.7. Functions S, $\tilde{S}$, S_+, S_-

These functions arise in the quantization of the spinorial field. They are obtained by applying the operator $(\hat{\partial} - m)$ to D_m, $\tilde{D}_m$, D_{m+}, and D_{m-}, specifically

$$S(x) = (\hat{\partial} - m)\,D_m(x)\ ; \quad \tilde{S}(x) = (\hat{\partial} - m)\,\tilde{D}_m(x), \qquad (A.93)$$

$$S_\pm(x) = (\hat{\partial} - m)\,D_{m\pm}(x). \qquad (A.94)$$

Since

$$(\hat{\partial} - m)e^{ikx} = (i\hat{k} - m)e^{ikx} = i(\hat{k} + im)e^{ikx}, \qquad (A.95)$$

it follows from (A.60) and (A.51) that the Fourier transforms of the new functions are

$$\begin{cases} S(k) = \frac{1}{2\pi}(\hat{k} + im)\,\epsilon(k_0)\,\delta(k^2 + m^2)\ ; \\ \tilde{S}(k) = \frac{1}{2\pi i}(\hat{k} + im)\,\delta(k^2 + m^2)\ ; \\ S_\pm(k) = \pm\frac{1}{2\pi}(\hat{k} + im)\,\theta(\pm k_0)\,\delta(k^2 + m^2). \end{cases} \qquad (A.96)$$

But, at the same time,

$$(\hat{\partial} - m)e^{-ikx} = (-i\hat{k} - m)e^{-ikx} = -i(\hat{k} - im)e^{-ikx}, \qquad (A.97)$$

so that we can write

$$(\hat{\partial} - m)e^{\pm ikx} = \pm i(\hat{k} \pm im)e^{\pm ikx}, \qquad (A.98)$$

and (A.57) gives

$$S_{\pm}(x) = \frac{1}{(2\pi)^3} \int \frac{(\hat{k} \pm im)}{2\omega} e^{\pm ikx} \, d\vec{k}, \qquad (A.99)$$

where $k_0 = \omega$ (see (A.52)).

According to (A.58), (A.93), and (A.94), we have

$$\begin{cases} S(x) = S_{+}(x) + S_{-}(x) \; ; \\ \tilde{S}(x) = -i[S_{+}(x)S_{-}(x)] \; , \end{cases} \qquad (A.100)$$

and (A.99) yields

$$\begin{cases} S(x) = \frac{1}{(2\pi)^3} \int \frac{1}{2\omega} \left[(\hat{k} + im)e^{i\vec{k}\cdot\vec{x}}e^{-i\omega t} + (\hat{k} - im)e^{-i\vec{k}\cdot\vec{x}}e^{i\omega t} \right] d\vec{k} \; ; \\ \tilde{S}(k) = \frac{1}{(2\pi)^3 i} \int \frac{1}{2\omega} \left[(\hat{k} + im)e^{i\vec{k}\cdot\vec{x}}e^{-i\omega t} - (\hat{k} - im)e^{-i\vec{k}\cdot\vec{x}}e^{i\omega t} \right] d\vec{k}. \end{cases}$$

If we set $\vec{k} \to -\vec{k}$ in the factor $e^{-i\vec{k}\cdot\vec{x}}$, we have

$$\begin{cases} S(x) = \frac{1}{(2\pi)^3 i} \int \frac{\hat{k}+im}{\omega} e^{i\vec{k}\cdot\vec{x}} \sin \omega t \, d\vec{k} \; ; \\ \tilde{S}(x) = \frac{1}{(2\pi)^3 i} \int \frac{\hat{k}+im}{\omega} e^{i\vec{k}\cdot\vec{x}} \cos \omega t \, d\vec{k} \; . \end{cases} \qquad (A.101)$$

A.8. Retarded and advanced functions

The Green function of the scalar field is defined by

$$(\Box_x - m^2) \, G(x - x') = -\delta(x - x'), \qquad (A.102)$$

where the index of the d'Alembertian shows the variable on which it acts. Using the Fourier transform (A.50), we have

$$G(x - x') = \frac{1}{(2\pi)^2} \int e^{ik(x-x')} \, G(k) \, d\vec{k} \, dk_0, \qquad (A.103)$$

therefore (A.102) becomes

$$\frac{1}{(2\pi)^2} \int (k^2 + m^2) \, G(k) e^{ik(x-x')} \, d\vec{k} \, dk_0 = \delta(x - x'). \qquad (A.104)$$

Comparing (A.104) with (A.44), we get

$$G(k) = \frac{1}{(2\pi)^2} \frac{1}{k^2 + m^2} = \frac{1}{(2\pi)^2} \frac{1}{\vec{k}^2 - k_0^2 + m^2}. \qquad (A.105)$$

As shown in (A.105), $G(k)$ becomes infinite for $k_0 = \pm\omega$ (see (A.52)). The usual procedure is then to consider k_0 as a complex variable

$$k_0 = Re\, k_0 + i\, Im\, k_0, \qquad (A.106)$$

and $G(k)$ will have two simple poles at $\pm\omega_0$. To bypass the poles, the integration contour can be chosen in two ways, as shown in Fig.A.2:

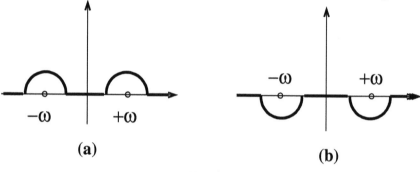

(a)   **(b)**

Fig.A.2

Let us focus our attention on the integration contour displayed in Fig.A.1(a). The Green function associated with this path is called *retarded Green function*, and is denoted by $D_m^{ret}(x - x')$. In view of (A.103) and (A.103), one can write

$$D_m^{ret}(x - x') = \frac{1}{(2\pi)^4} \int \frac{e^{ik(x-x')} \, d\vec{k} \, dk_0}{\vec{k}^2 - k_0^2 + m^2}. \qquad (A.107)$$

Using a standard procedure, we close the contour (a) by a semi-circle of radius $R \to \infty$ centered at the origin. The choice of the semi-circle (in the upper or lower half-plane) is dictated by the requirements that the integral over this contour cancel in the limit $R \to \infty$. This limit is insured by the timelike part of the exponential function in the integral in (A.107), which is

$$e^{-ik_0(t-t')} = e^{-i\, Re\, k_0(t-t')} \, e^{Im\, k_0(t-t')}.$$

Therefore we must have

$$Im\, k_0(t - t') < 0. \qquad (A.108)$$

Relation (A.108) shows that:

- If $(t - t') < 0$, then $Im\, k_0 > 0$, and we choose the semi-circle in the upper half-plane. But, in this case, our closed contour does not contain any pole, so that

$$D_m^{ret}(x - x') = 0 \quad \text{for} \quad t - t' < 0, \qquad (A.109)$$

which also justifies the chosen function name (*retarded* comes from the Latin *retardare*).

- If $(t - t') > 0$, we impose $Im\, k_0 < 0$, close the contour with the semi-circle in the lower half-plane, and integrate over k_0 in (A.107). Thus, we arrive at the integral

$$I = \int \frac{e^{-ik_0(t-t')}}{\omega^2 - k_0^2}\, dk_0.$$

Since

$$\frac{1}{\omega^2 - k_0^2} = -\frac{1}{2\omega}\left(\frac{1}{k_0 - \omega} - \frac{1}{k_0 + \omega}\right),$$

the two residues are

$$-\frac{e^{-i\omega(t-t')}}{2\omega} \quad \text{and} \quad +\frac{e^{i\omega(t-t')}}{2\omega},$$

and we are left with

$$I = \frac{2\pi i}{2\omega}\left[e^{i\omega(t-t')} - e^{-i\omega(t-t')}\right] = -\frac{2\pi}{\omega}\sin\omega(t - t').$$

The integral (A.107) is taken over the above chosen contour, and therefore writes

$$D_m^{ret}(x - x') = -\frac{1}{(2\pi)^3}\int e^{i\vec{k}\cdot(\vec{x}-\vec{x}')}\frac{\sin\omega(t - t')}{\omega}\, d\vec{k}\,; \quad (t - t') > 0.$$

$$(A.110)$$

Using the Heaviside step function (A.46) and comparing (A.110) with $(A.61)_1$, one can write $D_m^{ret}(x - x')$ for the whole interval spanned by the timelike variable as

$$D_m^{ret}(x - x') = \theta(t - t')\, D_m(x - x'). \qquad (A.111)$$

Sometimes it is more convenient to choose, rather then the contour shown in Fig.A.2(a), a straight line parallel to the real axis, in

Appendix. Distributions

the upper half-plane, this way replacing k_0 by $k_0 + i\alpha$ ($\alpha > 0$), and then take the limit for $\alpha \to 0$, obtaining

$$D_m^{ret}(x - x') = \frac{1}{(2\pi)^4} \lim_{\alpha \to 0} \int \frac{e^{ik(x-x')}}{\vec{k}^2 + m^2 - k_0^2 - 2i\alpha k_0} \, d\vec{k} \, dk_0, \quad (A.112)$$

where the term in α^2 has been omitted.

According to (A.112), the Fourier transform of $D_m^{ret}(x - x')$ is

$$D_m^{ret}(k) = \frac{1}{(2\pi)^2} \lim_{\alpha \to 0+} \frac{1}{k^2 + m^2 - 2i\alpha k_0}. \quad (A.113)$$

For the contour in Fig.A.2(b), a similar reasoning leads to the *advanced Green function*, defined as

$$D_m^{adv}(x - x') = \begin{cases} 0 & \text{for} \quad t - t' > 0 \, ; \\ \frac{1}{(2\pi)^3} \int e^{i\vec{k}\cdot(\vec{x}-\vec{x}')} \frac{\sin \omega(t-t')}{\omega} \, d\vec{k} & \text{for} \quad t - t' < 0 \, , \end{cases} \quad (A.114)$$

and we can write

$$D_m^{adv}(x - x') = -\theta(t' - t) \, D_m(x - x'). \quad (A.115)$$

In this case, k_0 is replaced by $k_0 - i\alpha$, so that

$$D_m^{adv} = D_m^{adv}(x - x') = \frac{1}{(2\pi)^4} \lim_{\alpha \to 0+} \int \frac{e^{ik(x-x')}}{\vec{k}^2 + m^2 - k_0^2 + 2i\alpha k_0} \, d\vec{k} \, dk_0, \quad (A.116)$$

and

$$D_m^{adv}(k) = \frac{1}{(2\pi)^2} \lim_{\alpha \to 0+} \frac{1}{k^2 + m^2 + 2i\alpha k_0}. \quad (A.117)$$

If $m = 0$, the retarded and advanced Green functions are

$$\begin{cases} D_0^{ret}(x - x') = \theta(t - t') \, D_0(x - x') \, ; \\ D_0^{adv}(x - x') = -\theta(t - t') D_0(x - x'), \end{cases} \quad (A.118)$$

and

$$\begin{cases} D_0^{ret}(k) = \frac{1}{(2\pi)^2} \lim_{\alpha \to 0} \frac{1}{k^2 - 2i\,\alpha\,k_0} \, ; \\ D_0^{adv}(k) = \frac{1}{(2\pi)^2} \lim_{\alpha \to 0} \frac{1}{k^2 + 2i\,\alpha\,k_0} . \end{cases} \quad (A.119)$$

An analogous theory applied to the Dirac spinorial field yields

$$\begin{cases} S^{ret}(x - x') = (\hat{\partial} - m) \, D_m^{ret}(x - x') \, ; \\ S^{adv}(x - x') = (\hat{\partial} - m) \, D_m^{adv}(x - x') \, , \end{cases} \quad (A.120)$$

as well as

$$\begin{cases} S^{ret}(k) = \frac{1}{(2\pi)^2} \lim_{\alpha\to 0+} \frac{i\hat{k}-m}{k^2+m^2-2i\alpha\,k_0} \; ; \\ S^{adv}(k) = \frac{1}{(2\pi)^2} \lim_{\alpha\to 0+} \frac{i\hat{k}-m}{k^2+m^2+2i\alpha\,k_0} \; . \end{cases} \tag{A.121}$$

A.9. Causal functions

A function f(t) is called *causal function* if it is zero when $t < 0$, because this reflects the constraint placed on the systems to insure that the effects do not precede the cause. Such a function was first introduced as (see (3.5.12))

$$D^c_m(x) = \theta(t)\, D_{m-}(x) - \theta(-t)\, D_{m+}(x). \tag{A.122}$$

A simple trick yields

$$D^c_m(x) = \theta(t)\,[D_{m-}(x) + D_{m+}(x)] - [\theta(t) + \theta(-t)]\, D_{m+}(x),$$

and, in view of $(A.47)_1$, $(A.58)_1$ and (A.111), we have

$$D^c_m(x) = D^{ret}_m(x) - D_{m+}(x). \tag{A.123}$$

Comparing (A.113) with (A.39) one can write

$$D^{ret}_m(k) = \frac{1}{(2\pi)^2} \begin{cases} \zeta^*(k^2 + m^2) & \text{for} \quad \alpha k_0 > 0 \;; \\ \zeta(k^2 + m^2) & \text{for} \quad \alpha k_0 < 0 \;. \end{cases} \tag{A.124}$$

Since in (A.113) $\alpha > 0$, using (A.113), (A.35), and (36) we also have

$$D^{ret}_m(k) = \frac{1}{(2\pi)^2} \left[\frac{P}{k^2 + m^2} + i\pi\,\epsilon(k_0)\,\delta(k^2 + m^2) \right]. \tag{A.125}$$

The Fourier transform of $D^c_m(x)$ is found by means of (A.123), (A.125), and (A.51):

$$D^c_m(k) = \frac{1}{(2\pi)^2} \left[\frac{P}{k^2 + m^2} + i\pi\,\epsilon(k_0)\,\delta(k^2+m^2) - 2\pi\,i\,\theta(k_0)\,\delta(k^2+m^2) \right]$$

$$= \frac{1}{(2\pi)^2} \left\{ \frac{P}{k^2 + m^2} - i\pi\,[2\theta(k_0) - \epsilon(k_0)]\,\delta(k^2 + m^2) \right\}.$$

Appendix. Distributions

Since $2\theta(k_0) - \epsilon(k_0) = 1$, we have

$$D_m^c(k) = \frac{1}{(2\pi)^2}\left[\frac{\mathcal{P}}{k^2 + m^2} - i\pi\,\delta(k^2 + m^2)\right]$$

$$= \frac{1}{(2\pi)^2}\,\zeta(k^2 + m^2) = \frac{1}{(2\pi)^2}\lim_{\alpha\to 0}\frac{1}{k^2 + m^2 + i\alpha}\,, \qquad (A.126)$$

where $(A.35)$ and $(A.39)_1$ have also been used.

If $D_m^c(k)$ given by $(A.126)$ is the Fourier transform of $D_m^c(x)$, then

$$D_m^c(x) = \frac{1}{(2\pi)^4}\lim_{\alpha\to 0}\int \frac{e^{ikx}}{k^2 + m^2 + i\alpha}\,d\vec{k}\,dk_0. \qquad (A.127)$$

Obviously, for $m = 0$ the causal function is

$$D_0^c(x) = \frac{1}{(2\pi)^4}\lim_{\alpha\to 0}\int \frac{e^{ikx}}{k^2 + i\alpha}\,d\vec{k}\,dk_0, \qquad (A.128)$$

and its Fourier transform is

$$D_0^c(k) = \frac{1}{(2\pi)^2}\,\zeta(k^2) = \frac{1}{(2\pi)^2}\lim_{\alpha\to 0}\frac{1}{k^2 + i\alpha}. \qquad (A.129)$$

In the case of a spinorial field, we have:

$$S(x) = (\hat{\partial} - m)D_0^c(x) = \frac{1}{(2\pi)^4}\lim_{\alpha\to 0}\int \frac{(i\hat{k} - m)\,e^{ikx}}{k^2 + m^2 + i\alpha}\,d\vec{k}\,dk_0, \quad (A.130)$$

as well as

$$S^c(k) = \frac{1}{(2\pi)^2}\lim_{\alpha\to 0}\frac{i\hat{k} - m}{k^2 + m^2 + i\alpha}. \qquad (A.131)$$

At the end of this Appendix we mention that, since $G(x - x')$ is a solution of the Gordon-Schrödinger equation, the functions $D_{m\pm}$, D_m, $\tilde{D}_m$, D_m^{ret}, D_m^{adv} and D_m^c are also solutions of this equation. Consequently, $D_{0\pm}$, D_0, $\tilde{D}_0$, D_0^{ret}, D_0^{adv} and D_0^c are solutions of the D'Alembert equation, while $S_\pm$, S, $\tilde{S}$, S^{ret}, S^{adv} and S^c satisfy the Dirac equation. Here is an example:

$$(\hat{\partial} + m)\,S(x) = (\hat{\partial} + m)(\hat{\partial} - m)D_m(x) = (\Box - m^2)D_m(x). \quad (A.132)$$

PROBLEMS with SOLUTIONS

This Chapter is a collection of problems assembled by the authors from a variety of sources, and adapted to the present theory layout. [1] [2] These problems have been chosen to help the reader with the learning the computational details, and to expand the range of applications given in the book. For further examples, the reader is referred to the books listed at the end, in particular the references [14], [24], and [30].

1. **Problem.** Show that the following Lagrangian density

$$\mathcal{L} = \frac{1}{2}[(\partial\phi_1)^2 + (\partial\phi_2)^2] - \frac{m^2}{2}(\phi_1^2 + \phi_2^2) - \frac{\lambda}{4}(\phi_1^2 + \phi_2^2)^2$$

is invariant under the rotation transformation

$$\phi_1 \rightarrow \phi_1' = \phi_1\cos\theta - \phi_2\sin\theta,$$
$$\phi_2 \rightarrow \phi_2' = \phi_1\sin\theta + \phi_2\cos\theta.$$

Find the Noether current and charge corresponding to this Lagrangian density.

Solution. The first term transforms as:

$$\frac{1}{2}[(\partial\phi_1)^2 + (\partial\phi_2)^2] \rightarrow \frac{1}{2}[(\partial\phi_1')^2 + (\partial\phi_2')^2] =$$

$$\frac{1}{2}\left[(\partial(\phi_1(\cos)\theta) - \partial(\phi_2(\sin)\theta))^2 + (\partial(\phi_1(\sin)\theta) + \partial(\phi_2(\cos)\theta))^2\right].$$

The terms cancel, and we are left with $1/2[(\partial\phi_1)^2 + (\partial\phi_2)^2]$, since $\sin^2\theta + \cos^2\theta = 1$. Similarly, one can easily show that the other two terms are also invariant. From (1.4.2), we can write

$$\delta\mathcal{L} = \frac{\partial}{\partial x_l}\left[\pi_{l(r)}^+\delta U_{(r)} + \pi_{l(r)}\delta U_{(r)}^+\right],$$

---

[1] Problem 10 has been adapted fom *Greiner, W., Reinhardt, J. - Quantum Electrodynamics, Springer, 4th ed. Edition (2008)*, with permission from the authors.

[2] Problems 1, 2, 8, 9, 10, 11, 12, 13, 17, 20, 21, and 22 have been adapted from *Radovanovic, V. - Problem Book in Quantum Field Theory, 2nd edition, Springer (2007)*, with permission from the author.

295

and, similar to the notation (1.8.5), we can define the current density as

$$s_l = \pi^+_{l(r)} \delta U_{(r)} + \pi_{l(r)} \delta U^+_{(r)},$$

or

$$s_l = \frac{\partial \mathcal{L}}{\partial U_{(r),l}} \delta U_{(r)} + \frac{\partial \mathcal{L}}{\partial U^+_{(r),l}} \delta U^+_{(r)}.$$

Since our fields are scalar and real, this expression becomes

$$s_l = \frac{\partial \mathcal{L}}{\partial \phi_{1,l}} \delta \phi_1 + \frac{\mathcal{L}}{\partial \phi_{2,l}} \delta \phi_2.$$

Using a Taylor expansion around the origin, the infinitesimal variations of the fields $\phi' - \phi$ are $\delta\phi_1 = -\theta\phi_2$ and $\delta\phi_2 = \theta\phi_1$, so that

$$s_l = \theta(\phi_1 \partial_l \phi_2 - \phi_2 \partial_l \phi_1)$$

The parameter θ can be dropped out since it is a constant. From (3.1.22), the Noether charge is $Q = 1/i \int d\vec{x} s_4$. Since $\partial_4 = 1/(ic)\partial_t$, the charge becomes

$$Q = -\frac{1}{c} \int d\vec{x} \, (\phi_1 \phi_{2,t} - \phi_2 \phi_{1,t}).$$

2. **Problem.** The Einstein-Hilbert gravitation action is defined as

$$S = \kappa \int d^4 x \sqrt{-g} R$$

where $g = g^\alpha_\alpha$ (sum over a; g_{ab} is the metric of the four-dimensional curved space-time, with a and b going from 1 to 4); R is the scalar curvature and κ is a constant. In the weak-field approximation, the metric is obtained by adding a small perturbation to the flat metric g^0_{ab}, i.e.

$$g_{ab}(x) = g^0_{ab} + h_{ab}(x),$$

where the perturbation $h_{ab}(x)$ is a symmetric second rank tensor field. In this approximation, the Einstein-Hilbert action reduces to an action in flat spacetime :

$$s = \int d^4 x \left(\frac{1}{2} \partial_c h_{ab} \partial^c h^{ab} - \partial_c h_{ab} \partial^b h^{ac} + \partial_c h^{ac} \partial_a h - \frac{1}{2} \partial_a h \partial^a h \right), \quad (1)$$

Problems with solutions

where $h = h^a_a$, and we used Einstein's convention that summation is performed over identical upper and lower indices.

(a) Derive the equations of motion for h_{ab}.

(b) Show that the linearized theory is invariant under the gauge symmetry transformation:

$$h_{ab} \to h_{ab} + \partial_a \Lambda_b + \partial_b \Lambda_a, \tag{2}$$

where $\Lambda_a(x)$ is any four-vector field.

Solution. (a) From (1), the Lagrangian density of the field is

$$\mathcal{L} = \frac{1}{2} h_{ab,c} h^{ab,c} - h_{ab,c} h^{ac,b} + h^{ac}_c h_{,a} - \frac{1}{2} h_{,a} h^{,a}. \tag{3}$$

We use Equation (1.1.15), where $U_{(r)}$ is replaced by h_{ab}. This equation becomes

$$\frac{\partial \mathcal{L}}{\partial h_{ab}} - \frac{\partial}{\partial x_c} \left(\frac{\partial \mathcal{L}}{\partial h_{ab,c}} \right) = 0. \tag{4}$$

Let us first show that

$$\frac{\partial \mathcal{L}}{\partial (h_{ab,c})} = h^{ab,c} - h^{ac,b} - h^{bc,a} + \frac{1}{2} g^{ac} h^{,b} + \frac{1}{2} g^{bc} h^{,a} + g^{ab} h^{dc}_{,d} - g^{ab} h^{,c}. \tag{5}$$

Here we make use of the relations $h_{ab} = g_{ac} g_{bd} h^{cd}$ and $\partial_c = g_{ac} \partial^a$ for raising and lowering indices, and keep in mind that the metric tensor is symmetric. Accordingly, the first term in the Lagrangian density writes $1/2 h_{ab,c} g^{am} g^{an} g^{aq} h_{mn,q}$, and its derivative with respect to $h_{ab,c}$ is $1/2 h^{ab,c} + 1/2 h^{mn,q} \delta_{am} \delta_{bn} \delta_{cq} = h^{ab,c}$. Similarly, the second term gives $-h^{ac,b} - h^{bc,a}$. The third term is $g^{ab} g^{cn} g^{dm} h_{mn,c} h_{ab,d}$, and since h_{mn} is symmetric, it's derivative can be written as

$$1/2 g^{ab} g^{cn} g^{dm} h_{ab,d} \delta_{ma} \delta_{nb} + 1/2 g^{ab} g^{cn} g^{dm} h_{ab,d} \delta_{mb} \delta_{na} +$$
$$g^{ab} g^{cn} g^{dm} h_{mn,c} \delta_{dc},$$

and therefore it becomes $1/2 g^{ac} h^{,b} + 1/2 g^{bc} h^{,a} + g^{ab} h^{mc}_{,m}$. Similarly, the last term will be $1/2 g^{cd} g^{am} g^{bn} h_{am,c} h_{bn,d}$, with the derivative $-g^{ab} h^{,c}$. Since the Lagrangian density does not depend directly on h_{ab}, the equations of motions result directly from (4) by taking the derivative of (5) with respect to x_c:

$$\Box h^{ab} - g^{ab} \Box h - h^{ac,b}_{,c} - h^{bc,a}_{,c} + g^{ab} h^{mn}_{,mn} + h^{,ab} = 0, \tag{6}$$

where $\Box = \partial^a \partial_a$ is the d'Alembert operator.

297

Problems with solutions

(b) The gauge invariance can be proven by showing that the Lagrangian density is changed only by a four-divergence term. However, since gauge invariance means that the equations of motion are left unchanged by the given transformation, it is straightforward to show this by replacing h^{ab} by (2) in (6). In this case, all terms containing *Lambda* cancel out.

3. **Problem.** Find the matrix elements of the operators c_i and c_i^+ for bosons in the occupation number space.

Solution. Since the state function ψ is orthonormalized, we can write

(a)
$$< \psi(n_1', n_2', ..., n_i', ...)|c_i|\psi(n_1, n_2, ..., n_i, ...) >$$

$$= \sqrt{n_i}(\psi(n_1', n_2', ..., n_i', ...), \psi(n_1, n_2, ..., n_i - 1, ...))$$

$$= \sqrt{n_i}\, \delta_{n_1, n_1'} ...\delta_{n_i - 1, n_i'} ...$$

(b)
$$< \psi(n_1' n_2'...n_i'...)|c_i^+|\psi(n_1 n_2...n_i...) >$$

$$= \sqrt{n_i + 1}(\psi(n_1', n_2', ..., n_i', ...), \psi(n_1, n_2, ..., n_i + 1, ...))$$

$$= \sqrt{n_i + 1}\, \delta_{n_1', n_1} ...\delta_{n_i', n_{i+1}} ...$$

4. **Problem.** Find the matrix elements of the operators c_i and c_i^+ for fermions in the occupation number space.

Solution. In this case, we have:

(a)
$$< \psi(n_1', n_2', ..., n_i', ...)|c_i|\psi(n_1, n_2, ..., n_i, ...) >$$

$$= (-1)^{\sum_{j=1}^{i-1} n_j} n_i(\psi(n_1', n_2', ..., n_i', ...), \psi(n_1, n_2, ...(1 - n_i)...)$$

$$= (-1)^{\sum_{j=1}^{i-1} n_j} n_i\, \delta_{n_1, n_1'} ...\delta_{n_i', 1 - n_i} ...$$

(b)
$$< \psi(n_1', n_2', ..., n_i', ...)|c_i^+|\psi(n_1, n_2, ..., n_i, ...) >$$

$$= (-1)^{\sum_{j=1}^{i-1} n_j} (1 - n_i)(\psi(n_1', n_2', ..., n_i', ...), \psi(n_1, n_2, ...(1 - n_i)...)$$

$$= (-1)^{\sum_{j=1}^{i-1} n_j} (1 - n_i) \, \delta_{n_1', n_1} ... \delta_{n_i', 1-n_i}$$

5. Problem. Derive the equation

$$c_i^+ c_i \psi(n_1, n_2, ... n_i, ...) = n_i \psi(n_1, n_2, ... n_i, ...),$$

using the relations

$$c_i \psi(n_1, n_2, .., n_i, ..) = \sqrt{n_i} \psi(n_1, n_2, .., n_{i-1}, ..) :$$

$$c_i^+ \psi(n_1, n_2, .., n_i, ..) = \sqrt{n_i + 1} \psi(n_1, n_2, .., n_i + 1, ..)$$

for bosons, and

$$c_i \psi(n_1, n_2, .., n_i, ..) = (-1)^{\sum_{j=1}^{i-1} n_j} n_i \psi(n_1, n_2, .., 1 - n_i, ..);$$

$$c_i^+ \psi(n_1, n_2, .., n_i, ..) = (-1)^{\sum_{j=1}^{i-1} n_j} (1 - n_i) \psi(n_1, n_2, .., 1 - n_i, ..)$$

for fermions.

Solution. For bosons, we have:

$$c_i^+ c_i \psi(n_1, n_2, ... n_i, ..) = c_i^+ \sqrt{n_i} \psi(n_1, n_2, .., n_i - 1, ..)$$

$$= n_i \psi(n_1, n_2, ... n_i, ..),$$

and for fermions:

$$c_i^+ c_i \psi(n_1, n_2, ... n_i, ..) = c_i^+ (-1)^{\sum_{j=1}^{i-1} n_j} n_i \psi(n_1, ... (1 - n_i), ..)$$

$$= n_i^2 \psi(n_1, n_2, ... n_i, ..) = n_i \psi(n_1, n_2, ... n_i, ..).$$

Since n_i equals either 0 or 1, the result is $n_i^2 = n_i$.

6. Problem. Derive the relations

$$(a) \qquad [c_i, c_j] = [c_i^+, c_j^+] = 0 ; \quad [c_i, c_j^+] = \delta_{ij}$$

for bosons, and

$$(b) \qquad \{c_i, c_j\} = \{c_i^+, c_j^+\} = 0 ; \quad \{c_i, c_j^+\} = \delta_{ij}$$

for fermions.

Solution

(a) Expanding the commutator, we have:

$$[c_i, c_j]|\psi(n_1, n_2, ...n_i, ..) >= (c_i c_j - c_j c_i)|\psi(n_1, n_2, ...n_i, ..) >= 0,$$

because:

$$c_i|\psi(n_1, n_2, ...n_i, ..) >= \sqrt{n_i}|\psi(n_1, n_2, ...n_i - 1, ..) >;$$

$$c_j c_i|\psi(n_1, .., n_i, .., n_j, ..) >= \sqrt{n_i n_j}|\psi(n_1, ...n_i - 1, .., n_j - 1, ..) >;$$

$$c_i c_j|\psi(n_1, .., n_i, .., n_j, ..) >= \sqrt{n_j n_i}|\psi(n_1, ...n_i - 1, .., n_j - 1, ..) > .$$

Similarly, one obtains

$$[c_i^+, c_j^+] = 0.$$

Next, let us show that $[c_i, c_j^+] = \delta_{ij}$. For $i \neq j$, we have:

$$[c_i, c_j^+]|\psi(n_1, .., n_i, ..) >= (c_i c_j^+ - c_j^+ c_i)|\psi(n_1, .., n_i, ..) >$$

$$= c_i\sqrt{c_j + 1}|\psi(n_1, .., n_j + 1, ..) > -c_j^+\sqrt{n_i}|\psi(n_1, .., n_i - 1, ..) >$$

$$= \sqrt{n_i(n_j + 1)}|\psi(n_1, .., n_j + 1, .., n_i - 1..) >$$

$$-\sqrt{(n_j + 1)n_i}|\psi(n_1, .., n_j + 1, .., n_i - 1..) >= 0.$$

In the case $i = j$, a similar calculation yields

$$[c_i, c_j^+]|\psi(n_1, .., n_i, ..) >= (c_i c_i^+ - c_i^+ c_i)|\psi(n_1, .., n_i, ..) >$$

$$= (n_i + 1)|\psi(n_1, .., n_i, ..) > -n_i|\psi(n_1, .., n_i, ..) >= |\psi(n_1, .., n_i, ..) > .$$

Therefore, $[c_i, c_i^+] = 1$.

(b) For $i = j$, we have:

$$c_i c_i \psi(n_1, .., n_i, ..) = c_i(-1)^{\sum_{j=1}^{i-1} n_j} n_i \psi(n_1, .., 1 - n_i, ..)$$

$$= n_i(1 - n_i)\psi(n_1, .., n_i, ..) \equiv 0,$$

since $n_i(1 - n_i) = 0$ for $n_i = 0, 1$. Therefore $\{c_i, c_i\} = 0$.

For $i \neq j$, with $j < i$, we have:

$$c_i c_j \psi(n_1, .., n_j, .., n_i, ..)$$

$$= (-1)^{\sum_{k=1}^{j-1} n_k} n_j (-1)^{\sum_{k=1}^{j-1} n_k + (1-n_j) + \sum_{k=j+1}^{i-1} n_k} n_i$$

300

$$\times \psi(n_1, .., (1 - n_j), .., (1 - n_i)..)$$

$$= n_i n_j (-1)^{1 - n_j + \sum_{k=j+1}^{i-1} n_k} \psi(n_1, .., (1 - n_j), .., (1 - n_i)..),$$

as well as

$$c_j c_i \psi(n_1, .., n_j, .., n_i, ..)$$

$$= n_i n_j (-1)^{\sum_{k=1}^{j-1} n_k + \sum_{k=1}^{i-1} n_k} \psi(n_1, .., (1 - n_j), .., (1 - n_i)..)$$

$$= n_i n_j (-1)^{n_j + \sum_{k=j+1}^{i-1} n_k} \psi(n_1, .., (1 - n_j), .., (1 - n_i)..).$$

As one can see, there appears the parenthesis $[(-1)^{1-n_j} + (-1)^{n_j}]$, which is zero for $n_j = 0, 1$. Similarly, it can be shown that $\{c_i^+, c_j^+\} = 0$.

Let us finally show that $\{c_i, c_j^+\} = \delta_{ij}$. If $i \neq j$, with $i > j$, since $\{c_i, c_j\} = 0$, we easily find $\{c_i, c_j^+\} = 0$. If $i = j$, using the result from the previous problem, we can write

$$c_i^+ c_i \psi(n_1, .., n_i, ..) = n_i^2 \psi(n_1, .., n_i, ..);$$

$$c_i c_i^+ \psi(n_1, .., n_i, ..) = c_i (-1)^{1 \cdot \sum_{k=1}^{i-1} n_k} (1 - n_i) \psi(n_1, .., 1 - n_i, ..)$$

$$= (-1)^{2 \cdot \sum_{k=1}^{n-1} n_k} (1 - n_i)^2 \psi(n_1, .., n_i, ..) = (1 - n_i^2) \psi(n_1, .., n_i, ..),$$

equivalently to $c_i c_i^+ = (1 - n_i)^2$. Therefore

$$\{c_i, c_j^+\} = (1 - n_i)^2 + n_i^2 = 1 - 2n_i + 2n_i^2 = 1,$$

since $n_i = n_i^2$ for $n_i = 0, 1$.

7. **Problem.** Prove the relativistic invariance of the Lagrangian density associated with the real scalar field

$$\mathcal{L} = -\frac{1}{2}(\psi_{,i}\psi_{,i} + m^2 \psi\psi)$$

Solution. Since

$$\psi(x) = \psi'(x') \ ; \quad \frac{\partial \psi(x)}{\partial x_i} = \frac{\partial \psi'(x')}{\partial x_j} a_{ji},$$

where a_{ji} is the Lorentz transformation matrix, we have:

$$\frac{\partial \psi(x)}{\partial x_i}\frac{\partial \psi(x)}{\partial x_i} = \frac{\partial \psi'(x')}{\partial x_j'}\frac{\partial \psi'(x')}{\partial x_k'} a_{js} a_{ks}$$

$$= \frac{\partial \psi'(x')}{\partial x'_j} \frac{\partial \psi'(x')}{\partial x'_k} \delta_{jk} = \frac{\partial \psi'(x')}{\partial x'_j} \frac{\partial \psi'(x')}{\partial x'_j},$$

therefore

$$\mathcal{L}' = -\frac{1}{2} \left[\left(\frac{\partial \psi'}{\partial x'} \right)^2 + m^2 \psi'^2 \right] = -\frac{1}{2} \left[\left(\frac{\partial \psi}{\partial x} \right)^2 + m^2 \psi^2 \right] = \mathcal{L}.$$

8. **Problem.** The action of time reversal is defined as

$$T\psi(x)T^{-1} = \eta\psi(\vec{x}, -t),$$

where T is an anti-unitary operator (i.e. $T i T^{-1} = -i$), and η is a phase factor.

(a)   Prove that

$$Tc(\vec{p})T^{-1} = \eta c(\vec{p}),$$

$$Tc^+(\vec{p})T^{-1} = \eta c^+(-\vec{p})$$

(b)   Derive the transformation rules for the Hamiltonian and momentum of the field under the action of time reversal.

Solution. (a) Following (3.3.38), and (3.2.24), we can write

$$\psi(x) = \frac{1}{(2\pi)^{3/2}} \int \frac{d\vec{p}}{\sqrt{2E_p}} \left(\phi_-(\vec{p}) e^{-ipx} + \phi_+(\vec{p}) e^{ipx} \right).$$

From (3.4.7), ϕ_- is a creation operator, and can be renamed $c^+(\vec{p})$, while ϕ_+ is an annihilation operator, which can also be written as $a(\vec{p})$. Therefore, $\psi(x)$ becomes

$$\psi(x) = \frac{1}{(2\pi)^{3/2}} \int \frac{d\vec{p}}{\sqrt{2E_p}} \left(c^+(\vec{p}) e^{-ipx} + a(\vec{p}) e^{ipx} \right).$$

Since T is anti-unitary, we have that $T a T^{-1} = a^*$, and therefore the time reversal of $\psi(x)$ becomes

$$T\psi(x)T^{-1} = \frac{1}{(2\pi)^{3/2}}$$

$$\int \frac{d\vec{p}}{\sqrt{2E_p}} \left(Tc^+(\vec{p})T^{-1} e^{i(\vec{p}\cdot\vec{x} - E_p t)} + T a(\vec{p})T^{-1} e^{-i(\vec{p}\cdot\vec{x} - E_p t)} \right).$$

Problems with solutions

On the other hand, we have

$$\eta\psi(\vec{x}, -t) = \frac{1}{(2\pi)^{3/2}}$$

$$\int \frac{d\vec{p}}{\sqrt{2E_p}} \left(\eta c^+(\vec{p}) e^{-i(\vec{p}\cdot\vec{x}+E_p t)} + \eta a(\vec{p}) e^{i(\vec{p}\cdot\vec{x}+E_p t)} \right).$$

Equating the two expressions, we observe that the equality of the exponents is obtained when we exchange $\vec{p}$ for $-\vec{p}$ in the second relation. It follows that the coefficients must also be equal, such that

$$\tau c^+(\vec{p})\tau^{-1} = \eta c^+(-\vec{p}),$$

and

$$\tau a(\vec{p})\tau^{-1} = \eta a(-\vec{p}).$$

(b) From (3.1.18) and (3.1.20) we have that

$$H = \int \left(\psi^+_{,\alpha}\psi_{,\alpha} + 1/c^2 \partial_t\psi^+ \partial_t\psi + m^2\psi^+\psi \right) d\vec{x},$$

and

$$P_\alpha = i/c \int \left(\psi^+_{,\alpha}\partial_t\psi + \partial_t\psi^+\psi_\alpha \right) d\vec{x}.$$

Applying the time reversal operator on H, we have

$$\tau\partial_t\psi^+\partial_t\psi\tau^{-1} = \tau\partial_t\psi^+\tau^{-1}\tau\partial_t\psi\tau^{-1}$$
$$= (-\partial_t\tau\psi^+\tau^{-1})(-\partial_t\tau\psi\tau^{-1})$$
$$= \eta^2\partial_{-t}\psi^+(-t)\partial_{-t}\psi(-t),$$

and

$$\tau\partial_\alpha\psi^+\partial_\alpha\psi\tau^{-1} = \tau\partial_\alpha\psi^+\tau^{-1}\tau\partial_\alpha\psi\tau^{-1}$$
$$= (\partial_\alpha\tau\psi^+\tau^{-1})(\partial_\alpha\tau\psi\tau^{-1})$$
$$= \eta^2\partial_\alpha\psi^+(-t)\partial_\alpha\psi(-t),$$

The last term also gives $\tau\psi^+(t)\psi(t)\tau^{-1} = \eta^2\psi^+(-t)\psi(-t)$, where we omitted writing the $\vec{x}$ dependence. Since the factor η is arbitrary, and we can make the change $-t \to t$ everywhere, then it follows that in this case

$$\tau H\tau^{-1} = H.$$

Problems with solutions

For the momentum, we have the product

$$\tau i \partial_\alpha \psi^+ \partial_t \psi \tau^{-1} = \tau \partial_\alpha \psi^+ \tau^{-1} \tau i \partial_t \psi \tau^{-1}$$
$$= \partial_\alpha (\tau \psi^+ \tau^{-1}) i \partial_t (\tau \psi \tau^{-1})$$
$$= \eta^2 \partial_\alpha \psi^+ (-t)(-i) \partial_{-t} \psi(-t).$$

Similarly,

$$\tau i \partial_t \psi^+ \partial_\alpha \psi \tau^{-1} = \eta^2 (-i) \partial_{-t} \psi^+ (-t) \partial_\alpha \psi(-t).$$

Therefore, after the substitution $-t \to ts$ we are left with a negative sign, and

$$\tau \vec{p} \tau^{-1} = -\vec{P}.$$

9. **Problem** For a scalar field, we define the parity operator as

$$P = \exp\left[-\frac{i\pi}{2} \int d\vec{k} \left(c^+(\vec{k}) c(\vec{k}) - \eta_p c^+(\vec{k}) c(-\vec{k}) \right) \right], \qquad (1)$$

where $\eta_p = \pm 1$ represents the intrinsic parity of the field. Using the relation

$$e^A H e^{-A} = H + [A, H] + \frac{1}{2}[A, [A, H]] + ..., \qquad (2)$$

prove that P commutes with the Hamiltonian.

Solution. From (3.3.40), the Hamiltonian of the scalar field can be written as

$$H = \int E_p \left[c(\vec{p}) c^+(\vec{p}) + a(\vec{p}) a^+(\vec{p}) \right] d\vec{p}.$$

From (3.4.6) we have that a and c commute, so that we only have to consider the part of the Hamiltonian containing c. From (1) and (2), it follows that we need to calculate the commutator

$$[(c^+(\vec{k}) c(\vec{k}) - \eta_p c^+(\vec{k}) c(-\vec{k})), c(\vec{p}) c^+(\vec{p})] =$$
$$[c^+(\vec{k}) c(\vec{k}), c(\vec{p}) c^+(\vec{p})] - \eta_p [c^+(\vec{k}) c(-\vec{k}), c(\vec{p}) c^+(\vec{p})]. \qquad (3)$$

From (3.4.20) we have that $[c(\vec{k}), c^+(\vec{p})] = \delta(\vec{p} - \vec{k})$, all other commutators being 0. The first term in (3) becomes $\delta(\vec{p} - \vec{k})(c^+(\vec{k}) c(\vec{p}) - c^+(\vec{p}) c(\vec{k})) = 0$. The second term can be expanded as

$$c(\vec{k}) c^+(\vec{k}) c^+(\vec{p}) c(-\vec{p}) - [\delta(\vec{k} - \vec{p}) + c(\vec{k}) c^+(\vec{p})][c^+(\vec{k}) c(-\vec{p}) - \delta(\vec{k} + \vec{p})]$$
$$= -c^+(\vec{p}) c(-\vec{p}) + c(-\vec{p}) c^+(\vec{p}) - \delta(\vec{k} - \vec{p}) \delta(\vec{k} + \vec{p}) = 0,$$

304

since the two δ-functions are non-zero at different positions. This means that $[A, H] = 0$, and all other commutators in (2) will also vanish, so that $PHP^{-1} = H$ and $[P, H] = 0$

10. **Problem.** Let us define the charge conjugation for the scalar field as
$$C\psi(x)C^{-1} = \eta_c \psi^+(x), \tag{1}$$
where η_c is a phase factor, $\eta_c \eta_c^* = 1$, and C is unitary $C^+ = C^{-1}$. Prove that $CQC^{-1} = -Q$, where Q stands for the charge operator.

Solution. Taking the adjoint of (1), and using the properties of unitary operators, we have
$$(C\psi C^{-1})^+ = C\psi^+ C^{-1},$$
and
$$(\eta_c \psi^+)^+ = \eta_c^* \psi,$$
so that $C\psi^+ C^{-1} = \eta_c^* \psi$. Using (3.1.22), we have
$$CQC^{-1} = e \int (C\psi^+ \psi_{,4} C^{-1} - C\psi_{,4}^+ \psi C^{-1}) d\vec{x}. \tag{2}$$

Since $C\psi_{,4} C^+ = i\psi_{,0}^+ = \psi_{,4}^+$, relation (2) becomes
$$CQC^{-1} = e \int (\psi \psi_{,4}^+ - \psi_{,4} \psi^+) d\vec{x},$$

and using the commutation relations (3.4.12), we arrive at the desired result.

11. **Problem.** Prove the following relation:
$$(\Box + m^2)\langle 0|T(\psi(x)\psi(y))|0\rangle = -i\delta(x - y). \tag{1}$$

Solution. From (3.5.7), we can write
$$T(\psi(x)\psi(y)) = \theta(x_0 - y_0)\psi(x)\psi(y) + \theta(y_0 - x_0)\psi(y)\psi(x). \tag{2}$$

We only need to calculate $\partial^2(\langle 0|T(\psi(x)\psi(y))|0\rangle)/\partial x_0{}^2$, since the other operators do not act on θ. The first derivative reads
$$\langle 0|\frac{\partial T(\psi(x)\psi(y))}{\partial x_0}|0\rangle = \delta(x_0 - y_0)\langle 0|[\psi(x), \psi(y)]|0\rangle +$$
$$\langle 0|\theta(x_0 - y_0)\psi(x)_{,0}\psi(y)|0\rangle + \langle 0|\theta(y_0 - x_0)\psi(y)\psi(x)_{,0}|0\rangle,$$

Problems with solutions

where the first term cancels due to (3.4.43). The second derivative will be

$$\langle 0|\frac{\partial T(\psi(x)\psi(y))}{\partial x_0}|0\rangle = \delta(x_0 - y_0)\langle 0|[\psi(x)_{,0}, \psi(y)]|0\rangle +$$
$$\langle 0|\theta(x_0 - y_0)\psi(x)_{,00}\psi(y)|0\rangle + \langle 0|\theta(y_0 - x_0)\psi(y)\psi(x)_{,00}|0\rangle.$$

Form the commutation relation (3.4.12) and $\langle 0|0\rangle = 1$, the first term becomes $-i\delta(x - y)$, so that we can write

$$(\Box + m^2)\langle 0|T(\psi(x)\psi(y))|0\rangle = -i\delta(x - y)$$
$$+ \langle 0|\theta(x_0 - y_0)(\Box + m^2)\psi(x)\psi(y)|0\rangle$$
$$+ \langle 0|\theta(y_0 - x_0)\psi(y)(\Box + m^2)\psi(x)|0\rangle,$$

where, according to the Klein-Gordon equation, the last two terms vanish.

12. **Problem.** Calculate the following commutators between the components of the electric and magnetic fields:

$$[E_\alpha(x), E_\beta(y)],$$
$$[B_\alpha(x), B_\beta(y)],$$
$$[E_\alpha(x), B_\beta(y)],$$

Also consider the case of equal time, $x_0 = y_0$.
    Solution. From (1.1.5) we have

$$[E_\alpha(x), E_\beta(y)] =$$
$$= \partial_{x_\alpha}\partial_{y_\beta}[A_0(x), A_0(y)] + \partial_{x_0}\partial_{y_0}[A_\alpha(x), A_\beta(y)],$$

and using (4.4.14),

$$[E_\alpha(x), E_\beta(y)] = i(\partial_{x_\alpha}\partial_{y_\beta} + \partial_{x_0}\partial_{y_0}\delta_{\alpha\beta})D_0(x - y).$$

Similarly, we have

$$[B_\alpha(x), B_\beta(y)] = i(\delta_{\alpha\beta}\partial_{x_\mu}\partial_{y_\mu} - \partial_{x_\beta}\partial_{y_\alpha})D_0(x - y),$$

and, using again (4.4.14), the last commutator yields

$$[E_\alpha(x), B_\beta(y)] = i\epsilon_{\beta\mu\alpha}\partial_{y_\mu}\partial_{x_0}D_0(x - y).$$

306

Problems with solutions

For calculating these commutators for equal time, we start with the relations (A.89) and (A.92). We can write

$$D_0(x - y) = -\frac{i}{(2\pi)^3} \int \left(e^{ik(x-y)} - e^{-ik(x-y)} \right) \frac{d\vec{k}}{2\omega}.$$

Taking the derivatives with respect to x_0, and then y_β and y_0, and imposing the equal time condition $x_0 = y_0$, we have

$$\partial_{x_0} D_0(x - y) = -\delta(\vec{x} - \vec{y}),$$

$$\partial_{x_\alpha} D_0(x - y) = 0,$$

$$\partial_{y_\beta} \partial_{x_\alpha} D_0(x - y) = 0,$$

$$\partial_{y_\alpha} \partial_{x_0} D_0(x - y) = -\partial_{y_\alpha} \delta(\vec{x} - \vec{y}),$$

where we have also used the relation (A.16), and the fact that the δ-function is symmetric. Therefore, for $x_0 = y_0$ we have

$$[E_\alpha(x), E_\beta(y)] = [B_\alpha(x), B_\beta(y)] = 0,$$

$$[E_\alpha(x), B_\beta(y)] = -i\epsilon_{\beta\mu\alpha} \partial_{y_\mu} \delta(\vec{x} - \vec{y}).$$

13. **Problem.** The *Casimir Effect*. Let us consider the electromagnetic field in space between two parallel uncharged square plates located at $z = 0$ and $z = a$. The two square plates have sides of length L, and are perfect conductors.

(a) Find the general form of the electromagnetic potential inside the capacitor.

(b) Write down the canonical commutation relations for the electromagnetic field in this region.

(c) Find the Hamiltonian H of the field, and show that the vacuum energy is given by

$$E = \frac{1}{2}L^2 \int \frac{d^2k}{(2\pi)^2} \left[2\sum_{n=1}^{\infty} \sqrt{k_x^2 + k_y^2 + \left(\frac{n\pi}{a}\right)^2} + \sqrt{k_x^2 + k_y^2} \right]. \qquad (1)$$

(d) Let us define the quantity ϵ, representing the difference between the vacuum energy per unit area in the presence and in the absence of plates, namely

$$\epsilon = \frac{E - E_0}{L^2}. \qquad (2)$$

This quantity is divergent, but the integral can be regularized using the function

$$f(k) = \begin{cases} 1, & k < \Lambda \;; \\ 0, & k > \Lambda \;, \end{cases} \tag{3}$$

where Λ is a cutoff parameter. Calculate ϵ, and show that an attractive force acts between the plates. Hint: Use the Euler-Maclaurin expansion formula.

Solution.

(a) Given the geometry of the problem, a natural approach is to take the vector potential $\vec{A}$ with a component perpendicular to the capacitor plates, along the z direction, and a component in the plane parallel to the plates, the sum of the x and y components:

$$\vec{A} = \vec{A}_\perp + \vec{A}_\| . \tag{4}$$

Moreover, since by (4.1.4) the vector potential is not uniquely determined, we work in the Coulomb gauge, characterized by $A_4 = iA_0 = iV = 0$, and $\text{div } \vec{A} = 0$. From (1.1.5) we then have

$$\vec{E} = -\frac{\partial \vec{A}}{\partial t} . \tag{5}$$

The vector potential $\vec{A}$ satisfies the wave equation (4.1.5), namely

$$\Box \vec{A} = \left(\Delta - \frac{\partial^2}{\partial t^2} \right) \vec{A} = 0.$$

This differential equation is solved by separation of variables. Let us take the solution to be

$$\vec{A} = F(x, y, t)[G_1(z)\vec{e}_1 + G_2(z)\vec{e}_2 + G_3(z)\vec{e}_3], \tag{6}$$

where $\vec{e}_i$ are the unit vectors. For each component we then have

$$G_j \left(\partial_t^2 - \partial_x^2 - \partial_y^2 \right) F - F \partial_z^2 G_j = 0.$$

This is satisfied only if we take $\partial_z^2 G = \alpha G$. Let $\alpha = k_z^2$. Then, the equation separates as

$$d_z^2 G_j + k_z^2 G_j = 0 \;; \quad (j = 1, 2, 3)$$

$$\left(\partial_t^2 - \partial_x^2 - \partial_y^2 + k_z^2 \right) F = 0. \tag{7}$$

308

Problems with solutions

The first equation has the well-known solution

$$G_j = a_j \sin k_z z + b_j \cos k_z z. \tag{8}$$

We now impose the boundary conditions. Since the plates are ideal conductors, the parallel component of the electric field, as well as the normal component of magnetic field will vanish on the plates, so that

$$\left.\frac{\partial \vec{A}_\|}{\partial t}\right|_{z=0} = \left.\frac{\partial \vec{A}_\|}{\partial t}\right|_{z=a} = 0, \tag{9}$$

$$B_z|_{z=0} = B_z|_{z=a} = 0. \tag{10}$$

From the first boundary condition we obtain $b_1 = b_2 = 0$, while the second one gives $k_z = n\pi/a$, with n integer.

For the second equation (7) we take a particular solution of the form $F = e^{i(-\omega t + k_x x + k_y y)}$. This yields

$$\omega = \pm\omega_{kn} = \pm\sqrt{k_x^2 + k_y^2 + \left(\frac{n\pi}{a}\right)^2}. \tag{11}$$

Imposing the Coulomb gauge condition, and equating to 0 the coefficients of $\sin k_z z$ and $\cos k_z z$, respectively, we obtain $a_3 = 0$, and

$$ia_1 k_1 + ia_2 k_2 - \frac{n\pi}{a} b_3 = 0. \tag{12}$$

Note that there are two polarization states for $n \neq 0$, and only one for $n = 0$. For $a_1 = a_2 = 1$, a particular solution for $\vec{A}$ can be written as

$$\vec{A} = F\left(\sin k_z z \vec{e}_\| + b_3 \cos k_z z \vec{e}_z\right),$$

where $k_z = n\pi/a$, and $\vec{e}_\|$ belongs to the xy-plane. Taking into account all possible values of n, as well as the 2 possible polarizations, we integrate over k and sum over polarizations, obtaining the general solution

$$\vec{A} = \sum_{n=1}^{\infty} \int \frac{d^2 k}{2\pi} \frac{1}{\sqrt{2\omega_{kn}}} \sum_{\lambda=1}^{2} \left\{ a_\lambda(k_x, k_y, n) \right.$$

$$\times e^{i(-\omega_{kn}t + k_x x + k_y y)} \left[\sin k_z z \vec{e}_\|(\vec{k}, n, \lambda) + \cos \frac{n\pi z}{a} \vec{e}_z \right]$$

$$+ a_\lambda^+(k_x, k_y, n) e^{i(\omega_{kn}t - k_x x - k_y y)}$$

$$\times \left[\sin k_z z \vec{e}_{\parallel}{}^*(\vec{k}, n, \lambda) + \cos k_z z \vec{e}_z \right] \}$$

$$+ \int \frac{d^2 k}{2\pi} \frac{1}{\sqrt{2\omega_k}} \left[a(k_x, k_y) e^{i(-\omega_k t + k_x x + k_y y)} \right.$$

$$\left. + a^+(k_x, k_y) e^{i(\omega_k t - k_x x - k_y y)} \right] \vec{e}_z, \tag{13}$$

where ω_{kn} is given by (11), and we denoted $\omega_k = \sqrt{k_1^2 + k_2^2}$, corresponding to $n = 0$.

(b) By analogy with (4.4.3) and (4.4.4), the non-zero canonical commutation relations are

$$[a_\lambda(k_x, k_y, n), a_\lambda^{'+}(k'_x, k'_y, m)] = \delta_{mn} \delta_{\lambda\lambda'} \delta(k_x - k'_x) \delta(k_y - k'_y) ;$$

$$[a(k_x, k_y), a^+(k'_x, k'_y)] = \delta(k_x - k'_x) \delta(k_y - k'_y). \tag{14}$$

(c) Similar to (4.3.29), we can write the Hamiltonian as

$$H = \int d^2 k \sum_{n=1}^{\infty} \frac{1}{2} \omega_{kn} \sum_{\lambda=1}^{2} [a_\lambda^+(k_x, k_y, n) a_\lambda(k_x, k_y, n)$$

$$+ a_\lambda(k_x, k_y, n) a_\lambda^+(k_x, k_y, n)]$$

$$+ \frac{1}{2} \int d^2 k \, \omega_k [a^+(k_x, k_y) a(k_x, k_y) + a(k_x, k_y) a^+(k_x, k_y)]. \tag{15}$$

For the ground state $|0>$, the energy is computed as

$$< 0|H|0 > = \sum_{n=1}^{\infty} \sum_{\lambda=1}^{2} \int d^2 k \, \frac{1}{2} \omega_{kn} < 0|a_\lambda(k_x, k_y, n) a_\lambda^+(k_x, k_y, n)|0 >$$

$$+ \int d^2 k \, \frac{1}{2} \omega_k < 0|a(k_x, k_y) a^+(k_x, k_y)|0 > . \tag{16}$$

Following (3.5.3), since the vacuum expectation value of the normal product is 0, then the vacuum expectation value of $a_\lambda(k_x, k_y, n) a_\lambda^+(k_x, k_y, n)$ is given by the commutator (14) for $k = k'$, and (16) becomes

$$< 0|H|0 > = \sum_{n=1}^{\infty} \frac{1}{2} \int d^2 k \, \omega_{kn} \, 2\delta^{(2)}(0) + \frac{1}{2} \int d^2 k \, \omega_k \delta^{(2)}(0),$$

where $\delta^{(2)}(0)$ is the 2D δ-function. Since

$$\delta^{(2)}(0) = \int \frac{dx\,dy}{(2\pi)^2} e^{ik_1 x + ik_2 y}\Big|_{\vec{k}_\| = 0} = \frac{L^2}{(2\pi)^2},$$

which is proportional to the area of the capacitor plate, the vacuum energy in this state becomes

$$E = \frac{L^2}{2(2\pi)^2} \int d^2k \left[2\sum_{n=1}^{\infty} \sqrt{k_x^2 + k_y^2 + \left(\frac{n\pi}{a}\right)^2} + \sqrt{k_x^2 + k_y^2} \right].$$

(d) In the absence of plates, k_z is no longer quantized as $n\pi/a$, and the sum is replaced by an integral over n. In this case, the vacuum energy becomes

$$E_0 = \frac{L^2}{2(2\pi)^2} \int d^2k \int \frac{a}{2\pi} dn\, 2\sqrt{k_x^2 + k_y^2 + \left(\frac{n\pi}{a}\right)^2}, \qquad (17)$$

where the $n = 0$ term also drops out.

Then ϵ defined by (2) takes the form

$$\epsilon = \frac{1}{2(2\pi)^2} \int d^2k \left[2\sum_{n=1}^{\infty} \sqrt{k_x^2 + k_y^2 + \left(\frac{n\pi}{a}\right)^2} \right.$$

$$\left. + \sqrt{k_x^2 + k_y^2} - 2\int_0^{\infty} \sqrt{k_x^2 + k_y^2 \left(\frac{n\pi}{a}\right)^2} \right],$$

which can be written in polar coordinates, after integrating over all angles, as

$$\epsilon = \frac{1}{2(2\pi)} \int_0^{\infty} k\,dk \left[k + 2\sum_{n=1}^{\infty} \sqrt{k^2 + \left(\frac{n\pi}{a}\right)^2} \right.$$

$$\left. - 2\int_0^{\infty} dn \sqrt{k^2 + \left(\frac{n\pi}{a}\right)^2} \right].$$

Making the change of variables $u = a^2 k^2/\pi^2$, this expression becomes

$$\epsilon = \frac{\pi^2}{8a^3} \int_0^{\infty} du \left(\sqrt{u} + 2\sum_{n=1}^{\infty} \sqrt{u + n^2} - 2\int_0^{\infty} dn \sqrt{u + n^2} \right).$$

Since these integrals are divergent, we introduce the function $f(x)$ given by (3), and we have

$$\epsilon = \frac{\pi^2}{8a^3} \int_0^\infty du \left[\sqrt{u} f\left(\frac{\pi\sqrt{u}}{a}\right) + 2\sum_{n=1}^\infty \sqrt{u+n^2} f\left(\frac{\pi\sqrt{u+n^2}}{a}\right) \right.$$

$$\left. -2\int_0^\infty dn \sqrt{u+n^2} f\left(\frac{\pi\sqrt{u+n^2}}{a}\right) \right], \tag{18}$$

which is finite.

Defining

$$F(n) = \int_0^\infty du \sqrt{u+n^2} \, f\left(\frac{\pi\sqrt{u+n^2}}{a}\right),$$

ϵ then takes the form

$$\epsilon = \frac{\pi^2}{8a^3} \left(F(0) + 2\sum_{n=1}^\infty F(n) - 2\int_0^\infty dn \, F(n) \right). \tag{19}$$

This expression can immediately be written in terms of the Bernoulli numbers B_α, using the Euler-Maclaurin expansion formula:

$$\sum_{n=1}^\infty F(n) - \int_0^\infty dn \, F(n) + \frac{1}{2}F(0) = -\frac{1}{2!}B_2 F'(0) - \frac{1}{4!}B_4 F'''(0) + ...,$$

where B_2, B_4, ... are defined as

$$\frac{x}{e^x - 1} = \sum_{\alpha=0}^\infty B_\alpha \frac{x^\alpha}{\alpha!}.$$

Substituting this expression into (19), we have

$$\epsilon = \frac{\pi^2}{4a^3} \left(-\frac{1}{2!}B_2 F'(0) - \frac{1}{4!}B_4 F'''(0) + ... \right).$$

One can easily compute $F'(0) = 0$, and $F'''(0) = -4$, and since $B_4 = 1/30$, the final form of the energy per unit surface will be

$$\epsilon = -\frac{1}{720}\frac{\pi^2}{a^3}.$$

Problems with solutions

By definition, the force per unit area between the conducting plates is computed as

$$f = -\frac{\partial \epsilon}{\partial a} = -\frac{1}{240}\frac{\pi^2}{a^4},$$

which is an attractive force. The total force between two plates with $L = 1$cm and separated by $a = 1\mu$m will be $F = 10^{-8}$N.

We can conclude that the vacuum energy of the electromagnetic field between the two *uncharged* conducting plates produces a weak attractive force between them. The effect is due to quantum vacuum fluctuations of the electromagnetic field, and is a striking example of a purely quantum field effect. This effect was predicted by the Dutch physicist *Hendrick Casimir* in 1948 and measured experimentally in 1997 by *Steven Lamoreaux*; his results were in agreement with the theory to within the experimental uncertainty of 5%.

14. **Problem.** Prove that there is no second rank matrix anti-commuting with the Pauli matrices σ_i.

Solution. In the space of the second-rank matrices there are only four linearly independent matrices. Let us first show that the Pauli matrices and the unit matrix I are linearly independent. To this end, we shall assume that the opposite is true, namely

$$a_i\sigma_i + a\,I = 0,$$

or

$$a_1 - ia_2 + 0 + 0 = 0,$$
$$a_1 + ia_2 + 0 + 0 = 0,$$
$$0 + 0 + a_3 + a = 0,$$
$$0 + 0 - a_3 + a = 0.$$

The coefficients a_i, a would be different from zero only if the determinant of the above system is zero, which is not true. Consequently, any matrix σ can be expressed as

$$\sigma = a_i\sigma_i + bI.$$

Next, let's show that σ and σ_k anti-commute. Using the well-known relation $\{\sigma_i, \sigma_k\} = 2.I.\delta_{ik}$, we have

$$a_i(\sigma_i\sigma_k + \sigma_k\sigma_i) + 2b\sigma_k = 0.$$

If $b = 0$, then $a_k = 0$, and therefore $\sigma = 0$. If $b \neq 0$, then

$$\sigma_k = -\frac{a_k}{b}I,$$

313

in contradiction with the initial assumption. We conclude that a fourth non-zero matrix, anti-commuting with the σ_i's, does not exist.

15. Problem. Show that the rank of the Dirac matrices is even.
Solution.
Since $\alpha_i\beta + \beta\alpha_i = 0$ $(i = 1, 2, 3)$, we have

$$\alpha_i\beta = (-I)\beta\alpha_i.$$

We can write

$$\det(\alpha_i\beta) = \det\alpha_i . \det\beta = det(-I)\det\alpha_i \det\beta.$$

Assuming that $\det\alpha_i$, $\det\beta \neq 0$, we have

$$\det(-I) = (-1)^n = 1$$

which is satisfied for $n = 2k$, where k is an integer.
Let us now show that $\det\alpha_i$, $\det\beta \neq 0$. Since

$$\alpha_i^2 = \beta^2 = I$$

we can write

$$\det\alpha_i^2 = \det\alpha_i \det\alpha_i = 1,$$

or

$$\det\alpha_i = \pm1 \ \ \det\beta = \pm1.$$

16. Problem. Prove the following relations involving the Dirac gamma matrices:
(a) $\gamma_n\hat{A} + \hat{A}\gamma_n = 2A_n$; $\gamma_n\hat{A} - \hat{A}\gamma_n = 2A_n - 2\hat{A}\gamma_n$; $\hat{A}\hat{B}$ $+\hat{A}\hat{B} = 2(AB).$
(b) $\gamma_i a\gamma_i = 4a$; $\gamma_i\hat{A}\gamma_i = -2\hat{A}$; $\gamma_i(\hat{A} + a)\gamma_i = -2\hat{A}$ $+4a$; $\gamma_i\gamma_n\gamma_i = -2\gamma_n.$
(c) $\gamma_i\hat{A}\hat{B}\gamma_i = 4(AB)$; $\gamma_i\gamma_m\gamma_n\gamma_i = 4\delta_{mn}$; $\gamma_i\hat{A}\gamma_n\gamma_i = 4A_n$ $= 2\gamma_n\hat{A} + 2\hat{A}\gamma_n$; $\gamma_i(\hat{A} + a)\gamma_n\gamma_i = 4A_n - 2a\gamma_n$; $\gamma_i(\hat{A} + a)$ $\times (\hat{B} + b)\gamma_i = 4(AB + ab) - 2(a\hat{B} + b\hat{A}).$
(d) $\gamma_i\hat{A}\hat{B}\hat{C}\gamma_i = -2\hat{C}\hat{B}\hat{A}$; $\gamma_i\hat{A}\gamma_n\hat{B}\gamma_i = 2\gamma_n\hat{B}\hat{A} - 4\hat{A}B_n$ $= -2\hat{B}\gamma_n\hat{A} = 2\hat{B}\hat{A}\gamma_n - 4\hat{B}A_n$; $\gamma_i(\hat{A} + a)\gamma_n(\hat{B} + b)\gamma_i$ $= 2\hat{B}\hat{A}\gamma_n - 4\hat{B}A_n + 4aB_n + 4bA_n - 2ab\gamma_n,$
where a is an arbitrary complex number, and A_i, B_i, and C_i are four-vectors.

Problems with solutions

Solution.
(a_1) $\gamma_n \hat{A} = \gamma_n \gamma_i A_i = A_i(2\delta_{ni} - \gamma_i \gamma_n) = 2A_n - A_i \gamma_i \gamma_n$
$\qquad = 2A_n - \hat{A}\gamma_n \; ; \quad \gamma_n \hat{A} + \hat{A}\gamma_n = 2A_n.$
(a_2) $\gamma_n \hat{A} = 2A_n - \hat{A}\gamma_n, \; \gamma_n \hat{A} - \hat{A}\gamma_n = 2A_n - 2\hat{A}\gamma_n.$
(a_3) $\qquad \hat{A}\hat{B} + \hat{B}\hat{A} = A_i B_k \gamma_i \gamma_k + B_k A_i \gamma_k \gamma_i$
$\qquad = A_i B_k(2\delta_{ik} - \gamma_k \gamma_i) + B_k A_i \gamma_k \gamma_i = 2A_i B_i = 2(AB).$

(b_1) $\qquad$ follows immediately as a result of $\gamma_i \gamma_i = 4.$
(b_2) $\qquad \gamma_i \hat{A}\gamma_i =$ (according to a_1) $= \gamma_i(2A_i - \gamma_i \hat{A})$
$\qquad\qquad = 2\hat{A} - 4\hat{A} = -2\hat{A}.$
(b_4) Consider the case (b_2) with $A_n = 1$, while the remaining components of the four-vector A_i are zero

(c_1) $\qquad \gamma_i \hat{A}\hat{B}\gamma_i =$ (according to (a_1)) $= \gamma_i \hat{A}.(2B_i - \gamma_i \hat{B}) =$
(according to (b_2)) $= 2\hat{B}\hat{A} + 2\hat{A}\hat{B} =$ (according to (a_3)) $= 4(AB).$
(c_2) $\qquad$ A particular case of (c_1), when the only non-zero components of A_i and B_i equal one.
(c_3) $\qquad$ A particular case of (c_1), when the only non-zero component B_n equals one:
$\qquad\qquad \gamma_i \hat{A}\gamma_n \gamma_i =$ (according to (a_1))
$\qquad\qquad = 2\gamma_n A + 2\hat{A}\gamma_n$
(c_4) $\qquad$ Follows as a consequence of (c_3) and (b_4)
(c_5) $\qquad$ Follows as a consequence of (c_1), (b_2), and (b_1)

(d_1) $\qquad \gamma_i \hat{A}\hat{B}\hat{C}\gamma_i =$ (according to (a_1))$=\gamma_i \hat{A}\hat{B}(2C_i - \gamma_i \hat{C})$
$=$ (according to (c_1)) $= 2\hat{C}\hat{A}\hat{B} - 4(AB)\hat{C} =$ (according to (a_3))$=$
$\qquad 2\hat{C}[2(AB) - \hat{B}\hat{A}] - 4(AB)\hat{C} = -2\hat{C}\hat{B}\hat{A}.$
(d_2) $\qquad$ Follows as a consequence of (d_1) and (a_1)
(d_3) Follows as a consequence of $(d_2), (c_2), (a_1)$, and (b_4)

17. **Problem.** Consider the Lagrangian density
$$\mathcal{L} = i\bar{\psi}\gamma_i \partial_i \psi - gx^2\bar{\psi}\psi,$$
where g is a constant. (a) Derive the expression for the energy-momentum tensor of the field, T_{ij}, and its divergence, $\partial_j T_{ij}$. (b) Calculate the commutator between the momentum components, $[P_0(t), P_\alpha(t)]$.
Solution. (a) From (1.5.4) and (1.1.17) we have
$$T_{ij} = i\bar{\psi}\gamma_i \partial_j \psi - (i\bar{\psi}\gamma_l \partial_l \psi - gx^2\bar{\psi}\psi)\delta_{ij}.$$
Form (1.1.15), the equations of motion are
$$i\partial_j \bar{\psi}\gamma_j + gx^2\bar{\psi} = 0,$$

315

$$i\gamma_j \partial_j \psi - g x^2 \psi = 0.$$

Using these equations, the divergence of the energy-momentum tensor becomes

$$\partial_i T_{ij} = 2 g x_j \bar\psi \psi.$$

(b) First prove that

$$[\psi^+(x)\psi(x), \psi^+(y)\partial_\alpha\psi(y)] =$$
$$= \psi^+(x)\partial_\alpha\psi(y)\delta(\vec{x}-\vec{y}) - \psi^+(y)\psi(x)\partial_{y_\alpha}\delta\vec{x}-\vec{y},$$
$$[\psi^+(x)\partial_\beta\psi x, \psi^+ y\partial_\alpha\psi y] =$$
$$= \psi^+(x)\partial_\alpha\psi(y)\partial_{x_\beta}\delta(\vec{x}-\vec{y}) - \psi^+(y)\partial_\beta\psi(x)\partial_{y_\alpha}\delta\vec{x}-\vec{y},$$
$$[\bar\psi(x)\gamma_\beta\partial_\beta\psi(x), \psi^+(y)\partial_\alpha\psi(y)] =$$
$$= \bar\psi(x)\gamma_\beta\partial_\alpha\psi(y)\partial_{x_\beta}\delta(\vec{x}-\vec{y}) - \bar\psi(y)\gamma_\beta\partial_\beta\psi(x)\partial_{y_\alpha}\delta\vec{x}-\vec{y}.$$

Using the expressions (5.2.12) and (5.2.13), and taking into account the Dirac equation, the final result is

$$[P_0(t), P_\alpha(t)] = -2ig \int x_\alpha \bar\psi\psi d\vec{x}.$$

The surface integrals are not shown, as they cancel at infinity.

18. **Problem.** Let us consider the coupled system consisting of a Maxwell field, and an electrically charged Dirac field, describing quantum electrodynamics. The coupling in this system is driven by the Dirac current $j_k = ie\bar\psi\gamma_k\psi$. Derive the equations of motion, and show that the energy-momentum vector is gauge invariant, and that the energy and momentum are conserved in this system.

Solution. For this system, the Lagrangian density can be written as the sum

$$\mathcal{L} = \mathcal{L}_{Dirac} + \mathcal{L}_{EM} + \mathcal{L}_{INT}$$

$$= -\bar\psi(\hat\partial + m)\psi - \frac{1}{4}F_{ik}F_{ik} - ie\bar\psi\hat A\psi, \tag{1}$$

where the interaction Lagrangian can also be found in (6.3.3). Using the Euler-Lagrange equations (1.1.15) for the fields $\bar\psi$, ψ, and A_k, we obtain the following equations of motion:

$$[\hat\partial + m + ie\hat A]\psi = 0,$$

$$\bar\psi[\overleftarrow{\hat\partial} - m - ie\hat A] = 0, \tag{2}$$

Problems with solutions

$$\Box A_i - A_{k,ki} = ie\overline{\psi}\gamma_i\psi.$$

Let us consider the following local gauge transformations

$$A'_k = A_k + \partial_k\Lambda,$$

$$\psi' = exp[-ie\Lambda]\psi , \qquad (3)$$

$$\overline{\psi}' = exp[ie\Lambda]\overline{\psi}.$$

One can easily show that the total Lagrangian density (1) is invariant under these transformations. In fact, the extra terms resulting from the expansion of the Dirac Lagrangian density are exactly cancelled by the terms form the interaction Lagrangian density. The same effect can be obtained if we drop the interaction term, and instead replace ∂_k with the so-called gauge-covariant derivative, defined as $D_k = \partial_k + ieA_k$. This procedure is also called *minimal coupling* prescription.

The total energy-momentum tensor can be defined in the same manner as the sum

$$T^{(c)}_{jk} = T^{(c)Dirac}_{jk} + T^{(c)EM}_{jk} + T^{(c)INT}_{jk}. \qquad (4)$$

From the definition (1.5.4), we have

$$T^{(c)Dirac}_{jk} = -\overline{\psi}\gamma_j\psi_{,k} + \delta_{jk}\overline{\psi}(\hat{\partial} + m)\psi,$$

$$T^{(c)EM}_{jk} = F_{jl}A_{l,k} + \delta_{jk}\frac{1}{4}F_{st}F_{st}, \qquad (5)$$

$$T^{(c)INT}_{jk} = ie\delta_{jk}\overline{\psi}\hat{A}\psi.$$

After applying the gauge transformations (3), we are left with the extra terms:

$$\delta T^{(c)Dirac}_{jk} = ie\overline{\psi}\gamma_j\psi\Lambda_{,k} - \delta_{jk}ie\overline{\psi}\gamma_l\psi\Lambda_{,l},$$

$$\delta T^{(c)EM}_{jk} = -F_{jl}\Lambda_{,kl},$$

$$\delta T^{(c)INT}_{jk} = ie\delta_{jk}\overline{\psi}\gamma_l\psi\Lambda_{,l}.$$

The sum of these contributions reduces to

$$\delta T_{jk} = ie\overline{\psi}\gamma_j\psi\Lambda_{,k} - F_{jl}\Lambda_{,kl}$$

$$= (ie\overline{\psi}\gamma_j\psi - F_{jl,l})\Lambda_{,k} - \partial_l(F_{jl}\Lambda_{,k}) = -\partial_l(F_{jl}\Lambda_{,k}),$$

Problems with solutions

where we have used the third equation of motion. The energy-momentum four-vector is given by (5.2.12) such that

$$\delta P_j = i \int \delta T_{j4}^{(c)} d\vec{x} = -i \int \partial_l (F_{l4}\Lambda_{,j}) d\vec{x},$$

which reduces to a 3D surface integral, since F_{ij} is antisymmetric. This integral cancels at infinity, proving that the energy-momentum 4-vector is gauge invariant.

In order to prove that the energy and momentum are conserved, let us compute the 4-divergence of the energy-momentum tensor (4). Using the equations of motion (2), we obtain the three terms as follows:

$$T_{jk,j}^{(c)Dirac} = -ie\partial_k(\overline{\psi}\gamma_l\psi)A_l,$$

$$T_{jk,j}^{(c)EM} = -ie(\overline{\psi}\gamma_l\psi)A_{l,k},$$

$$T_{jk,j}^{(c)INT} = ie\partial_k(\overline{\psi}\gamma_l\psi)A_l + ie(\overline{\psi}\gamma_l\psi)A_{l,k}.$$

Adding up these three contributions, we obtain the energy-momentum conservation, since $T^{(c)}_{jk,j} = 0$. Prove that this conservation equation also holds for the symmetrized energy-momentum tensor T_{jk}.

19. **Problem.** Calculate the differential cross section for the scattering of an electron by the Coulomb field of an external nucleus of charge Ze.

Solution. Using Fourier series, and the convenient normalization factor $(2\pi)^3 \to V$, where V is the volume, the quantized state functions ψ and $\overline{\psi}$ write

$$\psi(x) = \frac{1}{\sqrt{V}} \sum_{\vec{p}} \left(\frac{m}{p_0}\right)^{1/2}$$

$$\times \sum_{s=1}^{2} \left[a_s(\vec{p})u^{(s)}(\vec{p})e^{ipx} + b_s^+(\vec{p})v^{(s)}(\vec{p})e^{-ipx}\right];$$

$$\overline{\psi}(x) = \frac{1}{\sqrt{V}} \sum_{\vec{p}} \left(\frac{m}{p_0}\right)^{1/2}$$

$$\times \sum_{s=1}^{2} \left[a_s^+(\vec{p})\overline{u}^{(s)}(\vec{p})e^{-ipx} + b_s(\vec{p})\overline{v}^{(s)}(\vec{p})e^{ipx}\right].$$

The potential four-vector for the electromagnetic field of the nucleus is

$$A_j^{ext} = \begin{cases} 0, & j = 1,2,3 : \\ i\frac{Ze}{4\pi|\mathbf{x}|}, & j = 4 \end{cases} \qquad (1)$$

This is a first order process. The associated Feynman diagram is

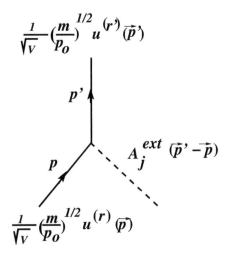

The corresponding matrix element for the scattering process $i \to f$, is

$$M_{fi} = 2\pi\delta(E_{\vec{p}} - E_{\vec{p}'})\left(\frac{e}{V}\right)\frac{m}{\partial_0}A_4^{ext}(\vec{p}' - \vec{p})u^{(r')} + (\vec{p}')u^{(r)}(\vec{p}), \qquad (2)$$

and the transition probability is

$$W = |M_{fi}|^2. \qquad (3).$$

We are interested in the transition probability per unit time. To this end, let us write

$$\left[2\pi\delta(E_{\vec{p}} - E_{\vec{p}'})\right]^2 = 2\pi\delta(E_{\vec{p}} - E_{\vec{p}'})\int e^{i(E_{\vec{p}} - E_{\vec{p}'})t}\,dt$$

$$= 2\pi\delta(E_{\vec{p}} - E_{\vec{p}'})\int dt.$$

The transition probability per unit time is then

$$w_{fi} = \frac{|M_{fi}|^2}{\int dt} = 2\pi\delta(E_{\vec{p}} - E_{\vec{p}'})\frac{e^2}{V^2}\left(\frac{m}{E_{\vec{p}}}\right)^2$$

$$\times \left|A_4^{ext}(\vec{p}\,' - \vec{p})\right|^2 \left|u^{(r')+}(\vec{p}\,')u^{(r)}(\vec{p})\right|^2. \tag{4}$$

For this system, the density of the final states is

$$\rho_f = \frac{V\,d^3p'}{(2\pi)^3} = \frac{V}{(2\pi)^3}p'^2\,dp'\,d\Omega$$

$$= \frac{V}{(2\pi)^3}p'\,E_{\vec{p}\,'}\,dE_{\vec{p}\,'}\,d\Omega. \tag{5}$$

Using the incident flux density

$$J = \frac{v}{V} = \frac{p}{V E_{\vec{p}}}, \tag{6}$$

where $p = |\vec{p}|$, and integrating with respect to energy, one obtains the differential cross section over the solid angle $d\Omega = 2\pi \sin\theta\,d\theta$, where θ is the angle between $\vec{p}$ and $\vec{p}\,'$:

$$\frac{d\sigma}{d\Omega} = \left(\frac{V E_{\vec{p}}}{p}\right)\frac{V}{(2\pi)^3}\int dE_{\vec{p}\,'}\,p'\,E_{\vec{p}\,'}\,2\pi$$

$$\times \delta(E_{\vec{p}} - E_{\vec{p}\,'})\frac{e^2}{V^2}\left(\frac{m}{E_{\vec{p}}}\right)^2$$

$$\left|A_4^{ext}(\vec{p}\,' - \vec{p})\right|^2 \cdot \left|u^{(r')+}(\vec{p}\,')\gamma_4 u^{(r)}(\vec{p})\right|^2$$

$$= \left(\frac{m\,e}{2\pi}\right)^2 \left|A_4^{ext}(\vec{p}\,' - \vec{p})\right|^2 \cdot \left|u^{(r')+}(\vec{p}\,')\gamma_4 u^{(r)}(\vec{p})\right|^2 \tag{7}$$

We have obtained the differential cross section for a given polarization of the incoming and scattered particles. If we do not know the polarization state of the incoming particles, and we are not interested in the polarization of the scattered particles, we have to average over the incoming particle polarizations, and sum over the polarizations of the scattered particles:

$$\frac{1}{2}\sum_{r=1}^{2}\sum_{r'=1}^{2}\left|\bar{u}^{(r')}(\vec{p}\,')\gamma_4 u^{(r)}(\vec{p})\right|^2 = \frac{1}{2}Tr\left[\gamma_4\Lambda^+(\vec{p}\,')\gamma_4\Lambda^+(\vec{p})\right]$$

$$= \frac{1}{2m^2}(m^2 + E_{\vec{p}}^2 + \vec{p}\cdot\vec{p}\,') = \left(\frac{E_{\vec{p}}}{m}\right)^2\left(1 - v^2\sin^2\frac{\theta}{2}\right), \tag{8}$$

where $\Lambda^+ = \frac{\hat{p}+im}{2\,i\,m}$.

We then obtain:

$$\frac{d\sigma}{d\Omega} = \left(\frac{eE_{\vec{p}}}{2\pi}\right)^2 \left|A_4^{ext}(\vec{p}' - \vec{p})\right|^2 \left(1 - v^2 \sin^2 \frac{\theta}{2}\right). \tag{9}$$

However,

$$A_4^{ext}(\vec{p}' - \vec{p}) = \frac{ize}{|\vec{p}' - \vec{p}|^2} = \frac{ize}{\left(2p\sin\frac{\theta}{2}\right)^2}, \tag{10}$$

and then we get

$$\frac{d\sigma}{d\Omega} = \left(\frac{ze^2}{8\pi}\right)^2 \left(\frac{E_{\vec{p}}}{p^2 \sin^2 \frac{\theta}{2}}\right)^2 \left(1 - v^2 \sin^2 \frac{\theta}{2}\right). \tag{11}$$

In the non-relativistic approximation ($v \ll 1$, $E_{\vec{p}} \to m$, $p \to mv$), relation (11) becomes

$$\frac{d\sigma}{d\Omega} = \left(\frac{ze^2}{8\pi} \frac{1}{mv^2} \frac{1}{\sin^2 \frac{\theta}{2}}\right)^2, \tag{12}$$

which is known as the *Rutherford formula*.

20. **Problem.** Using the Feynman rules, write the expressions for the amplitudes corresponding to the following QED processes:
(a) $\mu^- \mu^+ \to e^- e^+$,
(b) $e^- \mu^+ \to e^- \mu^+$.

Calculate the quantity $|M|^2$ by averaging over all initial polarizations and summing over the final polarizations of the particles. In the ultra-relativistic limit, calculate the differential cross sections in the center-of-mass frame.

Solution. This problem is similar to Section 7.7, only now the particles have different masses, and we use the indices 1 and 2 to denote the momenta of the two particles. (a) The diagram for this process is represented in the figure below, similar to the second graph in Figure 7.9.

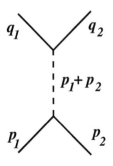

Problems with solutions

Following the GRAPH TABLE, the matrix element for this process will be

$$S_{12} = -\frac{e^2}{2(2\pi)^2} \int \bar{v}_-^{(2)}(\vec{q_1})\gamma_j v_-^{(1)}(\vec{q_2})\delta_{jl}\frac{1}{k^2}\bar{v}_+^{(1)}(\vec{p_1})\gamma_l v_+^{(2)}(\vec{p_2})$$
$$\times \delta(q_1 + q_2 - k)\delta(k - p_1 - p_2)dk$$
$$= -\frac{e^2}{2(2\pi)^2}\frac{1}{(p_1 + p_2)^2}(\bar{v}_-^{(2)}(\vec{q_1})\gamma_j v_-^{(1)}(\vec{q_2}))(\bar{v}_+^{(1)}(\vec{p_1})\gamma_j v_+^{(2)}(\vec{p_2})).$$

Since this case is also similar to the Möller scattering in Section 7.5, we can follow the same steps and write (see 7.4.22 and 7.4.23)

$$|M|^2 = \frac{e^4}{64(p_1 + p_2)^4(2\pi)^4 E_\mu^+ E_\mu^- E_e^+ E_e^-}$$

$$\times Tr[\gamma_j(\hat{q_2} - m_e)\gamma_l(\hat{q_1} + m_e)]Tr[\gamma_j(\hat{p_1} + m_\mu)\gamma_l(\hat{p_2} - m_\mu)].$$

Following (7.7.10), we have

$$|M|^2 = \frac{e^4}{64(p_1 + p_2)^4(2\pi)^4 E_\mu^+ E_\mu^- E_e^+ E_e^-}X, \qquad (1)$$

where

$$X = \frac{1}{64}\frac{B}{(p_1 + p_2)^2} = \frac{m_e^2 m_\mu^2}{2}\frac{\kappa^2 + \lambda^2 - 2\mu + 2}{(p_1 + p_2)^4},$$

and we took into account that now the particles have different masses, such that (7.7.18) and (7.7.22) become

$$\kappa = -\gamma_e\gamma_\mu(1 + \beta_e\beta_\mu cos\theta),$$
$$\lambda = \gamma_e\gamma_\mu(1 - \beta_e\beta_\mu cos\theta),$$
$$\mu = -(2\gamma_e\gamma_\mu - 1),$$
$$(p_1 + p_2)^2 = -m_e^2(2\gamma_e)^2.$$

Note that here the index μ stands for the muon particle/anti-particle, i.e. the particles in the initial state. Making the substitutions, we obtain

$$X = \frac{m_\mu m u^2}{16\gamma_e^4 m_e^2}(\gamma_e^2\gamma_\mu^2 + \gamma_e^2\gamma_\mu^2\beta_e^2\beta_\mu^2 cos^2\theta - 2\gamma_e\gamma_\mu + 1),$$

which can then be replaced in Equation (1) for the scattering amplitude. For the scattering cross-section, we use the general expression

322

(7.7.14), where now β'_- is β_e^-, E'_- is E_e^-, v_1 is the relative velocity of the two incident particles $v_1 = 2\beta_m u$ (in CMF), E_+ and E_- are $E_\mu^+ = E_\mu^- = E_\mu$, respectively, and $E = E_\mu^+ + E_\mu^- = 2E_\mu$. In we also have CMF $|\vec{P}| = 0$, so that

$$\frac{d\sigma}{d\Omega} = \frac{e^4}{(4\pi)^2} \frac{\beta_e \gamma_e m_e}{\beta_m u \gamma_m u m_m u (\gamma_m u m_m u)^2} X.$$

In the high-energy limit, we have the approximations $\beta \approx 1$, $E \approx p$, $\gamma_\mu m_\mu \approx \gamma_e m_e$, and $1/\gamma^2 \rightarrow 1$. In this limit, we are left with

$$\frac{d\sigma}{d\Omega} = \frac{e^4}{256\pi^2 E_\mu^2}(1 + \cos^2\theta).$$

(b)   For this process the Feynman diagram is

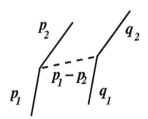

with the corresponding matrix element

$$S_{12} = -\frac{e^2}{2(2\pi)^2} \int \bar{v}_+^{(1)}(\vec{q_1})\gamma_j v_-^{(1)}(\vec{q_2})\delta_{jl}\frac{1}{k^2}\bar{v}_-^{(2)}(\vec{p_2})\gamma_l v_+^{(2)}(\vec{p_1})$$
$$\times \delta(q_2 - q_1 - k)\delta(p_2 - p_1 + k)$$
$$= -\frac{e^2}{2(2\pi)^2}\frac{1}{(p_1 - p_2)^2}(\bar{v}_+^{(1)}(\vec{q_1})\gamma_j v_-^{(1)}(\vec{q_2}))(\bar{v}_-^{(2)}(\vec{p_2})\gamma_j v_+^{(2)}(\vec{p_1})).$$

This is similar to the first diagram in Figure 7.9. Following (7.4.22), after averaging over the spin states of the initial particles and summing over the spin states of the final particles, we obtain the squared amplitude

$$|M|^2 = \frac{e^4}{64(p_1 - p_2)^4(2\pi)^4 E_\mu^+ E_\mu^- E_e^+ E_e^-}$$
$$\times Tr[\gamma_j(\hat{p}_1 + m_e)\gamma_l(\hat{p}_2 + m_e)]Tr[\gamma_j(\hat{q}_2 - m_\mu)\gamma_l(\hat{q}_2 - m_\mu)].$$

Note that here p_1 and p_2 still correspond to the electrons, even though now one is in the initial state and the other one in the final state. Using the same steps as above, we can write

$$|M|^2 = \frac{e^4}{64(p_1 - p_2)^4(2\pi)^4 E_\mu^+ E_\mu^- E_e^+ E_e^-}X, \qquad (1)$$

where

$$X = \frac{1}{64} \frac{A}{(p_1 - p_2)^2} = \frac{m_e^2 m_\mu^2}{2} \frac{\kappa^2 + \mu^2 - 2\lambda + 2}{(p_1 - p_2)^4.}$$

From (7.7.21), we have

$$(p_1 - p_2)^2 = 2m_e^2(\gamma_e^2 - 1)(1 - costheta).$$

Making the substitutions and passing to the high-energy limit, we are left with

$$X = \frac{1}{8} \frac{4 + (1 + cos\theta)^2}{(1 - cos\theta)^2}.$$

For the scattering cross section, we start with the same expression (7.7.14) as above, working in the CMF, where now $v_r = \beta_e + \beta_m u$, and $E = E_\mu + E_e$. After cancellations, in the high-energy limit, this becomes

$$\frac{d\sigma}{d\Omega} = \frac{e^4}{128\pi^2 E_i^2} \frac{4 + (1 + cos\theta)^2}{(1 - cos\theta)^2},$$

where now

$$E_e \approx E_\mu = E_i$$

.

21. **Problem.** Write the expressions for the matrix elements corresponding to the following Feynman diagrams:

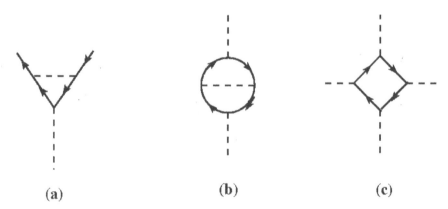

(a)　　　　　　　　(b)　　　　　　　　(c)

Solution. (a) We derive the expressions for the matrix elements by applying the Feynman rules given in the GRAPH TABLE, and following the same steps as in Chapter 8.

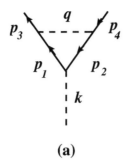

(a)

The relevant quantities are labeled in the figure. Following Section 8.1, we have

$$S = -\frac{e^3}{(2\pi)^9/2}\frac{1}{\sqrt{2\omega}}\bar{v}_-(p_4)\bar{v}_+(p_3)\int \gamma_l \frac{i\hat{p}_1-m}{p_1^2+m^2}\gamma_j\frac{i\hat{p}_2-m}{p_2^2+m^2}$$

$$\times \gamma_l \frac{1}{q^2}\delta(p_4-p_2-q)\delta(p_3-p_1+q)\delta(p_1+p_2-k)dp_1dp_2dq(e_1)_j,$$

or, integrating with respect to p_2, and renaming $p_3 = p$ we finally obtain

$$S = -\frac{e^3}{(2\pi)^1/2}\frac{1}{\sqrt{2\omega}}\bar{v}_-(p_4)\bar{v}_+(p)\int \gamma_l \frac{i(\hat{p}-\hat{q})-m}{(p-q)^2+m^2}\gamma_j$$

$$\times \frac{i(\hat{p}+\hat{k}-\hat{q})-m}{(p+k-q)^2+m^2}\gamma_l\frac{1}{q^2}\frac{dp}{(2\pi)^4}\frac{dq}{(2\pi)^4}(e_1)_j\delta(p+p_4-k).$$

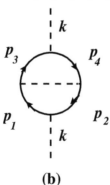

(b)

(b) Following Section 8.2, we take the trace over the fermionic propagators, following the graph top-down, and the matrix element reads

$$S = -\frac{1}{(2\pi)^7}\frac{e^4}{4\omega}(e_2)_j\int\frac{1}{q^2}Tr\left[\gamma_j\frac{i\hat{p}_3-m}{p_3^2+m^2}\gamma_i\frac{i\hat{p}_1-m}{p_1^2+m^2}\right.$$

$$\left.\times\gamma_l\frac{i\hat{p}_2-m}{p_2^2+m^2}\gamma_i\frac{(i\hat{p}_4-m)}{p_4^2+m^2}\right]\delta(p_3+p_4-k)\delta(p_3-q-p_1)$$

$$\times \delta(p_2-p_4-q)\delta(p_1+p_2-k)dp_1dp_2dp_3dp_4dq(e_1)_l,$$

where the vectors p_{1-4}, q, and k are as shown in the figure. After integrating with respect to p_2, p_3, and p_4, and renaming $p_1 = p$, we are left with

$$S = -2\pi \frac{e^4}{4\omega}(e_2)_j \int \frac{1}{q^2} Tr\left[\gamma_j \frac{i(\hat{p} - \hat{q}) - m}{(p - q)^2 + m^2}\gamma_i \frac{(i\hat{p} - m)}{p^2 + m^2}\right.$$

$$\times \gamma_l \frac{i(\hat{p} - \hat{k}) - m}{(p - k)^2 + m^2}\gamma_i \frac{i(\hat{p} - \hat{k} + \hat{q}) - m}{(p - k + q)^2 + m^2}\left.\right] \frac{dp}{(2\pi)^4}\frac{dq}{(2\pi)^4}(e_1)_l.$$

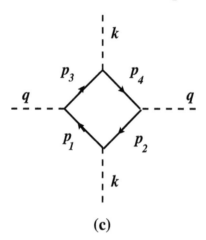

(c)

(c) We follow the same steps as before, except now we have two extra external photonic lines instead of the internal photonic propoagator, and we will not integrate over q. The final formula is

$$S = -\frac{1}{(2\pi)^2}\frac{e^4}{4\omega_k\omega_q}(e_2)_j(e_2)_l \int Tr\left[\gamma_j \frac{i(\hat{p} - \hat{q}) - m}{(p - q)^2 + m^2}\gamma_i \frac{(i\hat{p} - m)}{p^2 + m^2}\right.$$

$$\times \gamma_g \frac{i(\hat{p} - \hat{k}) - m}{(p - k)^2 + m^2}\gamma_h \frac{i(\hat{p} - \hat{k} - \hat{q}) - m}{(p - k - q)^2 + m^2}\left.\right] \frac{dp}{(2\pi)^4}(e_1)_g(e_1)_h.$$

22. **Problem.** Find the divergent part of the following diagram:

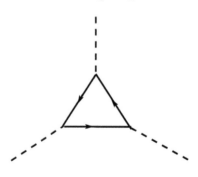

Show that this diagram cancels with the diagram that has the reverse orientation inside the fermion loop.

Solution. Following the GRAPH TABLE, for the diagram

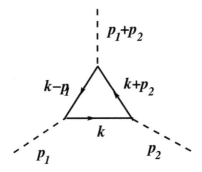

we obtain the matrix element

$$S_{ijk} = \frac{1}{2\sqrt{2}} \frac{e^3}{(2\pi)^{5/2}} \frac{1}{\sqrt{\omega_1 \omega_2 \omega_3}} (e_i)^1 (e_j)^2 (e_k)^3$$

$$\times \int \mathrm{Tr}\left[\gamma_i \frac{i(\hat{k} - \hat{p}_1) - m}{(k - p_1)^2 + m^2} \gamma_j \frac{i\hat{k} - m}{k^2 + m^2} \right.$$

$$\left. \times \gamma_k \frac{i(\hat{k} + \hat{p}_2) - m}{(k + p_2)^2 + m^2} \right] \frac{dk}{(2\pi)^4}.$$

Expanding the integral in this expression, the terms containing an odd number of internal momenta k vanish, while the terms proportional to m^2 and $\mathrm{Tr}[\gamma_i(p_1 z - p_2 x - p_1)\gamma_j(p_1 z - p_2 x + p_2)\gamma_k(p_1 z - p_2 x)]$ are finite. The first of the remaining integrals is

$$I_1 = 8e^3 \int_0^1 \int_0^{1-x} \int \left[\frac{2k_j(k_i G_l - \delta_{il} G \cdot k + k_l G_i)}{(k^2 - A)^3} \right.$$

$$\left. - \frac{k^2(\delta_{ij} G_l - \delta_{il} G_j + \delta_{jl} G_i)}{(k^2 - A)^3} \right] dx dz \frac{d^4 k}{(2\pi)^4}$$

$$= \frac{-4ie^3}{(4\pi)^2} \Gamma(\epsilon/2) \int_0^1 \int_0^{1-x} \left[1 - \frac{\epsilon}{2} \ln A + \mathcal{O}(\epsilon^2) \right]$$

$$\times (\delta_{ij} G_l - \delta_{il} G_j + \delta_{jl} G_i) dx dz,$$

where $G = (p_1 z - p_2 x)$, and $A = (p_2 x - p_1 z)^2 - p_2^2 x - p_1^2 z + m^2$. The divergent part of this integral is

$$I_1^{div} = \frac{-ie^3}{2\pi^2 \epsilon} \int_0^1 \int_0^{1-x} (\delta_{ij} G_l - \delta_{il} G_j + \delta_{jl} G_i) dx dz.$$

Problems with solutions

Calculating all other terms in the same manner, we obtain the final result

$$I^{div} = \frac{-ie^3}{2\pi^2\epsilon}\left[\frac{1}{6}\left(\delta_{ij}(p_1 - p_2)_l + \delta_{il}(p_1 - p_2)_j + \delta_{jl}(p_1 - p_2)_i\right)\right.$$
$$\left. + \frac{1}{2}\left(\delta_{ij}(p_1 + p_2)_l + \delta_{il}(p_2 - p_1)_j - \delta_{jl}(p_1 + p_2)_i\right)\right].$$

In reversing the diagram, we also reverse the signs of the momenta. In this case, the expression for the matrix element is the same, except for the trace term, which should be replaced by

$$\text{Tr}\left[\gamma_k \frac{i(-\hat{k} - \hat{p}_2) - m}{(-k - p_2)^2 + m^2}\gamma_j \frac{i(-\hat{k}) - m}{(-k)^2 + m^2}\right.$$
$$\left. \times \gamma_i \frac{i(\hat{p}_1 - \hat{k}) - m}{(p_1 - k)^2 + m^2}\right].$$

Using the charge conjugation operators (Problem 10), we can show that this expression is also equal to

$$(-1)^3\text{Tr}\left[\gamma_k \frac{i(\hat{k} + \hat{p}_2) - m}{(k + p_2)^2 + m^2}\gamma_j \frac{i\hat{k} - m}{k^2 + m^2}\right.$$
$$\left. \times \gamma_i \frac{i(\hat{k} - \hat{p}_1) - m}{(k - p_1)^2 + m^2}\right],$$

quantity which cancels exactly the first diagram. This result is also valid for all such diagrams with an odd number of vertices, and is known as the *Furry theorem*.

REFERENCES

1. Akhiezer, A.I., Berestetzkij, V.B.: Quantum electrodynamics, Wiley, New York (1965)

2. Araki, H.: Mathematical Theory of Quantum Fields, Oxford University Press (1999)

3. Auyang, Y.: How is Quantum Field Theory Possible? Oxford University Press, New York (1995)

4. Berestetskii, V.B., Pitaevskii, L. P., Lifshitz, E.M.: Quantum Electrodynamics, Second. Ed., Butterworth-Heinemann (1982)

5. Bogoliubov, N.N., Shirkov, D.V.: Introduction to the Theory of Quantized Fields, 3rd ed., Wiley, New York (1980)

6. Brown, L. S.: Quantum Field Theory, Cambridge University Press (1992)

7. Chaichian, M., and Demichev, A.: Path Integrals in Physics, Volume II, Quantum Field Theory, Statistical Physics and other Modern Applications, Institute of Physics Publishing Ltd., Bristol and Philadelphia (2001)

8. Cohen-Tannoudji, C., Dupont-Roc, J., Grynberg, G.: Photons and Atoms: Introduction to Quantum Electrodynamics, Wiley, New York (1989)

9. D'Auria, R., and Trigiante, M.: From Special Relativity to Feynman Diagrams, Springer (2011)

10. Dimock, J.: Quantum Mechanics and Quantum Field Theory: A Mathematical Primer, Cambridge University Press (2011)

11. Feynman, R.P.: Quantum Electrodynamics, Addison-Wesley, Reading, MA (1998)

12. Feynman, R.P.: The Development of the Space-Time View of Quantum Electrodynamics (Nobel lecture), Stockholm: Nobel Foundation (1966).

329

13. Fujimotu, M.: Physics of Classical Electromagnetism, Springer (2007)

14. Greiner, W., Reinhardt, J. : Quantum Electrodynamics, Springer, 4th ed. Edition (2008)

15. Itzykson, C., Zuber, J-B.: Quantum Field Theory, Dover Publications (2006)

16. Jauch, J.M., Rohrlich, F.: The Theory of Photons and Electrons, Springer-Verlag (1980)

17. Kinoshita, T.: Quantum Electrodynamics, World Scientific, Singapore (1990)

18. Maggiore, M.: A Modern Introduction to Quantum Field Theory, Oxford Master Series in Statistical, Computational, and Theoretical Physics, Oxford University Press (2005)

19. McMahon, D.: Quantum Field Theory Demystified, McGraw-Hill Professional (2008)

20. Miller, A.I.: Early Quantum Electrodynamics, A Sourcebook, Cambridge University Press (1995)

21. Milonni, P.W.: The quantum vacuum - an introduction to quantum electrodynamics, Academic Press (1994)

22. Ohlsson, T.: Relativistic Quantum Physics: From Advanced Quantum Mechanics to Introductory Quantum Field Theory, Cambridge University Press (2011)

23. Peskin, M.E., and Schroeder, D.V.: An Introduction to Quantum Field Theory, Addison-Wesley, Reading, Mass. (1995)

24. Radovanovic, V.: Problem Book in Quantum Field Theory, 2nd edition, Springer (2007)

25. Reinhardt, J., Greiner, W.: Field Quantization, Springer (2008)

26. Ryder, L. H.: Quantum Field Theory, 2nd ed., Cambridge University Press (1996)

27. Synge, J.L.: Relativity: The Special Theory, 2nd ed., Elsevier (1980).

28. Schweber, S.S.: QED and the Men Who Made It, Princeton University Press (1994)

29. Scharf, G.: Finite Quantum Electrodynamics: The Casual Approach, 2nd ed., Springer-Verlag, New York (1995)

30. Schechter, A.S.: Basics of Quantum Field Theory (in Russian), Saratov State University (1965)

31. Schwinger, J.S. (Ed): Selected Papers on Quantum Electrodynamics, Dover Publications, New York (1958)

32. Srednicki, M.A.: Quantum Field Theory, Cambridge University Press (2007)

33. Sterman, G.: An Introduction to Quantum Field Theory, Cambridge University Press (1993)

34. Tannoudji-Cohen, C., Dupont-Roc, J., Grynberg, G.: Photons and Atoms: Introduction to Quantum Electrodynamics, Wiley-Interscience (1997)

35. Thaller, B.: The Dirac Equation, Springer-Verlag, Berlin (1992)

36. Thirring, W.E.: Principles of Quantum Electrodynamics, Academic Press (1962)

37. Weinberg, S.: The Quantum Theory of Fields, Vol. 1: Foundations, Cambridge University Press (1995)

38. Weinberg, S.: The Quantum Theory of Fields, Vol. 2: Modern Applications, Cambridge University Press (1996)

39. Williams, L.P.: The Origins of Field Theory, Random House, New York (1966)

SUBJECT INDEX

Infinitesimal Lorentz transformation, 5, 6, 12
Infrared divergence, 219
Interaction
    Hamiltonian density, 147, 151, 153, 253
    Lagrangian density, 143, 149, 151
    representation, 143

K

Klein-Gordon-Schrödinger equation, 48, 49, 54
Klein-Nishina-Tamm formula, 187

L

Laboratory reference frame (LF), 193, 195, 215
Lagrangian density, 1, 2, 3, 12, 13, 20, 24, 50, 64, 72, 85,
    104, 115, 143, 269
Larmor invariant, 2
Longitudinal photons, 88, 106
Lorentz-Fermi condition, 102
Lorentz
    invariance, 6, 13
    gauge condition, 83, 85, 104

M

Mass renormalization, 254, 272
Matrix representation, 36, 40, 43, 144
Maxwell's equations, 83, 84
Meson, 49, 51, 53, 70, 89, 150
Methods of field quantization, 43, 46, 64
Minkowski metric, 46
Möller scattering, 198, 210, 240
    transition probability, 198
    cross section, 210
Möller's formula, 218
Momentum four-vector, 14, 38, 42, 45, 50, 52, 86, 110, 117,
    190, 215, 281
Momentum representation, 53, 57, 60, 86, 89, 93, 104, 130,
    155, 168, 199, 254, 267
Momentum spacial components, 57
Monochromatic, 118, 148

N

Necessity of field quantization, 27, 32, 43
Nobel Prize, 152, 220
Noether's theorem, 20
Non-observable vacuum process, 164
Normal product, 66, 71, 72, 76, 79, 81, 91, 140, 151

O

Operator,
 annihilation, 33, 39, 66, 71, 75, 96, 106,
  137, 153, 190, 220, 233
 antihermitian, 100, 102
 creation, 33, 66, 69, 71, 75, 77,
  137, 153, 168, 174, 220, 242
 Hermitian, 37, 100, 101, 102, 111, 127, 177
 idempotent, 128

P

Pauli exclusion principle, 29, 218, 232
Phase
 invariance, 24
 transformations, 24
Photon
 polarization, 88, 185, 224
 spin, 93
Photon-photon scattering, 220
Potentials, 2, 267
Principal part, 278, 279
Probability, 28, 148, 164, 170, 173, 198, 224, 237
Product of operators
 time-ordered, 72, 73, 76, 80, 81, 99,
  140, 146, 151, 158
 normal, 66, 71, 72, 75, 77, 81, 140, 151
Projection operator, 128
Propagator, 153, 155, 164, 165, 255, 271
Pure state vectors, 160

Q

Quantization condition, 65, 84, 97
Quantum statistics

Bose-Einstein, 29, 36, 39
Fermi-Dirac, 29, 41, 43

R

Renormalization, 63, 241, 255, 269, 272
Retarded and advanced functions, 289
Rotation operator, 19
Rutherfords formula, 220

S

Scalar field, 2, 7, 20, 49, 53, 60, 70, 85, 99, 150, 289
Scattering
 cross section, 167, 184, 210, 230, 233, 239
 matrix, 144, 156, 234
Second-order processes
 non-divergent, 173
 divergent, 241
Self-energy diagram, 241, 242, 267,
S-matrix, 144, 148, 154, 164, 192, 220, 270
Spin, 18, 20, 86, 89, 93, 98, 117, 130, 167, 202
Spinorial field, 1, 2, 8, 13, 109, 117, 131,141, 149,
 154, 288, 292, 294
Strong interactions, 150
Symmetry transformations, 20

T

Tensor
 angular momentum 15, 17, 18, 21, 40,
  51, 93, 116, 134
 energy-momentum, 13, 14, 15, 19, 45, 116
 metric, 17
Thomsons formula, 188
Time-ordered contraction, 76, 153
Total angular momentum density, 17, 18, 19
Transformation
 infinitesimal, 5, 20
 phase 24
Transition probability, 149, 164, 166, 173, 198, 224
Transversal photons, 88, 105

U
Unitary matrix, 30, 113
Units used in QFT, 46, 47, 48

V
Vacuum
    contraction, 76, 99, 141
    expectation value, 75, 76, 99, 141
    polarization, 256
    state, 38, 39, 74, 75, 154, 167, 256
Variables
    canonical conjugate, 44
    dependent, 2
    field, 2, 4, 6, 9, 21, 50, 53, 67, 84
    independent, 2, 53, 248
Virtual photons, 88, 103, 105, 108, 242

W
Weak interactions, 150
Wick's theorems, 77, 142, 153

Milton Keynes UK
Ingram Content Group UK Ltd.
UKHW020315111024
449327UK00040B/1249